Cyanobacterial Biotechnology in the 21st Century

Brett Neilan •
Michel Rodrigo Zambrano Passarini •
Prashant Kumar Singh • Ajay Kumar
Editors

Cyanobacterial Biotechnology in the 21st Century

Springer

Editors
Brett Neilan
School of Environmental & Life Sciences
University of Newcastle
Callaghan, NSW, Australia

Michel Rodrigo Zambrano Passarini
Laboratory of Environmental Biotechnology
Federal University of Latin-American Integration (UNILA)
Foz do Iguaçu, Paraná, Brazil

Prashant Kumar Singh
Department of Biotechnology
Mizoram University
Aizawl, India

Ajay Kumar
Agriculture Research Organization
Volcani Centre, Israel

ISBN 978-981-99-0180-7 ISBN 978-981-99-0181-4 (eBook)
https://doi.org/10.1007/978-981-99-0181-4

This Springer imprint is published by the registered company Springer Nature Singapore Pte Ltd.
The registered company address is: 152 Beach Road, #21-01/04 Gateway East, Singapore 189721, Singapore

Preface

The book *Cyanobacterial Biotechnology in the 21st Century* explores in detail the ecology, diversity, and biotechnological applications, including the potential for bioremediation, of the Cyanobacteria, addressing the diverse platform of technological tools used in their study. Some subjects addressed in recent years, including cyanotoxins, xenobiotics, secondary metabolites, pesticides, nanoparticles, biofertilizers, water pollution indicators, anthropogenic impact, and hydrogen production, are covered in this book. These microorganisms display significant advantages over other life forms, including resistance to desiccation, fixing nitrogen, and readily switching between phototrophic and mixotrophic nutrition. They have proven to be a very promising group of microbes for their application to environmental and industrial biotechnology.

Callaghan, Australia	Brett Neilan
Foz do Iguaçu, Brazil	Michel Rodrigo Zambrano Passarini
Aizawl, India	Prashant Kumar Singh
Volcani Centre, Israel	Ajay Kumar

Contents

Editors and Contributors

About the Editors

Brett Neilan is a molecular biologist who studies microbes and their chemistries, including the genetics of toxic cyanobacteria. The research has led to an understanding of the biochemical pathways responsible for producing their toxins. He also studies natural microbial products from novel environments, including symbioses, and the mechanisms of complex biosynthesis of these bioactive compounds.

Michel Rodrigo Zambrano Passarini's research areas are microbial ecology and microbial bioprospecting of extreme environments in the search for enzymes and metabolites of industrial interest applicable to the production of biofuel and degradation processes of environmental pollutants, as well as metagenomics and metabarcoding.

Prashant Kumar Singh is currently working at the Department of Biotechnology, Mizoram University, PUC campus, India. His research focuses on plant and cyanobacterial stress biology around the omics approaches.

Ajay Kumar is a microbiologist, and his research area includes microbiome, postharvest management, microbial biocontrol, plant-microbe interactions, cyanobacteria, and endophytes.

Contributors

Ravi K. Asthana R. N. Singh Memorial Laboratory, Centre of Advanced Study in Botany, Institute of Science, Banaras Hindu University, Varanasi, Uttar Pradesh, India

Sandra Barca Departamento de Zooloxía, Xenética e Antropoloxía Física, Universidade de Santiago de Compostela, Santiago de Compostela, Spain

Gabriela Sosa Benegas Latin American Institute for the Science of Life and Nature, Federal University of Latin American Integration (UNILA), Foz do Iguaçu, Brazil

Fernando Cobo Departamento de Zooloxía, Xenética e Antropoloxía Física, Universidade de Santiago de Compostela, Santiago de Compostela, Spain

M. Carmen Cobo Department of Biological Sciences, University of Alabama, Tuscaloosa, AL, USA

Caroline da Costa Silva Gonçalves Laboratory of Environmental Biotechnology, Federal University of Latin-American Integration (UNILA), Foz do Iguaçu, Parana, Brazil

Alka Devi Centre for Conservation and Utilisation of Blue Green Algae (CCUBGA), Division of Microbiology, ICAR-Indian Agricultural Research Institute, New Delhi, India

Luis Fernando Romanholo Ferreira Institute of Research and Technology, University of Tiradentes – UNIT, Aracaju, Sergipe, Brazil

Rajan Kumar Gupta Laboratory of Algal Research, Centre of Advanced Study in Botany, Institute of Science, Banaras Hindu University, Varanasi, India

Mohneesh Kalwani Centre for Conservation and Utilisation of Blue Green Algae (CCUBGA), Division of Microbiology, ICAR-Indian Agricultural Research Institute, New Delhi, India
School of Biotechnology, Banaras Hindu University, Varanasi, Uttar Pradesh, India

Kaushalendra Department of Zoology, Mizoram University, Aizawl, India

Surbhi Kharwar Department of Botany, University of Lucknow, Lucknow, Uttar Pradesh, India

Ajay Kumar Department of Botany, Centre of Advanced Studies in Botany, Institute of Science, Banaras Hindu University, Varanasi, India
Agriculture Research Organization, Volcani Center, Department of Postharvest Science, Rishon Lezzion, Israel
Department of Postharvest Science, Agriculture Research Organization, Volcani Center, Rishon LeZion, Israel

Indrajeet Kumar Laboratory of Algal Research, Centre of Advanced Study in Botany, Institute of Science, Banaras Hindu University, Varanasi, India

Vinod Kumar Department of Botany, Pandit Prithi Nath College, Kanpur, Uttar Pradesh, India

Arti Kumari Division of Biochemistry, ICAR-Indian Agricultural Research Institute, New Delhi, India

Esther Lalnunmawii Department of Biotechnology, Mizoram University, Tanhril, Mizoram, India

Lalrokimi Department of Biotechnology, Mizoram University, Tanhril, Mizoram, India

Laxmi R. N. Singh Memorial Laboratory, Centre of Advanced Study in Botany, Institute of Science, Banaras Hindu University, Varanasi, Uttar Pradesh, India

Stella T. Lima Department of Chemistry and Biochemistry, University of North Carolina at Greensboro, Greensboro, NC, USA

Keith Dayane Leite Lira Laboratory of Health and Environment – LABMSMA, Department of Pharmaceutical Sciences, Federal University of São Paulo (UNIFESP), Diadema, São Paulo, Brazil

Gajanan Mehetre Department of Biotechnology, Mizoram University, Tanhril, Mizoram, India

Sergio Mendez-Gaona Polytechnic Faculty, University of Asuncion (UNA), San Lorenzo, Paraguay

Arpan Mukherjee Institute of Environment and Sustainable Development, Banaras Hindu University, Varanasi, Uttar Pradesh, India

Júlia Ronzella Ottoni Laboratory of Environmental Biotechnology, Federal University of Latin-American Integration (UNILA), Foz do Iguaçu, Parana, Brazil

Sunil Pabbi Centre for Conservation and Utilisation of Blue Green Algae (CCUBGA), Division of Microbiology, ICAR-Indian Agricultural Research Institute, New Delhi, India

Michel Rodrigo Zambrano Passarini Laboratory of Environmental Biotechnology, Federal University of Latin-American Integration (UNILA), Foz do Iguaçu, Parana, Brazil

Krutika Patil Centre for Conservation and Utilisation of Blue Green Algae (CCUBGA), Division of Microbiology, ICAR-Indian Agricultural Research Institute, New Delhi, India

Cecilio Correa Perez CEMIT Multidisciplinary and Technological Center of Studies and Research of the University of Asuncion (UNA), San Lorenzo, Paraguay

Kodandoor Sharathchandra Department of Biosciences, Mangalore University, Mangaluru, Karnataka, India

Ekta Shukla Sunbeam College for Women Bhagwanpur, Varanasi, Uttar Pradesh, India

Livleen Shukla Division of Microbiology, Indian Agricultural Research Institute, New Delhi, India

Pratyoosh Shukla School of Biotechnology, Banaras Hindu University, Varanasi, Uttar Pradesh, India

Amit Kumar Singh Department of Botany, Magadh University, Bodh-Gaya, India

Avinash Singh R. N. Singh Memorial Laboratory, Centre of Advanced Study in Botany, Institute of Science, Banaras Hindu University, Varanasi, Uttar Pradesh, India

Bhim Pratap Singh Department of Agriculture and Environmental Sciences (AES), National Institute of Food Technology Entrepreneurship and Management (NIFTEM), Sonepat, Haryana, India

Prashant Kumar Singh Department of Biotechnology, Mizoram University, Aizawl, India
Department of Biotechnology, Mizoram University (A Central University), Pachhunga University College Campus, Aizawl, India

Pratika Singh Department of Life Science, School of Earth, Biological and Environmental Sciences, Central University of South Bihar, Gaya, Bihar, India

Rahul Prasad Singh Laboratory of Algal Research, Centre of Advanced Study in Botany, Institute of Science, Banaras Hindu University, Varanasi, India

Sandeep Kumar Singh Division of Microbiology, Indian Agricultural Research Institute, Pusa, New Delhi, India

Shiv Mohan Singh Department of Botany, Centre of Advanced Studies in Botany, Institute of Science, Banaras Hindu University, Varanasi, India

Sweksha Singh R. N. Singh Memorial Laboratory, Centre of Advanced Study in Botany, Institute of Science, Banaras Hindu University, Varanasi, Uttar Pradesh, India

Kandikere R. Sridhar Department of Biosciences, Mangalore University, Mangaluru, Karnataka, India

Kandikere Ramaiah Sridhar Department of Biosciences, Mangalore University, Mangalore, Karnataka, India

Amrita Srivastava Department of Life Science, School of Earth, Biological and Environmental Sciences, Central University of South Bihar, Gaya, Bihar, India

Aruna Tyagi Division of Biochemistry, ICAR-Indian Agricultural Research Institute, New Delhi, India

Suzan Pantarotto de Vasconcellos Laboratory of Health and Environment – LABMSMA, Department of Pharmaceutical Sciences, Federal University of São Paulo (UNIFESP), Diadema, São Paulo, Brazil

Ekta Verma Department of Botany, Magadh University, Bodh-Gaya, India

Rufino Vieira-Lanero Departamento de Zooloxía, Xenética e Antropoloxía Física, Universidade de Santiago de Compostela, Santiago de Compostela, Spain

Mukesh Kumar Yadav Department of Biotechnology, Mizoram University, Aizawl, India

Nisha Yadav Division of Agricultural Extension, ICAR-Indian Agricultural Research Institute, New Delhi, India

Priya Yadav Laboratory of Algal Research, Centre of Advanced Study in Botany, Institute of Science, Banaras Hindu University, Varanasi, India

Shivam Yadav Department of Botany, T.P.S. College, Patna, Bihar, India

Zothanpuia Department of Biotechnology, Mizoram University, Pachhunga University College Campus, Aizawl, Mizoram, India

Cyanobacteria and Cyanotoxins in Underground Water and the New Perspectives in a Climate Breakdown Scenario

1

Stella T. Lima

Abstract

Biospheric environmental perturbations, such as nutrient loading and climatic changes, strongly affect cyanobacterial growth and blooms' potential in freshwater and marine ecosystems. Harmful cyanobacterial blooms (cyanoHABs) cause recurrent toxic events in global watersheds and in underground water wells, highlighting a significant public health issue in a moment of climate and water crisis. Cyanobacteria are a prolific source of natural products some of them incredibly toxic for humans, animals, and plants. Consequently, in global conditions where the frequency and intensity of cyanoHABs might be increased, harmful blooms should be more than a casual public health issue and be part of a complex and multidisciplinary agenda for water management agencies all over the world.

Keywords

Cyanotoxins · Water supply · Cyanobacteria · Climate crisis

1.1 Freshwater and Cyanobacteria Harmful Blooms

Water is an essential resource for life. The global distribution of water is extremely uneven on the surface of Earth. Only 3% of the water is fresh, 69% of it resides in glacier and icecaps, 30% resides underground, and just over 1% is in rivers, lakes, and swamps (Gleick 2011). Whether we change our perspective of freshwater

S. T. Lima (✉)
Department of Chemistry and Biochemistry, University of North Carolina at Greensboro, Greensboro, NC, USA
e-mail: s_delimacam@uncg.edu

B. Neilan et al. (eds.), *Cyanobacterial Biotechnology in the 21st Century*,
https://doi.org/10.1007/978-981-99-0181-4_1

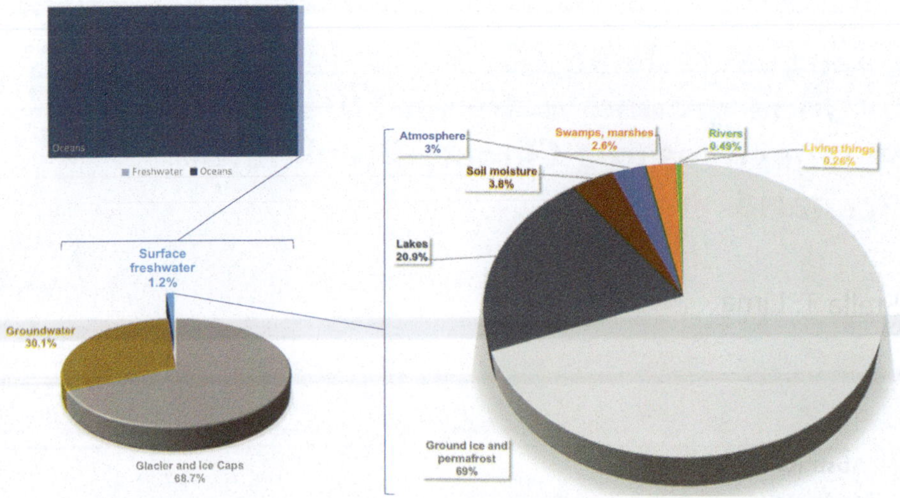

Fig. 1.1 Distribution of freshwater sources worldwide

readily accessible, only a relatively small portion is available to sustain human, plant, and animal life (Fig. 1.1).

Over the decades to meet the ongoing growth of the world's population, increasing need for high-quality freshwater will be demanded in domestic, agriculture, and industry. Additionally, contamination of surface water and groundwater increases the difficulties to manage water distribution and guarantee that communities have access to the stablished quality and quantity regulated by the World Health Organization (WHO). Global heating also plays its role in causing water shortage, since rainfall patterns, glaciers' collapse, as well as intense droughts and floods have been registered and defined as a new pattern in the climate breakdown (Zareian et al. 2019). Currently, we are globally facing a water crisis which requires fast development of strategies and attention of the scientific community.

The surface blooms are one of the most prominent visual expressions of cyanobacterial growth. To proliferate, cyanobacteria must supply its demand for light energy and nutrient, mostly phosphorus (P) and nitrogen (N). Competing in the environment for these resources with other phytoplankton and each other, some genus of cyanobacteria has a number of specific strategies that favor them over other grazers in the bloom community, such as fixation of atmospheric N and/or buoyancy regulation. Under eutrophic conditions, water is turbid, and low level of oxygen (hypoxia) can kill fishes and aquatic plants, reducing essential fish habitats (National Oceanic and Atmospheric Administration (NOAA), USA). Photosynthesis is restricted to mostly upper water layer of a few meters or even less than 1 m, when blooms are denser, exposing phytoplankton cells to limited time of light of the day which may then limit their growth rates. Once again, a massive growth of cyanobacteria indicates that these photoautotrophic, and mostly toxic, microorganisms could survive and/or dominate both natural and anthropogenically altered environments.

The complexity of the global climate crisis and ecosystem interactions, such as rising temperatures, nutrient over-enrichment, and hydrological alterations of waterways, are major drivers of cyanobacteria harmful blooms (cyanoHABs) and harmful algae blooms (HABs), which have been increasingly reported worldwide, causing significant environmental effects. These changes may work together to promote cyanobacteria dominance and remain (Paerl and Paul 2012), assembling a synergy that might be a significant challenge to water quality and water supply management agencies. From the time when blooms intensities, potential and control might be based in the contemporaneous climate changes and hydrological new regimes, well understanding of the interactions between these drivers and cyanobacterial can pave the path for next research interest in the field.

1.2 Cyanotoxins

Arsenal of freshwater cyanobacteria toxic compounds include the well-known microcystin, cylindrospermopsin, saxitoxin, and anatoxin-a, as well as guanitoxin (formerly known as anatoxin-a(s)) (Fiore et al. 2020), and aetokthonotoxin (Breinlinger et al. 2021) (Table 1.1). The physiological function of these molecules for cyanobacteria is not so clear, but it has been believed that this collection of chemically rich compounds benefits these microorganisms, at least in part, through allelopathic interactions with other primary producers, grazers, and competitors, increasing the ability of cyanobacteria to stay within the complex community of blooms' microenvironments, such as regulation of colony formation, floatage, nutrient acquisition, and tolerance of extremes in radiation and salinity (Welker and Von Döhren 2006; Zhang et al. 2020).

Microcystin (MC) is the most prevalent, well-studied, and structurally diverse group of cyanobacterial toxins, comprising around 250 isoforms with different toxicity and molecular weight range of 800–1100 Da. MC is an intracellular peptide which belongs to the class of hepatotoxins which inhibit eukaryotic protein phosphatases types 1 and 2A, being able to penetrate liver cells through active transport (Pearson et al. 2016). With LD_{50} of approximately 25–150 μg/kg (intraperitoneal route) of body weight in mice (World Health Organization 2006), MC has been responsible for fatal intoxications of humans and animals in the last few decades. The most significant intoxication attributed to MC occurred at a hemodialysis clinic in Caruaru, Brazil, where at least 50 adults' patients showed symptoms of acute liver damage after contact with contaminated water (Frazier et al. 1998; Jochimsen et al. 1998). Nodularins are also cyclic peptides structurally similar to microcystins and similar mode of action (Table 1.1). Nodularin has ten variants currently known (Chorus and Welker 2021).

Cylindrospermopsin (CYN) is one of the most diversified cyanotoxins in terms of cyanobacteria producers. CYN is an extracellular polyketide originated alkaloid with hepatotoxic, nephrotoxic, and general cytotoxic action, as well as a potent protein

Table 1.1 Overview of freshwater cyanotoxins (Huisman et al. 2018)

Toxin	Mode of action	Toxic effects	Structure
Microcystins	Inhibition of phosphatase in eukaryotes	Liver and kidney damage, gastroenteritis, carcinogenic effect, reduced DNA repair, and reproductive toxicity	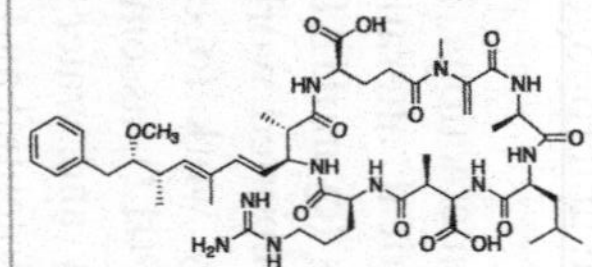
Nodularin	Inhibition of phosphatase in eukaryotes	Same effects as microcystins; however, weak carcinogenicity	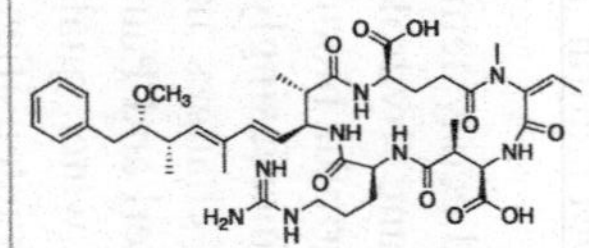
Cylindrospermopsin	Inhibition of protein synthesis, DNA damage, and cell death	Multiple damage in organs, gastroenteritis, and genotoxicity	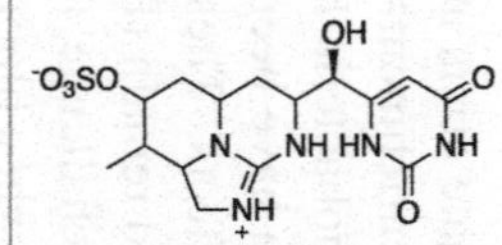
Saxitoxin	Block voltage-gated sodium channels of neurons	Paraesthesia, numbness, paralysis, and respiratory failure	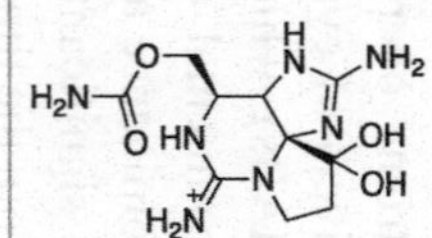
Anatoxin-a	Agonist of the nicotinic acetylcholine receptors	Loss of coordination, muscle tremors, and respiratory failure	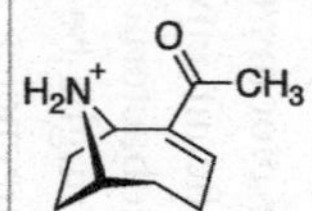

Guanitoxin	Irreversible inhibitors of acetylcholinesterase	Salivation, incontinence, muscle tremors, and respiratory failure	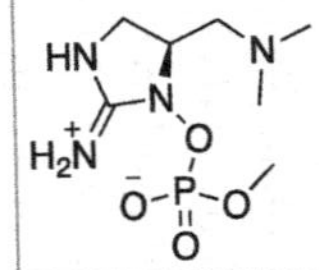
Aetokthonotoxin	Vascular myelinopathy (VM) brain lesions	Neurological behaviors in fishes and chickens	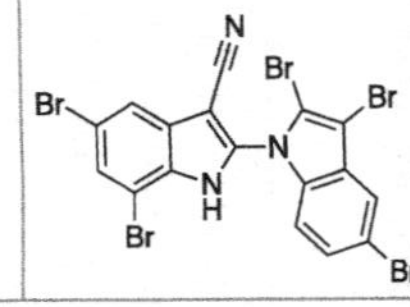

synthesis inhibitor (Pearson et al. 2016). Pure cylindrospermopsin has an LD_{50} in mice (i.p.) of 2.1 mg/kg at 24 h and 0.2 mg/kg at 5–6 days (Sivonen 2009). It was originally isolated from an outbreak in Australia in 1979, where many of the intoxicated cases were children that showed a serious hepatoenteritis (Hawkins et al. 1997).

Moving for neurotoxic class of toxins, cyanobacteria have a quite diverse resource for compounds that acts in the neuro system of mammals, birds, and fishes (Wang et al. 2016). Saxitoxin (STX) is a potent neurotoxic alkaloid produced by both marine dinoflagellates and freshwater cyanobacteria (Lukowski et al. 2019). Belonging to a family of 57 analogs, saxitoxin intoxication is caused by the reversible intermolecular interaction between the positive charge of the toxin in the guanidinium moiety and negative charge of carboxyl groups at Na^+ channels in an equimolar rate, blocking the channels (Wiese et al. 2010). Another neurotoxic alkaloid, anatoxin-a (ATX), is known to quickly intoxicate animals by its action against the cholinergic system of mammals (Méjean et al. 2009), acting as an agonist of the nicotinic acetylcholine receptor that upon binding induces the opening of the channel depolarizing the cell membrane (Valério et al. 2010). With a LD_{50} of 200–250 μg/kg (i.p.) of body weight (Carmichael 2001), ATX was isolated in the same bloom as anatoxin-a(s) (Carmichael and Gorham 1978), currently named as guanitoxin (Fiore et al. 2020).

Guanitoxin (GNT) is the only known natural organophosphate neurotoxin. Sharing the same mechanism of action with organophosphates as the chemical weapon agent sarin and the banned pesticide parathion, GNT is an irreversible inhibitor of the enzyme acetylcholinesterase in the cholinergic system of mammals (Valério et al. 2010). Through a covalent bound between guanitoxin and the active site of acetylcholinesterase, the enzyme cannot perform the recycling of acetylcholine, which induces acute neurological toxicity that can lead to rapid death. GNT shows a comparable lethality ($LD_{50} = 20$ μg/kg i.p.) (Carmichael 2001) to saxitoxin, the most powerful known cyanotoxin. The ordinary amino acid L-arginine serves as the starting point for the biosynthesis of guanitoxin, which involves nine biosynthetic steps (Lima et al. 2022). The GNT biosynthetic cassette includes enzymatic reactions of oxidation (*gntA*, *gntB*, and *gntD*), PLP-dependent biochemistry (*gntC*, *gntE*, and *gntG*), *S*-adenosylmethionine (SAM)-dependent reactions (*gntF* and *gntJ*), and a kinase (*gntI*) (Lima et al. 2022).

Recently, a novel toxin, aetokthonotoxin (AETX), was reported to be the responsible agent of a neurological impairment in birds that occurred over the winter of 1994 to 1995 (Breinlinger et al. 2021). This neurotoxic cyanotoxin is produced by the epiphytic cyanobacterium *Aetokthonos hydrillicola*, isolated from the invasive submerged aquatic plant *Hydrilla verticillata*. The chemical structure of AETX is unique among cyanotoxins composed for a pentabrominated biindole nitrile skeleton. The presence of bromides is notable to suggest the neurotoxicity of this molecule, as polyhalogenated compounds, such as bromethalin or hexachlorophene, are known to provoke vascular myelinopathy (VM) brain lesions in birds and mammals (Dorman et al. 1992; Thomas et al. 1998). The BGC of AETX has six biosynthetic genes organized in 9.2 kb length (Breinlinger et al. 2021). In a recent work, the enzymatic steps were elucidated showing the complete biosynthesis of AETX (Adak et al. 2022).

1.3 Toxic Genera of Cyanobacteria

Despite an incredible knowledge in natural products, toxicity, ecology, physiology, and taxonomy of cyanobacteria, an important gap about how to manage the methods of cyanotoxin detection and how to use them to predict toxic cyanoHABs, still have been a challenge.

Considering the toxic species, several genera are responsible for harmful blooms in freshwater bodies (Table 1.2). *Microcystis* is the dominant genus occurring harmfully around 79 countries in freshwater spots (Harke et al. 2016) and appears to be expanding. *Microcystis* is a single-celled cyanobacterium that can form buoyant colonies equipped with a set of genes responsible for microcystin's production, and even non-toxic blooms of *Microcystis* carry ecosystem and economic disturbing consequences. Their blooms are commonly reported to increase the pH of surface well above to 9, because of the rapid consumption of inorganic carbon for cyanobacteria (Wilhelm et al. 2020). Its conditions can give *Microcystis* advantage in the use of alternative sources of carbon (bicarbonate and urea, e.g.), as well as competitive strategies against phytoplankton community and other cyanobacteria present in the bloom (Krausfeldt et al. 2019). It makes this genus a promising candidate for global attention in a climate crisis scenario and the most studied one in different aspects, such as interactions, ecology, growth conditions, biochemistry, and biosynthesis of compounds (Wilhelm et al. 2020). Typically, *Microcystis* is the only cyanobacterium (and consequently its toxin) screened in municipal water treatment systems (Harke et al. 2016).

Since the mid-1990s, Lake Erie, Ohio, USA, has experienced the dominance of harmful blooms of *Microcystis* and *Planktothrix* (Huisman et al. 2018). *Planktothrix* genus has planktonic and benthonic strain, as well as a remarkable potential for

Table 1.2 Main cyanotoxin-producing genera (Huisman et al. 2018; Lima et al. 2022)

Main producing genera	Cyanotoxin – freshwater bodies
Microcystis, Anabaena, Dolichospermum, Leptolyngbya, Nostoc, Phormidium, Planktothrix, Synechococcus	Microcystins
Nodularia	Nodularins
Cylindrospermopsis, Anabaena, Aphanizomenon, Chrysosporum, Raphidia	Cylindrospermopsins
Anabaena, Aphanizomenon, Cuspidothrix, Dolichospermum, Oscillatoria, Phormidium	Anatoxin-a
Dolichospermum, Sphaerospermopsis, Cuspidothrix, Aphanizomenon	Guanitoxin
Aphanizomenon, Cuspidothrix, Cylindrospermopsis, Dolichospermum	Saxitoxin
Aetokthonos	Aetokthonotoxin

natural product production, which has been made its being frequently reported in blooms at temperate areas of the globe (Bothe et al. 2010; Briand et al. 2008; Kurmayer and Gumpenberger 2006; Pancrace et al. 2017).

Some other cyanobacteria genera have significant expansion of their geographical dominance, such as *Cylindrospermopsis* and *Leptolyngbya*. Initially, *Cylindrospermopsis* has been described as a tropical and subtropical dominant genus, especially due to toxic events in Brazil (Lagos et al. 1999) and Australia (Neilan et al. 2003; Saker et al. 2003). However, *C. raciborskii* has been progressively documented in Europe, specifically Greece, Hungary, France, Portugal, the Netherlands, and Germany (Antunes et al. 2015; Saker et al. 2003; Stüken et al. 2006). The United States also experiences an expressive number of cyanoHABs dominated by *Cylindrospermopsis* genus such as in Florida, Southeast, and Midwest reservoirs (Chapman and Schelske 1997; Paerl 2008). Furthermore, *Cylindrospermopsis* is known to perform good adaptation in water with temperatures above 20 °C, disperse in the water column, and survive in adverse conditions because of specialized resting cells named akinetes, which make the genus a suitable candidate for eutrophication on global climate crisis.

Filamentous toxin-producing genus *Lyngbya* has been also showing a remarkable invasive ability in a diverse of aquatic environments. *Lyngbya* cyanobacteria species often grow in periphytic or benthic mats; in addition, some species, like *L. birgei*, are planktonic. *Lyngbya* blooms have been associated with human health concern since the genus is a large producer of natural products and toxic compounds including lyngbyatoxin-a, kalkitoxin, and antillatoxin, of marine species *L. majuscula*, and paralytic shellfish-poisoning toxins from *L. wollei* in freshwater environments (Carmichael 2001). This genus has the ability to be the primer colonizer in freshwater environments and aggressive opportunist when conditions permit. Growing in blooms of dense floating mats that shade other primary producers, *Lyngbya* can take advantage of other cyanobacteria, such as *Microcystis* and *Cylindrospermopsis*, on both scenarios of eutrophication and global warming (Paerl and Huisman 2009).

Besides *Cylindrospermopsis*, others nitrogen-fixing filamentous cyanoHABs are commonly associated with member of the order *Nostocales*. Numerous of those members are characterized by their ability to generate differentiated cells enabling long-term dormancy (akinetes) and nitrogen fixation (heterocysts) (Driscoll et al. 2018). The genera *Anabaena*, *Aphanizomenon*, *Sphaerospermopsis*, *Dolichospermum*, *Nodularia*, and *Nostoc* are widely presented in the global harmful blooms associated with human health concerns due to their faculty to synthesizing toxin compounds (Dittmann et al. 2015).

1.4 Underground Waters

In 2020, while 74% (5.8 billion people) of the global population have access for a safety managed drinking water service, around 2 billion people live in different circumstances, consuming non-treated water for drinking, water livestock, or irrigation (WHO, March 2022). Allied with that, most freshwater source has been experiencing eutrophication events due to agricultural and industrial inputs of nutrients, and climate crisis, that intensify the formation of harmful cyanobacteria blooms.

Also, in arid environmental, both natural extreme conditions and anthropogenic actions make the access to water sources scarcity. It is quite common in these places the use of marine water desalination resources and underground water uses to supply the constant growth of populations, agriculture, and livestock. Desalinated marine water sustains some desert regions in the globe, such as the Arabian Gulf countries, which accounts over 20% of the total world's desalinated water (Le Quesne et al. 2021). Desalinated water is stored in dams, mega-reservoirs, and secondary house water tanks and can suffer cyanobacteria contamination during the storage and transport process. In the State of Qatar, cyanobacteria and cyanotoxins like microcystin-LR were found in water impoundments and groundwater wells destined for drinking and irrigation purposes (Chatziefthimiou et al. 2016).

Cyanobacteria are photoautotrophic organisms. Could they inhabit aphotic zones as groundwater wells? The answer is yes. Cyanobacteria also possess several pathways for light-independent energy generation, as the use of hydrogen gas to produce energy, what allow them to survive under very dark conditions like deep subsurface (Puente-Sánchez et al. 2018). However, most of the available data concerning groundwater cyanobacteria and cyanotoxins shows that the common well-known groups appear related to groundwater under the direct influence of surface water wells (Mohamed and Al Shehri 2009; Yang et al. 2016).

Surface water and groundwater systems are connected in landscapes (Liu et al. 2016). Surface water may interact with nearby groundwater through groundwater inflow, seepage loss, or combination of the two processes (Winter et al. 1998). When the stage of a surface water body is lower than that of groundwater, groundwater inflow occurs; however, when the stage of a surface water body is higher than that of groundwater, surface water moves toward groundwater (Abesh et al. 2019). Groundwater can also connect lakes across the landscape (Dodds and Whiles 2020), and thus toxins, as MCs in contaminated lakes, can migrate into groundwater (Eynard et al. 2000). The presence of MCs in groundwater was already reported previously in Italy, Latvia, and China (Eynard et al. 2000; Messineo et al. 2006; Ueno et al. 1996). In addition, Chen et al. 2006 demonstrated that MCs have high mobility in soil with low clay content. Zhang et al. 2021 detected microcystins in drinking water sourced from an artesian well and in the municipal drinking water supply across southcentral Quebec, Canada. A high density of *Oscillatoria limnetica* was described in well waters in Asir region, southwest of Saudi Arabia, producing microcystin (Mohamed and Al Shehri 2009). It was also demonstrated that concentrations of dissolved MCs in groundwater decreased with the well deepness (Mohamed et al. 2022).

A few studies have revealed the occurrence of cyanobacteria in deep subterranean habitats (Hubalek et al. 2016; Kormas et al. 2003; Rastogi et al. 2010); however, only Hubalek and co-workers have attempted to discuss their origin. They postulated that thousands of years ago, a bloom of aquatic cyanobacteria was trapped in a groundwater aquifer with no further connection to the surface. This strongly differs from the work executed by Puentes-Sanches and collaborators (2018), which analyzed rock sample instead of groundwater, finding out cyanobacteria lineages that are endolithic rather than aquatic, possibly responsible for the colonization of deep subsurface environments.

Groundwater is an important resource which should be protected from contamination and receives an extra attention for potential spots of cyanobacteria harmful formation. Groundwater source that is geographically close to water surface which has been often experiencing eutrophics events could be a rich tool to understand the transportation of certain biomass/molecules in the hydrological process. We cannot exclude the possibility that cyanobacteria produce cyanotoxins in agricultural soils, which are then transported through the vadose zone to subsurface drainage and groundwater receptors, as well as potentially to downstream aquatic ecosystems, but the transport processes and magnitude of the transfer are still unknown.

1.5 CyanoHABs' Control

In a scenario where the magnitude of cyanoHABs can become more frequently, the possibility to contamination of water surface as well as groundwater, dams, mega-reservoirs, and secondary house water tanks, causing a public health threat to communities around the world, especially the ones that experience a disordered growth, is real. Following this train of thought, novel approaches for the long-term control of cyanoHABs need to be developed including not only the macroscale control, such as overload nutrients, temperature, pH, or hydrology waterways, but also detailed understanding of microscale ongoings placed on quantifying the ecophysiological mechanisms involving cyanobacteria and the microbial community (Wells et al. 2020). Biotic associations that affect the bloom development, persistence, and toxin production will be crucial to better understanding the new scenario.

In terms of improvement of water treatment system, ideally, the containment of cyanobacteria might need to be the last line of defense from cyanotoxin, to ensure the provision of safe, clean, and quality water for population. For most of the cases, cyanotoxins are intracellular molecules, except cylindrospermopsin, that can occur in the solution. Thus, any physical process that would remove cyanobacterial cells keeping their integrity will offer an effective barrier to cyanotoxins. However, this is not in line with reality. Currently, the most popular method for cyanotoxin removal is pre-oxidation. The technique is applied for other treatment goals such as manganese removal or improvement of coagulation. It can have a range of potential risks of cell consequences on cyanobacteria cells, from minor wall cell damage to cell death and lysis, leading to cyanotoxin release, exposing populations even on a nanometric scale, to the most toxic levels (Chorus and Welker 2021; Pietsch et al. 2002).

Therefore, could we be consuming cyanotoxins in our water without knowing it? The recent work of guanitoxin biosynthesis confirmed that the answer for this question might be yes. Querying cyanobacteria bloom environmental samples using the genetic information of GNT biosynthetic genes, the authors showed that one of the most lethal cyanotoxins, yet long-term undetected, was being produced for multiple years in different locations in the United States and Latin America (Lima et al. 2022). And what are the consequences for a possible chronic exposure for cyanotoxins? This question remains unknown.

In a climate crisis scenario and intensive anthropogenic activity against the environment, harmful blooms' events will be increasingly frequent. Despite a long history of developing methods for cyanotoxin detections, an effective and rapid tool to evaluate and predict cyanobacteria risk in a harmful event is still a challenge for the scientific community and water management agencies. Currently, we understand best the effects of acute toxicity, but almost nothing is known about the risks of chronic subacute levels of toxin exposure.

References

Abesh BF, Liu G, Liu G, Vázquez-Ortega A, Vázquez-Ortega A, Gomezdelcampo E, Gomezdelcampo E (2019) Modeling and understanding groundwater contamination caused by cyanotoxins from harmful algal blooms in Lake Erie. https://doi.org/10.1130/abs/2019sc-326944

Adak S, Lukowski AL, Schäfer RJB, Moore BS (2022) From tryptophan to toxin: nature's convergent biosynthetic strategy to aetokthonotoxin. J Am Chem Soc 144:2861–2866. https://doi.org/10.1021/jacs.1c12778

Antunes JT, Leão PN, Vasconcelos VM (2015) Cylindrospermopsis raciborskii: review of the distribution, phylogeography, and ecophysiology of a global invasive species. Front Microbiol 6:1–13. https://doi.org/10.3389/fmicb.2015.00473

Bothe H, Tripp HJ, Zehr JP (2010) Unicellular cyanobacteria with a new mode of life: the lack of photosynthetic oxygen evolution allows nitrogen fixation to proceed. Arch Microbiol 192:783–790. https://doi.org/10.1007/s00203-010-0621-5

Breinlinger S, Phillips TJ, Haram BN, Mareš J, Martínez Yerena JA, Hrouzek P, Sobotka R, Henderson WM, Schmieder P, Williams SM, Lauderdale JD, Wilde HD, Gerrin W, Kust A, Washington JW, Wagner C, Geier B, Liebeke M, Enke H, Niedermeyer THJ, Wilde SB (2021) Hunting the eagle killer: a cyanobacterial neurotoxin causes vacuolar myelinopathy. Science 80:371. https://doi.org/10.1126/science.aax9050

Briand E, Gugger M, François JC, Bernard C, Humbert JF, Quiblier C (2008) Temporal variations in the dynamics of potentially microcystin-producing strains in a bloom-forming Planktothrix agardhii (Cyanobacterium) population. Appl Environ Microbiol 74:3839–3848. https://doi.org/10.1128/AEM.02343-07

Carmichael WW (2001) Health effects of toxin-producing cyanobacteria: "The CyanoHABs". Hum Ecol Risk Assess 7:1393–1407

Carmichael WW, Gorham PR (1978) Anatoxins from clones of Anabaena flos-aquae isolated from lakes of western Canada. Int Vereinigung Theor Angew Limnol 21:285–295. https://doi.org/10.1080/05384680.1978.11903972

Chapman AD, Schelske CL (1997) Recent appearance of Cylindrospermopsis (Cyanobacteria) in five hypereutrophic Florida lakes. J Phycol 33:191–195. https://doi.org/10.1111/j.0022-3646.1997.00191.x

Chatziefthimiou AD, Metcalf JS, Glover WB, Banack SA, Dargham SR, Richer RA (2016) Cyanobacteria and cyanotoxins are present in drinking water impoundments and groundwater wells in desert environments. Toxicon 114:75–84. https://doi.org/10.1016/j.toxicon.2016.02.016

Chen W, Song L, Gan N, Li L (2006) Sorption, degradation and mobility of microcystins in Chinese agriculture soils: risk assessment for groundwater protection. Environ Pollut 144:752–758. https://doi.org/10.1016/j.envpol.2006.02.023

Chorus I, Welker M (2021) Toxic cyanobacteria in water a guide to their public health consequences, monitoring and management, 2nd edn. CRC Press, Boca Raton

Dittmann E, Gugger M, Sivonen K, Fewer DP (2015) Natural product biosynthetic diversity and comparative genomics of the cyanobacteria. Trends Microbiol 23:642–652. https://doi.org/10.1016/j.tim.2015.07.008

Dodds WK, Whiles MR (2020) The hydrologic cycle and physiography of groundwater habitats. In: Freshwater ecology. Academic Press, San Diego, pp 73–93. https://doi.org/10.1016/b978-0-12-813255-5.00004-1

Dorman DC, Zachary JF, Buck WB (1992) Neuropathologic findings of bromethalin toxicosis in the cat. Vet Pathol 29:139–144. https://doi.org/10.1177/030098589202900206

Driscoll CB, Meyer KA, Šulčius S, Brown NM, Dick GJ, Cao H, Gasiūnas G, Timinskas A, Yin Y, Landry ZC, Otten TG, Davis TW, Watson SB, Dreher TW (2018) A closely-related clade of globally distributed bloom-forming cyanobacteria within the nostocales. Harmful Algae 77:93–107. https://doi.org/10.1016/j.hal.2018.05.009

Eynard F, Mez K, Walther JL (2000) Risk of cyanobacterial toxins in Riga waters (Latvia). Water Res 34:2979–2988. https://doi.org/10.1016/S0043-1354(00)00042-7

Fiore MF, de Lima ST, Carmichael WW, McKinnie SMK, Chekan JR, Moore BS (2020) Guanitoxin, re-naming a cyanobacterial organophosphate toxin. Harmful Algae 92:2019–2021. https://doi.org/10.1016/j.hal.2019.101737

Frazier K, Colvin B, Styer E, Hullinger G, Garcia R (1998) Microcystin toxicosis in cattle due to overgrowth of blue-green algae. Veterinary and human toxicology. Vet Hum Toxicol 40:23

Gleick PH (2011) Water resources. In: Encyclopedia of climate and weather, 2nd edn. Oxford University Press, New York

Harke MJ, Steffen MM, Gobler CJ, Otten TG, Wilhelm SW, Wood SA, Paerl HW (2016) A review of the global ecology, genomics, and biogeography of the toxic cyanobacterium, Microcystis spp. Harmful Algae 54:4–20. https://doi.org/10.1016/j.hal.2015.12.007

Hawkins PR, Chandrasena NR, Jones GJ, Humpage AR, Falconer IR (1997) Isolation and toxicity of Cylindrospermopsis raciborskii from an ornamental lake. Toxicon 35:341–346

Hubalek V, Wu X, Eiler A, Buck M, Heim C, Dopson M, Bertilsson S, Ionescu D (2016) Connectivity to the surface determines diversity patterns in subsurface aquifers of the Fennoscandian shield. ISME J 10:2447–2458. https://doi.org/10.1038/ismej.2016.36

Huisman J, Codd GA, Paerl HW, Ibelings BW, Verspagen JMH, Visser PM (2018) Cyanobacterial blooms. Nat Rev Microbiol 16:471–483. https://doi.org/10.1038/s41579-018-0040-1

Jochimsen EM, Carmichael WW, An J, Cardoso DM, Cookson ST, Holmes CEM, Antunes MB, de Melo Filho DA, Lyra TM, Barreto VST, Azevedo SMFO, Jarvis WR (1998) Liver failure and death after exposure to microcystins at hemodialysis center. Infect Control Hosp Epidemiol 19:465–465. https://doi.org/10.1017/s0195941700087890

Kormas KA, Smith DC, Edgcomb V, Teske A (2003) Molecular analysis of deep subsurface microbial communities in Nankai trough sediments (ODP leg 190, site 1176). FEMS Microbiol Ecol 45:115–125. https://doi.org/10.1016/S0168-6496(03)00128-4

Krausfeldt LE, Farmer AT, Castro Gonzalez HF, Zepernick BN, Campagna SR, Wilhelm SW (2019) Urea is both a carbon and nitrogen source for *Microcystis aeruginosa*: tracking 13C incorporation at bloom pH conditions. Front Microbiol 10:1–12. https://doi.org/10.3389/fmicb.2019.01064

Kurmayer R, Gumpenberger M (2006) Diversity of microcystin genotypes among populations of the filamentous cyanobacteria Planktothrix rubescens and Planktothrix agardhii. Mol Ecol 15:3849–3861. https://doi.org/10.1111/j.1365-294X.2006.03044.x

Lagos N, Onodera H, Zagatto PA, Andrinolo D, Azevedo SMFQ, Oshima Y (1999) The first evidence of paralytic shellfish toxins in the freshwater cyanobacterium Cylindrospermopsis raciborskii, isolated from Brazil. Toxicon 37:1359–1373. https://doi.org/10.1016/S0041-0101(99)00080-X

Le Quesne WJF, Fernand L, Ali TS, Andres O, Antonpoulou M, Burt JA, Dougherty WW, Edson PJ, El Kharraz J, Glavan J, Mamiit RJ, Reid KD, Sajwani A, Sheahan D (2021) Is the development of desalination compatible with sustainable development of the Arabian Gulf? Mar Pollut Bull 173:112940. https://doi.org/10.1016/j.marpolbul.2021.112940

Lima ST, Fallon TR, Cordoza JL, Chekan JR, Delbaje E, Hopiavuori AR, Alvarenga DO, Wood SM, Luhavaya H, Baumgartner JT, Dörr FA, Etchegaray A, Pinto E, McKinnie SMK, Fiore MF, Moore BS (2022) Biosynthesis of guanitoxin enables global environmental detection in freshwater cyanobacteria. J Am Chem Soc 144:9372–9379. https://doi.org/10.1021/jacs.2c01424

Liu G, Schwartz FW, Wright CK, McIntyre NE (2016) Characterizing the climate-driven collapses and expansions of wetland habitats with a fully integrated surface–subsurface hydrologic model. Wetlands 36:287–297. https://doi.org/10.1007/s13157-016-0817-9

Lukowski AL, Denomme N, Hinze ME, Hall S, Isom LL, Narayan ARH (2019) Biocatalytic detoxification of paralytic shellfish toxins. ACS Chem Biol 14:941–948. https://doi.org/10.1021/acschembio.9b00123

Méjean A, Mann S, Maldiney T, Vassiliadis G, Lequin O, Ploux O (2009) Evidence that biosynthesis of the neurotoxic alkaloids anatoxin-a and homoanatoxin-a in the cyanobacterium Oscillatoria PCC 6506 occurs on a modular polyketide synthase initiated by L-proline. J Am Chem Soc 131:7512–7513. https://doi.org/10.1021/ja9024353

Messineo V, Mattei D, Melchiorre S, Salvatore G, Bogialli S, Salzano R, Mazza R, Capelli G, Bruno M (2006) Microcystin diversity in a Planktothrix rubescens population from Lake Albano (Central Italy). Toxicon 48:160–174. https://doi.org/10.1016/j.toxicon.2006.04.006

Mohamed ZA, Al Shehri AM (2009) Microcystins in groundwater wells and their accumulation in vegetable plants irrigated with contaminated waters in Saudi Arabia. J Hazard Mater 172:310–315. https://doi.org/10.1016/j.jhazmat.2009.07.010

Mohamed ZA, Alamri S, Hashem M (2022) The link between microcystin levels in groundwater and surface Nile water, and assessing their potential risk to human health. J Contam Hydrol 244: 103921. https://doi.org/10.1016/j.jconhyd.2021.103921

Neilan BA, Saker ML, Fastner J, Törökné A, Burns BP (2003) Phylogeography of the invasive cyanobacterium Cylindrospermopsis raciborskii. Mol Ecol 12:133–140. https://doi.org/10.1046/j.1365-294X.2003.01709.x

Paerl H (2008) Nutrient and other environmental controls of harmful cyanobacterial blooms along the freshwater – marine continuum. Adv Exp Med Biol 619:216–241

Paerl HW, Huisman J (2009) Climate change: a catalyst for global expansion of harmful cyanobacterial blooms. Environ Microbiol Rep 1:27–37. https://doi.org/10.1111/j.1758-2229.2008.00004.x

Paerl HW, Paul VJ (2012) Climate change: links to global expansion of harmful cyanobacteria. Water Res 46:1349–1363. https://doi.org/10.1016/j.watres.2011.08.002

Pancrace C, Barny MA, Ueoka R, Calteau A, Scalvenzi T, Pédron J, Barbe V, Piel J, Humbert JF, Gugger M (2017) Insights into the Planktothrix genus: genomic and metabolic comparison of benthic and planktic strains. Sci Rep 7:1–10. https://doi.org/10.1038/srep41181

Pearson LA, Dittmann E, Mazmouz R, Ongley SE, D'Agostino PM, Neilan BA (2016) The genetics, biosynthesis and regulation of toxic specialized metabolites of cyanobacteria. Harmful Algae 54:98–111. https://doi.org/10.1016/j.hal.2015.11.002

Pietsch J, Bornmann K, Schmidt W (2002) Relevance of intra- and extracellular cyanotoxins for drinking water treatment. Acta Hydrochim Hydrobiol 30:7–15

Puente-Sánchez F, Arce-Rodríguez A, Oggerin M, García-Villadangos M, Moreno-Paz M, Blanco Y, Rodríguez N, Bird L, Lincoln SA, Tornos F, Prieto-Ballesteros O, Freeman KH, Pieper DH, Timmis KN, Amils R, Parro V (2018) Viable cyanobacteria in the deep continental subsurface. Proc Natl Acad Sci U S A 115:10702–10707. https://doi.org/10.1073/pnas.1808176115

Rastogi G, Osman S, Kukkadapu R, Engelhard M, Vaishampayan PA, Andersen GL, Sani RK (2010) Microbial and mineralogical characterizations of soils collected from the deep biosphere of the former Homestake gold mine, South Dakota. Microb Ecol 60:539–550. https://doi.org/10.1007/s00248-010-9657-y

Saker ML, Nogueira ICG, Vasconcelos VM, Neilan BA, Eaglesham GK, Pereira P (2003) First report and toxicological assessment of the cyanobacterium Cylindrospermopsis raciborskii from Portuguese freshwaters. Ecotoxicol Environ Saf 55:243–250. https://doi.org/10.1016/S0147-6513(02)00043-X

Sivonen K (2009) Encyclopedia of microbiology. Elsevier Scientific, Amsterdam

Stüken A, Rücker J, Endrulat T, Preussel K, Hemm M, Nixdorf B, Karsten U, Wiedner C (2006) Distribution of three alien cyanobacterial species (Nostocales) in northeast Germany: Cylindrospermopsis raciborskii, Anabaena bergii and Aphanizomenon aphanizomenoides. Phycologia 45:696–703. https://doi.org/10.2216/05-58.1

Thomas NJ, Meteyer CU, Sileo L (1998) Epizootic vacuolar myelinopathy of the central nervous system of bald eagles (Haliaeetus leucocephalus) and American coots (Fulica americana). Vet Pathol 35:479–487

Ueno Y, Nagata S, Tsutsumi T, Hasegawa A, Watanabe MF, Park HD, Chen GC, Chen G, Yu SZ (1996) Detection of microcystins, a blue-green algal hepatotoxin, in drinking water sampled in Haimen and Fusui, endemic areas of primary liver cancer in China, by highly sensitive immunoassay. Carcinogenesis 17:1317–1321. https://doi.org/10.1093/carcin/17.6.1317

Valério E, Chaves S, Tenreiro R (2010) Diversity and impact of prokaryotic toxins on aquatic environments: a review. Toxins (Basel) 2:2359–2410. https://doi.org/10.3390/toxins2102359

Wang DZ, Zhang SF, Zhang Y, Lin L (2016) Paralytic shellfish toxin biosynthesis in cyanobacteria and dinoflagellates: a molecular overview. J Proteome 135:132–140. https://doi.org/10.1016/j.jprot.2015.08.008

Welker M, Von Döhren H (2006) Cyanobacterial peptides - nature's own combinatorial biosynthesis. FEMS Microbiol Rev 30:530–563. https://doi.org/10.1111/j.1574-6976.2006.00022.x

Wells ML, Karlson B, Wulff A, Kudela R, Trick C, Asnaghi V, Berdalet E, Cochlan W, Davidson K, De Rijcke M, Dutkiewicz S, Hallegraeff G, Flynn KJ, Legrand C, Paerl H, Silke J, Suikkanen S, Thompson P, Trainer VL (2020) Future HAB science: directions and challenges in a changing climate. Harmful Algae 91:101632. https://doi.org/10.1016/j.hal.2019.101632

WHO (2006) Guidelines for drinking-water quality, 3rd edn. World Health Organization, Geneva

Wiese M, D'Agostino PM, Mihali TK, Moffitt MC, Neilan BA (2010) Neurotoxic alkaloids: saxitoxin and its analogs. Mar Drugs 8:2185–2211. https://doi.org/10.3390/md8072185

Wilhelm SW, Bullerjahn GS, McKay RML (2020) The complicated and confusing ecology of microcystis blooms. MBio 11:1–5. https://doi.org/10.1128/MBIO.00529-20

Winter TC, Harvey JW, Franke OL, Alley WM (1998) Ground water and surface water - a single resource - U.S. Geological Survey Circular 1139. USGS Publications Warehouse, vol 1, p 79

Yang Z, Kong F, Zhang M (2016) Groundwater contamination by microcystin from toxic cyanobacteria blooms in Lake Chaohu, China. Environ Monit Assess 188:280. https://doi.org/10.1007/s10661-016-5289-0

Zareian MJ, Eslamian S, Ostad-Ali-Askari K (2019) Global warming and sustainable development. In: Encyclopedia of sustainability in higher education, vol 1–13. Springer, Cham. https://doi.org/10.1007/978-3-319-63951-2_470-1

Zhang X, Ye X, Chen L, Zhao H, Shi Q, Xiao Y, Ma L, Hou X, Chen Y, Yang F (2020) Functional role of bloom-forming cyanobacterium Planktothrix in ecologically shaping aquatic environments. Sci Total Environ 710:136314. https://doi.org/10.1016/j.scitotenv.2019.136314

Zhang Y, Husk BR, Duy SV, Dinh QT, Sanchez JS, Sauvé S, Whalen JK (2021) Quantitative screening for cyanotoxins in soil and groundwater of agricultural watersheds in Quebec, Canada. Chemosphere 274:129781. https://doi.org/10.1016/j.chemosphere.2021.129781

On the Pigment Profile of 12 Cyanobacteria Isolated from Unpolluted and Polluted Habitats of Southwest India

2

Kodandoor Sharathchandra and Kandikere R. Sridhar

Abstract

Cultures of 12 species of cyanobacteria isolated from different habitats were assessed to quantify 17 pigments by high-performance liquid chromatography (13 pigments) and spectrophotometric (4 pigments) methods. Three pigments were common to all cyanobacteria (chlorophyll-a, lycopene, and β-carotene), while none of them produced four pigments (chlorophyll-b, chlorophyll-c, fucoxanthin, and peridinin). Based on the pigment profile of cyanobacteria, four species such as *Jaaginema pseudogeminatum*, *Leptolyngbya fragilis*, *Nostoc oryzae*, and *Planktolyngbya limnetica* have been considered as creative species owing to their efficiency of production of different pigments in substantial quantities. These species were isolated from stressed habitats (thermal spring or domestic sewage) except for *N. oryzae* (temple tank). Their pigment production potential could be enhanced by exposure to different stressed nutritional and physicochemical regimes.

Keywords

Carotene · Chlorophyll · Creative species · Lutein · Lycopene · Peridinin · Phaeophytin · Phycobilin · Xanthin · Xanthophyll

K. Sharathchandra · K. R. Sridhar (✉)
Department of Biosciences, Mangalore University, Mangaluru, Karnataka, India

B. Neilan et al. (eds.), *Cyanobacterial Biotechnology in the 21st Century*,
https://doi.org/10.1007/978-981-99-0181-4_2

2.1 Introduction

Cyanobacteria are Gram-negative, oxygenic, and photosynthetic prokaryotes widely distributed in different ecosystems (terrestrial, freshwaters, and marine habitats) owing to their capability to withstand extreme habitats like temperature, pH, and salinity. Morphologically, they are diverse (unicellular, multicellular, filamentous, heterocystous, and non-heterocystous forms) and live independently as well as in association with other biotrophs (bryophytes, pteridophytes, angiosperms, and lichens). They constitute the baseline of the food web in different habitats as they play important roles in primary production (Klemer 1990). Owing to their occupation and functions in diverse habitats, they possess valuable traits of applications in agriculture (food, feed, and fertilizer), biotechnology (fatty acids, vitamins, antioxidants, pigments, enzymes, and toxins), industries (biofuels, pharmaceuticals, and cosmeceuticals), and environmental (pollution control, abatement, and bioremediation) facets (Thajuddin and Subramanian 2005; Anupama 2000; Shetty and Krishnakumar 2020).

One of the important properties of cyanobacteria is the production of various pigments of biological and industrial significance (Duangsee et al. 2009; Prasanna et al. 2010; Kumar et al. 2011; Hifney et al. 2013; Hemlata et al. 2018). Photosynthetic pigments of cyanobacteria are accountable for photosynthetic activities (absorption, transfer, and transform light energy into biomass) (Eullaffroy and Vernet 2003). Pigments of cyanobacteria are useful in food, pharmaceutical, and cosmeceutical industries owing to their antioxidant and anti-inflammatory potential (Hemlata et al. 2018). Studies have been focused on the production of different pigments by individual species of cyanobacteria for extraction, purification, commercial exploitation, and cultivation, to assess the impact of abiotic conditions, and to control their blooms (Duangsee et al. 2009; Sarada et al. 1999; Hong et al. 2010; Hemlata et al. 2011; Jodłowsak and Latała 2011; Kumar et al. 2011; Hifney et al. 2013). Although cyanobacteria are known to produce value-added pigments, comprehensive information about their capability of production of a wide range of pigments is not available in a nutshell. Hence, the present study envisaged a quantitative analysis of 17 pigments of 12 cultured cyanobacteria isolated from unpolluted and polluted habitats of Southwest India in view of their future commercial utilization.

2.2 Cultures of Cyanobacteria

To assess the pigments, 12 species of cyanobacterial cultures (8 non-heterocystous and 4 heterocystous forms) isolated from freshwater and polluted habitats of Southwest India were selected (Table 2.1). They were isolated, cultured, and grown in the laboratory on dry agar followed by BG11 broth with nitrate (Stanier et al. 1971; Dor 1987). Their identity was confirmed based on morphological features (Desikachary 1959; Anagnostidis and Komarek 1998). The cultures were maintained at 26 ± 2 °C under 14:10 h light and dark cycles (illumination, 2000 lux). They were harvested

Table 2.1 Cyanobacteria assessed for pigment analysis (*, heterocystous form)

	Species code in Fig. 2.1	Geographic origin	Geographic coordinates
**Anabaena variabilis* Kützing ex Bornet & Flahault	Anva	Fishing tank, Gajanoor	13°50′N, 75°31′E
**Calothrix fusca* (Kutz.) B. & F.	Cafu	Thermal spring, Panekal	12°54′N, 75°17.5′E
Geitlerinema calcuttense (Biswas) Anagnostidis	Geca	Dairy effluent, Mangalore	12°53′N, 74°53′E
Jaaginema pseudogeminatum Anagnostidis & Komarek	Japs	Domestic sewage, Mangalore	12°52′N, 74°51′E
Leptolyngbya fragilis (Gomont) Anagnostidis & Komarek	Lefr	Thermal spring, Panekal	12°54′ N, 75°17.5′E
**Nostoc oryzae* (Fritsch) Komárek & Anagnostidis	Noor	Temple tank, Halebeedu	13°12′N, 75°59′E
Oxynema acuminatum (Gomont) Chatchawan, Komarek, Strunecky, Smarda & Peerapornpisal	Oxac	Tank, Malavalli	12°22′N, 77°4′E
Phormidium chalybeum (Mertens ex Gomont) Anagnostidis & Komarek	Phcha	Pharmaceutical effluent, Mangalore	12°51′N, 74°51′E
Phormidium chlorinum (Kützing ex Gomont) Umezaki & Watanabe	Phchl	Domestic sewage, Mangalore	12°52′N, 74°51′E
Phormidium lucidum Kützing ex Gomont	Phlu	Bhadra reservoir, Shimoga	13°42′N, 75°38′E
Planktolyngbya limnetica (Lemmermann) J. Komárková-Legnerová & G. Cronberg	Plli	Thermal spring, Panekal	12°54′N, 75°17.5′E
**Scytonema bohnerii* Schmidle	Scbo	Thermal spring, Panekal	12°54′N, 75°17.5′E

during the exponential phase (12 days) by centrifugation (3000 rpm, 10 min), and the wet biomass was subjected for pigment analysis.

2.3 Pigment Analysis

Total carotenoids present in the samples were extracted using 90% acetone at 4 °C in a refrigerator overnight. After extraction, samples were centrifuged (5000 rpm, 10 min), and the absorbance of the solution was assessed (480 and 510 nm) by UV-visible spectrophotometer (Implen, Schatzbogen 52, Germany), using 90% acetone as blank (Parsons and Strickland, 1963). The absorbance was also measured at 750 nm, which was subtracted from the values at 480 and 510 nm to minimize the error. The concentration of total carotenoids was estimated:

$$\text{Total carotenoids } (\mu g/mL) = 7.6\ (E480 - E750) - 1.49\ (E510 - E750)$$

Phycobilins (phycocyanin, allophycocyanin, and phycoerythrin) were determined by the spectrophotometric assay (Bennet and Bogorad, 1973; Abalde et al. 1998). Pellets obtained by centrifugation of a defined volume of cyanobacterial suspensions were suspended in phosphate buffer (in 5 ml, 50 mM with pH 7). The contents were subjected to freeze-thaw cycles (5–6 times) followed by centrifugation for extraction. The absorbance of pooled supernatant was assessed (565, 620, and 650 nm) with phosphate buffer as blank. The amount of phycocyanin, allophycocyanin, and phycoerythrin was estimated (De Marsac and Houmard 1988):

$$\text{Phycocyanin } (\mu g/mL) = \left(\frac{A620 - (0.7 \times A650)}{17.38}\right)$$

$$\text{Allophycocyanin } (\mu g/mL) = \left(\frac{A650 - (0.208 \times A620)}{15.09}\right)$$

$$\text{Phycoerythrin } (\mu g/mL) = \left(\frac{A565 - 2.41\ (\text{phycocyanin}) - 0.849\ (\text{allophyocyanin})}{19.62}\right)$$

Cultures which were grown in BG11 broth (100 mL) in replicates were filtered through Millipore filters (porosity, 1.2 μm) under low vacuum ($P < 70$ mm Hg) and kept at −80 °C until analysis to prevent the degradation of pigments. Pigments from filters were extracted (after attaining laboratory temperature) using 90% acetone (5 mL) and stored in dark (4 °C, 1 h). The consortium (filter + mucus + acetone) was agitated and vortexed prior to collecting the liquid phase by centrifugation (5000 rpm, 5 min). The supernatant was preserved at 4 °C in the dark for up to 6 h for extraction of pigment and clarified the extract by centrifugation under the same conditions. Injected the supernatant into a reverse-phase HPLC (model # HP 1100 series: Column Spherisorb ODS2–25 cm × 4.6 mm ID, 5 μm particle size and a loop 50 μL). The detection of pigment was carried out at 440 nm according to Wright et al. (1991) by maintaining the solvent, flow rate, and gradient. The calibration was accomplished using external standards except for phaeophytin-a (Sigma and the International Agency for ^{14}C determination, Denmark). Phaeophytin-a was assessed by the acidification (1 M HCl) of chlorophyll-a followed by the neutralization of the mixture using Na_2CO_3 and refiltration via Millipore filters (porosity, 1.2 μm) to clear up the extract.

2.4 Discussion

Among the 17 pigments assessed, chlorophyll-a, lycopene, β-carotene, carotenoids, phycocyanin, allophycocyanin, and phycoerythrin were present in all cyanobacterial species studied, while phaeophytin-a was present in all species except for *C. fusca* (Table 2.2, Fig. 2.1a–d). Four pigments were not represented by any of the species

Table 2.2 Pigment profile (μg/mL) of 12 cyanobacteria based on HPLC analysis ($n = 3$, mean ± SD; ND, not detectable; *, heterocystous form)

	Planktolyngbya limnetica	*Geitlerinema calcuttense*	**Nostoc oryzae*	*Leptolyngbya fragilis*	*Phormidium chlorinum*	*Jaaginema pseudogeminatum*
Chlorophyll-a	7.52 ± 0.05	6.82 ± 0.10	6.20 ± 0.02	7.20 ± 0.12	6.52 ± 0.02	8.50 ± 0.15
Lycopene	6.32 ± 0.02	5.45 ± 0.12	4.10 ± 0.20	5.68 ± 0.02	4.10 ± 0.22	7.20 ± 0.12
β-Carotene	8.68 ± 0.12	6.45 ± 0.05	7.42 ± 0.04	8.45 ± 0.05	5.50 ± 0.05	8.82 ± 0.06
Phaeophytin-a	2.45 ± 0.25	1.20 ± 0.30	3.22 ± 0.02	2.20 ± 0.03	2.12 ± 0.01	1.48 ± 0.02
Lutein	7.45 ± 0.25	2.28 ± 0.04	2.55 ± 0.15	6.50 ± 0.05	3.20 ± 0.01	4.75 ± 0.05
Zeaxanthin	2.82 ± 0.05	1.45 ± 0.05	1.80 ± 0.04	2.62 ± 0.02	1.50 ± 0.01	1.80 ± 0.04
Cis-β-carotene	0.58 ± 0.02	0.32 ± 0.02	0.95 ± 0.92	0.46 ± 0.08	0.45 ± 0.02	0.42 ± 0.02
Astaxanthin	0.88 ± 0.02	ND	0.78 ± 0.02	ND	0.55 ± 0.01	ND
Myxoxanthophyll	1.04 ± 0.02	1.20 ± 0.02	ND	0.80 ± 0.04	ND	ND
Chlorophyll-b	ND	ND	ND	ND	ND	ND
Chlorophyll-c	ND	ND	ND	ND	ND	ND
Fucoxanthin	ND	ND	ND	ND	ND	ND
Peridinin	ND	ND	ND	ND	ND	ND
Total number of pigments	9	8	8	8	8	7
Ratio of carotenoids-chlorophyll-a	1.15	1.57	1.08	1.47	2.23	1.48
	**Anabaena variabilis*	*Phormidium chalybeum*	**Scytonema bohnerii*	*Oxynema acuminatum*	*Phormidium lucidum*	**Calothrix fusca*
Chlorophyll-a	5.20 ± 0.05	4.40 ± 0.04	3.70 ± 0.01	4.20 ± 0.02	3.20 ± 0.01	2.82 ± 0.02
Lycopene	4.55 ± 0.05	2.40 ± 0.02	1.50 ± 0.02	3.50 ± 0.02	2.52 ± 0.05	1.90 ± 0.03
β-Carotene	6.80 ± 0.20	5.45 ± 0.25	4.25 ± 0.20	3.50 ± 0.05	2.80 ± 0.80	3.32 ± 0.10
Phaeophytin-a	1.92 ± 0.02	1.20 ± 0.03	0.72 ± 0.03	1.22 ± 0.02	0.50 ± 0.02	ND
Lutein	ND	3.45 ± 0.05	ND	0.75 ± 0.05	ND	ND
Zeaxanthin	1.45 ± 0.25	ND	0.52 ± 0.02	ND	ND	ND
Cis-β-carotene	ND	ND	ND	ND	ND	ND

(continued)

Table 2.2 (continued)

Astaxanthin	ND	ND	0.46 ± 0.01	ND	ND	ND
Myxoxanthophyll	0.45 ± 0.06	0.65 ± 0.02	ND	ND	0.22 ± 0.06	ND
Chlorophyll-b	ND	ND	ND	ND	ND	ND
Chlorophyll-c	ND	ND	ND	ND	ND	ND
Fucoxanthin	ND	ND	ND	ND	ND	ND
Peridinin	ND	ND	ND	ND	ND	ND
Total number of pigments	6	6	6	5	5	3
Ratio of carotenoids-chlorophyll-a	1.56	2.03	2.20	2.46	2.54	1.82

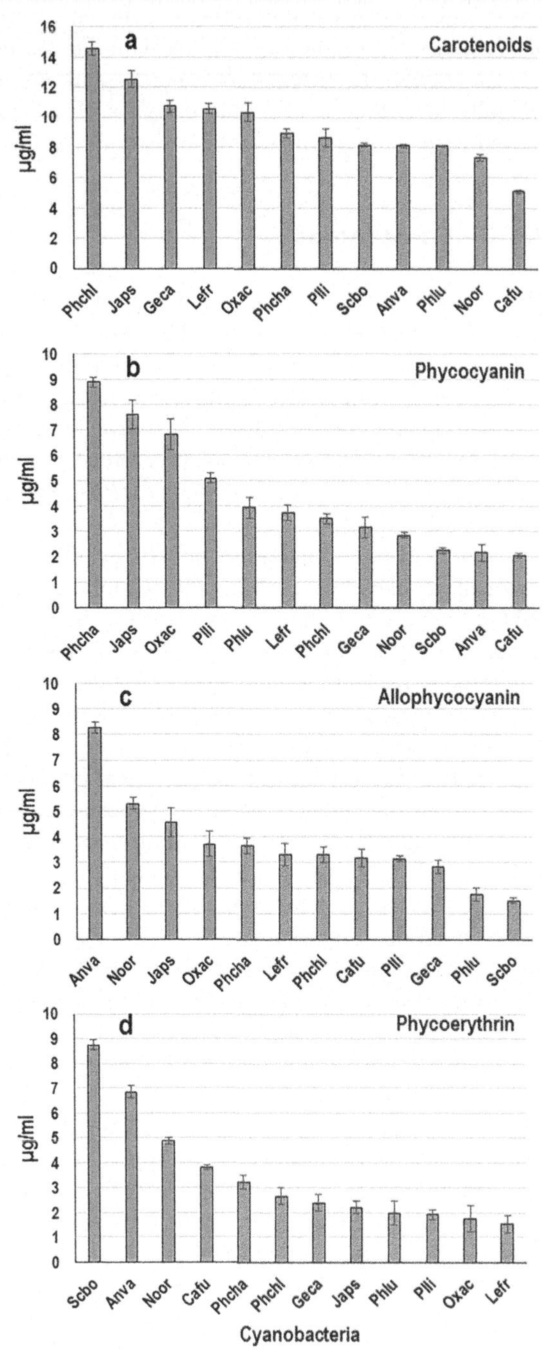

Fig. 2.1 Carotenoids (**a**), phycocyanin (**b**), allophycocyanin (**c**), and phycoerythrin (**d**) in 12 cyanobacteria assessed by the spectrophotometric method in decreasing order ($n = 3$, mean ± SD)

studied (chlorophyll-b, chlorophyll-c, fucoxanthin, and peridinin). The total number of pigments represented was highest in *Pl. limnetica* (13 pigments) followed by *G. calcuttense*, *L. fragilis*, *N. oryzae*, and *Ph. chalybeum* (12 pigments each). Overall, the carotenoids, chlorophyll-a, lycopene, β-carotene, and phycobilins were present in high quantity in at least six species studied.

2.4.1 Carotenoids

The overall carotenoid content in our study was higher in non-heterocystous forms rather than heterocystous forms. They were highest in *Ph. chlorinum* (14.6 μg/ml) followed by *J. pseudogeminatum*, *G. calcuttense*, *L. fragilis*, and *O. acuminatum* (range, 10.4–12.6 μg/mL), while it was lowest in *C. fusca* (5.1 μg/mL) (Fig. 2.1a). *J. pseudogeminatum*, *L. fragilis*, *N. oryzae*, and *Pl. limnetica* revealed the presence of carotenoid pigments especially β-carotene at high concentrations (7.4–8.8 μg/mL) (Table 2.1). These species as well as *Ph. chlorinum* showed a high quantity of chlorophyll-a.

Carotenoids are vital pigments in cyanobacteria as they serve as accessory pigments to elevate the ability of cyanobacteria to harness blue wavelengths of light, which are not directly absorbed by chlorophyll-a. They are known to protect against harmful photooxidation, especially prevent the oxidation of chlorophyll (Fresnedo et al. 1991; Linda and Lee 2000). Although carotenoids are not involved in photochemical reactions, their chief function is to protect the photosynthetic machinery from oxidative damages owing to their antioxidant capacities (Wada et al. 2013). The carotenoids have several commercial applications as natural food colorants, enhancement of color of the flesh of salmonids and improving the health/fertility of cattle (Emodi 1978). Among carotenoids, β-carotene is the most abundant and not only plays as a photosynthetic pigment but also serves as one of the important non-enzymatic antioxidants to defend cyanobacterial cells from destruction by external or internal hostile conditions (Bryant 1994). Chlorophyll-a involves the absorption of light, transmission and the majority of which participate in the transformation of light in cyanobacteria (Papageorgiou 1996; Eullaffroy and Vernet 2003).

Maximum carotenoid production is necessary to build up cyanobacterial biomass under optimum culture conditions (Olaizola and Duerr 1990). Increased carotenoid content at high light intensity can be explained by the protective roles of chlorophyll-a (Foyer et al. 1994; Chaneva et al. 2007). At the high light intensities, the quantity of chlorophyll-a dramatically decreases in cyanobacteria, whereas carotenoid content increases (Kopecky et al. 2000). In our study, all species showed a low quantity of chlorophyll-a than the carotenoids (Table 2.2). Carotenoid-to-chlorophyll-a ratio helps to understand the response of cyanobacteria to the changing light conditions (Jodłowsak and Latała 2011). In our study, this ratio was highest in *Ph. lucidum* (2.5) with the lowest chlorophyll-a content (3.2 μg/ml), while the ratio was least in *N. oryzae* (1.1) with 6.2 μg/ml chlorophyll-a content.

2.4.2 Phycobilins

The water-soluble photosynthetic phycobilins in cyanobacteria are mainly composed of three pigments (phycocyanin, allophycocyanin, and phycoerythrin). Among phycobilins in cyanobacteria, phycocyanin (range, 2.1–8.9 μg/mL) content was higher compared to phycoerythrin (range, 1.6–8.8 μg/mL) as well as allophycocyanin (range, 1.5–8.3 μg/mL) (Fig. 2.1b, c). Phycocyanin content was high in *Ph. chalybeum* (8.9 μg/mL) followed by *J. pseudogeminatum* (7.6 μg/mL), *O. acuminatum* (6.8 μg/mL), and *P. limnetica* (5.1 μg/mL), while it was least in *C. fusca* (2.1 μg/mL) (Fig. 2.1b). Allophycocyanin content was maximum in *A. variabilis* (8.3 μg/mL) followed by *N. oryzae* (5.3 μg/mL) and *J. pseudogeminatum* (4.6 μg/mL), while it was least in *S. bohnerii* (1.5 μg/mL) (Fig. 2.1c). Phycoerythrin content was highest in *S. bohnerii* (8.8 μg/mL) followed by *A. variabilis* (6.9 μg/mL) and *N. oryzae* (4.9 μg/mL), while it was least in *L. fragilis* (1.6 μg/mL) (Fig. 2.1d). Interestingly, the heterocystous forms in our study possess higher phycoerythrin than other species. Among the cyanobacteria studied, *Ph. chalybeum* is one of the top five phycobilin producers. Similarly, *A. variabilis* and *N. oryzae* were common among the top three producers of allophycocyanin as well as phycoerythrin.

The total phycobilins in phycobilisomes are attached to the surface of the thylakoid to facilitate photosynthesis (Hemlata et al. 2011). Many cultured species under optimum conditions consist 4% of total chlorophyll on a dry mass basis. The chlorophyll serves as a chelating agent, which is used in ointments and pharmaceutical products (e.g., treatment of the liver for recovery and ulcer) (El-Sayed et al. 2010). In addition, it also repairs cells, increases hemoglobin content in blood, and hastens the growth of cells. Tiwari et al. (2005) found maximum content of chlorophyll in *Phormidium* sp. (10.8 μg/mL), while minimum in *Nostoc linckia* (0.7 μg/mL). Our study corroborates with the results demonstrated by Tiwari et al. (2005).

Cyanobacterial phycobiliproteins are useful in the food, pharmaceutical, and cosmeceutical industries as a natural colorant. They also serve in diagnostics as fluorescent reagents. The phycofluors, mainly the allophycocyanin and phycoerythrin, are employed in fluorescence-activated cell sorting, histochemistry, and flow cytometry (Patterson 1996). In the current study, the phycobilins were found in better quantities, and such species could be used in specific pharmaceutical applications (e.g., *A. variabilis, J. pseudogeminatum, N. oryzae, O. acuminatum*, and *Ph. chalybeum*). Phycocyanin has many applications as a food colorant, a nutraceutical, and an immunomodulator and in immunodiagnostics in cancer therapy (Benedetti et al. 2004). Hifney et al. (2013) demonstrated the enhancement of phycobiliproteins by the induction of stress in *Spirulina* sp. (increased salt and decreased P, N, and S contents).

It has been suggested by Kobayashi et al. (1992) that cyanobacteria prefer low light intensities to stimulate the synthesis of phycobiliproteins owing to their low rate of specific maintenance energy and their pigment composition. In our study, phycoerythrin content was high in heterocystous nitrogen-fixing cyanobacteria than non-heterocystous forms. This result corroborates with the earlier reports by

Rodriguez et al. (1989) as well as Simeunovic et al. (2013). Phycoerythrin serves as a natural antioxidant in promoting human health and shows substantial antimicrobial activity (Hemlata et al. 2018).

2.4.3 Other Pigments

In the present study, chlorophyll-a, β-carotene, lycopene, and lutein contents were in higher quantities; phaeophytin-a and zeaxanthin were found in moderate quantities; and astaxanthin, cis-β-carotene, and myxoxanthophyll were found in a few species at low concentrations. The rest of the four pigments were not detectable (chlorophyll-b, chlorophyll-c, fucoxanthin, and peridinin). Astaxanthin was produced only in four species (*Ph. chlorinum*, *P. limnetica*, *N. oryzae*, and *S. bohnerii*). *P. limnetica*, *L. fragilis*, *J. pseudogeminatum*, *Ph. chlorinum*, and *N. oryzae* possess β-carotene, lycopene, and lutein in moderate quantities, whereas these pigments were in low quantities in *O. acuminatum*, *Ph. lucidum*, and *C. fusca*.

Significant quantities of β-carotene, lycopene, and lutein in cyanobacteria lead to considerable antioxidant potential (Subhashini et al. 2004; Velvizhi et al. 2011). These pigments were detected at moderate quantities in cyanobacteria isolated from a thermal spring (*P. limnetica* and *L. fragilis*) as well as sewage (*J. pseudogeminatum* and *Ph. chlorinum*). The β-carotene and astaxanthin are known to protect the living cells against oxidation (e.g., dissipating the energy as heat, quenching of singlet oxygen, and free radical scavenging). Zeaxanthin is an important pigment used to treat neurological diseases, allergies, and cancer (Bouyahya et al. 2021). *P. limnetica*, *J. pseudogeminatum*, and *L. fragilis* were the top three producers of zeaxanthin in our study. Astaxanthin serves as an extremely powerful antioxidant owing to its molecular structure. It very effectively quenches the singlet oxygen, scavenges free radicals, and effectively breaks the chain reaction of peroxide (Borowitzka 1995; Aboul-Enein et al. 2003). Among the top 3 producers of 13 pigments, *P. limnetica*, *L. fragilis*, *J. pseudogeminatum*, and *N. oryzae* could be considered as creative species as they are the top 3 producers of 9, 8, 7, and 6 pigments, respectively (Table 2.3). Interestingly, among these species, except for *N. oryzae*, the rest of them were isolated from stressed habitats (thermal spring and domestic sewage).

2.4.4 Conclusions

Cyanobacteria are well known for the production of a wide variety of pigments. Twelve cyanobacterial isolates from unpolluted and polluted habitats of Southwest India exhibited the production of 13 pigments. They produced substantial quantities of carotenoids, phycobilins, lycopene, and lutein. Based on the efficiency of pigment production, *Jaaginema pseudogeminatum*, *Leptolyngbya fragilis*, *Nostoc oryzae*, and *Planktolyngbya limnetica* could be considered as creative species. There is

Table 2.3 Top 3 pigment producers among 12 cyanobacteria

	Cyanobacteria
Carotenoids	*Phormidium chlorinum*, *Jaaginema pseudogeminatum*, and *Geitlerinema calcuttense*
Phycocyanin	*Phormidium chalybeum*, *Jaaginema pseudogeminatum*, and *Oxynema acuminatum*
Allophycocyanin	*Anabaena variabilis*, *Nostoc oryzae*, and *Oxynema acuminatum*
Phycoerythrin	*Scytonema bohnerii*, *Anabaena variabilis*, and *Nostoc oryzae*
Chlorophyll-a	*Jaaginema pseudogeminatum*, *Planktolyngbya limnetica*, and *Leptolyngbya fragilis*
Lycopene	*Jaaginema pseudogeminatum*, *Planktolyngbya limnetica*, and *Leptolyngbya fragilis*
β-Carotene	*Jaaginema pseudogeminatum*, *Planktolyngbya limnetica*, and *Leptolyngbya fragilis*
Phaeophytin-a	*Nostoc oryzae*, *Planktolyngbya limnetica*, and *Leptolyngbya fragilis*
Lutein	*Planktolyngbya limnetica*, *Leptolyngbya fragilis*, and *Jaaginema pseudogeminatum*
Zeaxanthin	*Planktolyngbya limnetica*, *Jaaginema pseudogeminatum*, and *Leptolyngbya fragilis*
Cis-β-carotene	*Nostoc oryzae*, *Planktolyngbya limnetica*, and *Leptolyngbya fragilis*
Astaxanthin	*Planktolyngbya limnetica*, *Nostoc oryzae*, and *Phormidium chlorinum*
Myxoxanthophyll	*Planktolyngbya limnetica*, *Nostoc oryzae*, and *Leptolyngbya fragilis*

ample scope to improve the production of industrially valued pigments by cyanobacteria by manipulation of nutrients, quality light, and light intensity.

Acknowledgments The authors acknowledge the support of Mangalore University and the Department of Biosciences for laboratory facilities. The authors are thankful to Dr. Mahadevakumar, Karnataka State Open University, Mysore, for constructive suggestions.

References

Abalde J, Betancourt L, Torres E, Cid A, Barwell C (1998) Purification and characterization of phycocyanin from the marine cyanobacterium *Synechococcus* sp. IO920. Plant Sci 136:109–120

Aboul-Enein AM, El-Baz FK, El-Baroty GS, Youssef AM, Abd El-Baky HH (2003) Antioxidant activity of algal extracts on lipid peroxidation. J Med Sci 3:87–98

Anagnostidis K, Komarek G (1998) Modern approach to the classification system of cyanobacteria. Archiv für Hydrobiol 80:372–470

Anupama PR (2000) Value added food: single cell protein. Biotechnol Adv 18:459–479

Benedetti S, Benvenutti F, Pagliarani S, Francogli S, Scoglio S, Canestrari F (2004) Antioxidant properties of a novel phycocyanin extract from the blue-green alga *Aphanizomenon flos-aquae*. Life Sci 75:2353–2362

Bennet A, Bogorad L (1973) Complementary chromatic adaptation in filamentous blue green algae. J Cell Biol 58:419–433

Borowitzka AM (1995) Microalgae as sources of pharmaceuticals and other biologically active compounds. J Appl Phycol 7:3–15

Bouyahya A, Omari NE, Hakkur M, Hachlafi NE, Sharfi S et al (2021) Sources, health benefits, and biological properties of zeaxanthin. Trends Food Sci Technol 118:519–538

Bryant DA (1994) The molecular biology of cyanobacteria. Kluwer Academic Publishers, Dordrecht, p 879

Chaneva G, Furnadzhieva S, Minkova K, Lukavsky J (2007) Effect of light and temperature on the cyanobacterium *Arthronema africanum* - a prospective phycobiliprotein-producing strain. J Appl Phycol 19:537–544

De Marsac NT, Houmard J (1988) Complementary chromatic adaptation: physiological conditions and action spectra. Meth Enzymol 167:318–328

Desikachary TV (1959) Cyanophyta. New Delhi, Indian Council of Agricultural Research, p 686

Dor I (1987) Preservation and microscopy of blue-green algae (cyanobacteria) on dry agar. Bot Mar 30:507–510

Duangsee R, Phoopat N, Ningsanond S (2009) Phycocyanin extraction from *Spirulina platensis* and extract stability under various pH and temperature. Asian J Food Ag-Ind 2:819–826

El-Sayed AB, El-Fouly MM, Abou El-Nour EAA (2010) Immobilized microalga *Scenedesmus* sp. for biological desalination of red sea water: I. Effect on growth. Nat Sci 8:69–76

Emodi A (1978) Carotenoids: properties and applications. Food Technol 32:38–42

Eullaffroy P, Vernet G (2003) The F684/F735 chlorophyll fluorescence ratio: a potential tool for rapid detection and determination of herbicide phytotoxicity in algae. Water Res 37:1983–1990

Foyer CH, Lelandais M, Kunert KJ (1994) Photooxidative stress in plants. Plant Physiol 92:696–717

Fresnedo O, Gomez R, Serra JC (1991) Carotenoid composition in the cyanobacterium *Phormidium laminosum* - effect of nitrogen starvation. FEBS Lett 282:300–304

Hemlata, Afreen S, Fatma T (2018) Extraction, purification and characterization of phycoerythrin from Michrochaete and its biological activities. Biocatal Agric Biotechnol 13:84–89

Hemlata, Pandey G, Bano F, Fatma T (2011) Studies on *Anabaena* sp. NCCU-9 with special reference to phycocyanin. J Algal Biomass Utln 2:30–51

Hifney AF, Issa AA, Fawazy MA (2013) Abiotic stress induced production of β-carotene, allophycocyanin and total lipids in *Spirulina* sp. J Biol Earth Sci 3:B54–B64

Hong Y, Hu HY, Sagehashi M (2010) Effects of allelochemical Gramine on photosynthetic pigments of cyanobacterium *Microcystis aeruginosa*. Int J Agric Biol Sci 1:10–14

Jodłowsak S, Latała A (2011) The comparison of spectrophotometric method and high-performance liquid chromatography in photosynthetic pigments analysis. J Biol Sci 11:63–69

Klemer AR (1990) Effects of nutritional status on cyanobacterial buoyancy, blooms and dominance, with special reference to inorganic carbon. Can J Bot 69:1133–1138

Kobayashi M, Van de Meent EJ, Oh-Oka H, Inoue K, Itoh S et al (1992) Pigment composition of heliobacteria and green sulfur bacteria. In: Murata N (ed) Research in photosynthesis. Kluwer Academic Publishers, Dordrecht, pp 393–396

Kopecky J, Schoeps B, Loest K, Stys D, Pulz O (2000) Microalgae as a source for secondary carotenoid production: a screening study. Algol Stud 36:153–168

Kumar M, Kulshreshtha J, Singh GP (2011) Growth and pigment profile of *Spirulina platensis* isolated from Rajasthan, India. Res J Agric Sci 2:83–86

Linda EG, Lee WW (2000) Algae. Prentice Hall, Upper Saddle River, p 640

Olaizola M, Duerr EO (1990) Effect of light intensity and quality on growth rate and photosynthetic pigment content of *Spirulina platensis*. J Appl Phycol 2:97–104

Papageorgiou GC (1996) The photosynthesis of cyanobacteria (blue bacteria) from the perspective of signal analysis of chlorophyll- a fluorescence. J Sci Ind Res 55:596–617

Parsons TR, Strickland JD (1963) Discussion of spectrophotometric determination of marine plant pigments with revised equations for ascertaining chlorophylls and carotenoids. J Mar Res 21: 155–163

Patterson GML (1996) Biotechnological applications of cyanobacteria. J Sci Ind Res 55:669–684

Prasanna R, Sood A, Jaiswal P, Nayak S, Gupta V et al (2010) Rediscovering cyanobacteria as valuable sources of bioactive compounds. Appl Biochem Microbiol 46:133–147

Rodriguez H, Rivas J, Guerrero MG, Losada M (1989) Nitrogen-fixing cyanobacterium with a high phycoerythrin content. Appl Environ Microbiol 55:758–760

Sarada R, Pillai MG, Ravishankar GA (1999) Phycocyanin from Spirulina sp: influence of processing of biomass on phycocyanin yield, analysis of efficacy of extraction methods and stability studies on phycocyanin. Proc Biochem 34:795–801

Shetty K, Krishnakumar G (2020) Algal and cyanobacterial biomass as potential dye biodecolorizing material: a review. Biotechnol Lett 42:2467–2488

Simeunovic J, Beslin K, Svireev Z, Kovac D, Babic O (2013) Impact of nitrogen and drought on phycobiliprotein content in terrestrial strains. J Appl Phycol 25:597–607

Stanier RY, Kunisawa R, Mandel M, Cohen-Bazire G (1971) Purification and properties of unicellular blue green algae (Order: Chroococcales). Bacteriol Rev 35:171–205

Subhashini J, Mahipal VK, Reddy MC, Reddy MM, Rachamallu A, Reddanna P (2004) Molecular mechanisms in C-phycocyanin induced apoptosis in human chronic myeloid leukemia cell line-K562. Biochem Pharmacol 68:453–462

Thajuddin N, Subramanian G (2005) Cyanobacterial biodiversity and potential applications in biotechnology. Curr Sci 89:47–57

Tiwari ON, Singh BV, Mishra U, Singh AK, Dhar DW, Singh PK (2005) Distribution and physiological characterization of cyanobacteria isolated from arid zones of Rajasthan. Trop Ecol 46:165–171

Velvizhi T, Varadharajan D, Babu R, Sundaramanickam A, Vijayalakshmi S, Balasubramanian T (2011) Studies on the effect of cyanobacteria on tobacco pasted albino rat (*Rattus norvegicus*). Adv Appl Sci Res 2:16–23

Wada N, Sakamoto T, Matsugo S (2013) Multiple roles of photosynthetic and sunscreen pigments in cyanobacteria focusing on the oxidative stress. Metabolites 3:463–483

Wright SW, Jeffrey SW, Mantoura RFC, Llewellyn CA, Bjornland T et al (1991) Improved HPLC method for the analysis of chlorophylls and carotenoids from marine phytoplankton. Mar Ecol Prog Ser 77:83–196

Cyanobacterial Stress and Its Omics Perspective

3

Surbhi Kharwar, Arpan Mukherjee, Vinod Kumar, and Ekta Shukla

Abstract

Cyanobacteria are prokaryotic oxygenic photoautotrophs exposed to various environmental stresses. Environmental stressors such as nutrient deficiency and high as well as low light conditions affect the growth and development of cyanobacteria. In order to overcome the effect of different stressors, they have evolved several adaptive mechanisms. Nutrients, like sulfur and iron, play a role in photosynthesis, respiration, nitrogen metabolism, and other processes. Different omics approaches show variations of different genes, transcripts, proteins, and metabolites of the affected organism. The present chapter discusses the impact of light and nutrient stress in cyanobacteria and the molecular mechanisms with the involvement of different omics approaches.

Keywords

Cyanobacteria · Light · Sulfur · Iron · Gene expression · Omics · Stress

S. Kharwar
Department of Botany, University of Lucknow, Lucknow, Uttar Pradesh, India

A. Mukherjee
Institute of Environment and Sustainable Development, Banaras Hindu University, Varanasi, Uttar Pradesh, India

V. Kumar
Department of Botany, Pandit Prithi Nath College, Kanpur, Uttar Pradesh, India

E. Shukla (✉)
Sunbeam College for Women Bhagwanpur, Varanasi, Uttar Pradesh, India

B. Neilan et al. (eds.), *Cyanobacterial Biotechnology in the 21st Century*,
https://doi.org/10.1007/978-981-99-0181-4_3

3.1 Introduction

Cyanobacteria are among the primitive organisms found to be present on Earth. They created the oxygenic environment required for all other organisms to survive. These organisms use solar radiation and nutrients from the soil and atmosphere to grow and develop. Cyanobacteria face stress condition such as light, nutrient, temperature, salt, and many more. In response to stress conditions, these organisms activate a cascade of gene expression and production of stress-specific metabolites which play a role in the acclimatization process. Cyanobacteria are the utmost research studied organism in response to various types of environmental stress conditions which regulated their mechanisms at the physiological and molecular levels. In the present chapter, we will describe the impacts of high and low light, along with sulfur and iron stress condition and the underlying acclimatory mechanisms using omics data. The effect of high and low light conditions was studied mainly in the model cyanobacteria.

3.2 "Omics" Data: An Overview of New Technologies

Omics approaches include genomics, transcriptomics, proteomics, and metabolomics. These omics approaches offer an outlook for the acquisition of new knowledge for wider insights of the organisms' cellular mechanism activated under stress-related conditions. Under the stress condition, changes in the gene expressions were correlated with modulation in the genetic makeup, proteome, and metabolomics of the organism. These changes that occurred at the transcriptional level are not necessary to coincide with changes that occurred at the proteome level. Hence, to provide a better understanding based on the molecular data of the cyanobacterial cell, a thorough investigation of the changes at different cellular levels is a prerequisite.

Kaneko et al. in 1996 firstly sequenced the complete genome of *Synechocystis* sp. PCC 6803. Later on, the entire genome sequences of *Anabaena* sp. PCC 7120, *Prochlorococcus marinus* SS 120, *Gloeobacter violaceus* PCC 7421, *Thermosynechococcus elongatus* BP 1, *Synechococcus* sp. WH8102, *P. marinus* MED4, and many more have also been submitted in the public database (Kaneko et al. 2001; Dufresne et al. 2003; Nakamura et al. 2002; Nakamura et al. 2003; Palenik et al. 2003). Subsequently in 1999, Takara Bio Co. (Japan) started the synthesis of genome-wide cDNA microarray in *Synechocystis* and revealed a total of 3165 genes (976%) are present on chromosomes. Analysis of the complete genome of the cyanobacteria decodes 5368 protein-coding genes on the main chromosome (Kaneko et al. 2001). Further, Sato et al. (2004) amplified a total of 2407 segments of 3–4 kb of *Anabaena* DNA by using polymerase chain reaction and spotted on the glass slides (used as microarrays). Recently, another type of microarray specific to the *Anabaena* genome has been developed by Ehira and Ohmori (2006) which involves spotted oligonucleotides. This microarray analysis covers a different set of regulatory genes present on the *Anabaena*. The original results of data analysis in terms of gene expression are accessible at the KEGG expression database (http://www.genome.jp/kegg/expression/). The present chapter deals with light stress-induced gene expression and its omics approaches in the model cyanobacteria, i.e., *Synechocystis* and *Synechococcus*. The studies based on DNA microarrays and modern methods for the identification of genes and its expression and regulation induced under stress condition, would not have been possible without the availability of entire genome sequences. Next, we will describe the environmental scenario of sulfur and iron, the physiological responses of sulfur and iron stress, and the mechanism of gene regulation in cyanobacteria.

3.3 Light Stress

3.3.1 Physiological Responses of Light Stress in Cyanobacteria

Light plays an important role with respect to the normal growth and survival; therefore, photoinhibition or photo-destruction is unavoidable in cyanobacteria. In order to acclimatize to light stress, cyanobacteria have adopted various mechanisms which in turn control the gene expression and other molecular functions of the cell (Rachedi et al. 2020; Biswas et al. 2022). The three most important strategies implemented by cyanobacteria are photoinhibition, non-photochemical quenching (NPQ), and phycobilisome decoupling (Adir et al. 2003; Kirilovsky 2007). Several modulations has been reported in the cyanobacterial cell during light stress apart from photo-destruction such as DNA destruction, protein degradation, and many more. The light sensor machinery of the cell plays a crucial part in the recognition of light signals, formation of unsaturated fatty acids by the destruction of membrane proteins, disturbance in metabolic equilibrium, activation of heat shock proteins and programmed cell death was described in the cyanobacteria upon light stress (Wase et al. 2012).

3.3.2 Expression and Regulation of Light-Responsive Genes in Cyanobacteria

Mechanisms of acclimation to the changing light field have been poorly understood in the organism. Schena et al. (1995) described that DNA microarray is used for the analysis of different gene expressions. Acclimatization of the cellular organisms to the changing light conditions, viz., from low light to high light, has been grouped into short-term and long-term processes (Anderson and Eiserling 1986; Anderson et al. 1995). The short-time process involves the conversion of state and degeneration of protective energy (Campbell et al. 1998; Niyogi 1999), and changes in the energy efficacy involve the allocation of the same to the photosystem II from the harvesting complex (Hassidim et al. 1997; Bissati et al. 2000) and produce a malfunctioned reaction center at photosystem II (Anderson et al. 1997), whereas long-term acclimation response to HL is a very slow process intricating structural, functional, and compositional variations in the photosynthetic apparatus. The physiological responses of the cell to the changing light intensity were reported by Neale and Melis (1989). Further, the underlying genetic machinery for the same has been described elaborately by Niyogi (1999). Kaneko et al. (2003) have worked on the cyanobacterium *Synechocystis* sp. PCC 6803 and have done extensive work on the cellular mechanism and time-dependent gene expression and its regulation from LL to HL condition.

Water acts as an electron donor mostly, in algae, blue-green algae, and higher plants, but all of them have different abilities to use water as an electron source. In the oxygenic photosynthetic organism, electron transfer occurred in thylakoid membrane-embedded proteins such as PS II, cytochrome b6f, and PS I which form complexes. These protein complexes together produce energy currency (chemical) as ATPs for the production of carbohydrates by using ATP synthase and NADPH. The core PS II pigment-protein complex consists of two core proteins, i.e., D1 and D2 (Mulo et al. 2009). The prolonged light stress condition reduces the number of D1 protein resulting decrease in the photosynthetic capacity of the cyanobacteria (Takahashi and Murata 2008). Protein D1 is encoded by gene *psbA* genes that vary from one to six depending upon the species. Multiple genes of *psbA* encode different types of D1 proteins in cyanobacteria and manifest the regulatory mechanisms responsible for maintaining PS II in the changing environmental conditions. Regulation of *psbA* in cyanobacteria depends on the two mechanistic principles. The first strategy is the replacement of D1 protein from the PS II under the normal condition with a form that is present mainly in reduced stress conditions, while in the second strategy, cyanobacteria elevate the turnover number of the same D1 protein under the stressed condition. These two important strategies were found in different cyanobacterial species. The third type of regulation process was documented in diverse groups of cyanobacteria where the silent *psbA* genes get induced low level of oxygen (Sicora et al. 2009).

Several studies have been focused on the cyanobacterial *psb*A gene and its regulation. Though a large number of scientific studies are still going on, the proper scientific explanations are at their initial. Transcript of *psb*A in cyanobacteria activates during the change in light intensity, but its regulation mechanism varies from one species to another. Here, in this chapter, we have discussed the basic and current study about the expression and regulation of *psb*A in *Synechococcus elongatus* sp. (PCC 7942) and *Synechocystis* sp. PCC 6803 and 6714 as model organism (Williams 2007).

In general, *psb*A gene expression was regulated at transcription and translation levels. In the case of the model organism cyanobacteria, during transcription, the initiation part is said to be the most detrimental phase of the gene expression and its regulation. The enzyme responsible for this is a holoenzyme RNA polymerase made up of catalytic active site and several sigma factors which initiates and regulates the transcription process. Sigma factors play a pivotal role in the promoter recognition and chromatin structure regulation which further controls/modulates the level of gene expression. Many studies exhibited that Group I sigma factor recognizes the hexameric region positioned between −35 and − 10 along the promoter region of *psbA* genes (Shibato et al. 1998). Another study suggests that the light-responsive/light-dependent expression of *psbA* requires SigB, SigD, and SigE (Pollari et al. 2009). Further, Willenbrock and Ussery (2004) reported that the DNA tertiary structure has some effects on gene expression. A = T repeats of DNA are often found between the −240 and −40 region from the transcriptional site which influences DNA double helix formation and modulates transcriptional response within the cell. In *Synechocystis* 6803 and *Microcystis aeruginosa* K-81, it is reported that in the upstream region of the *psbA* genes, an intrinsic curvature of A = T tracts is present (Asayama et al. 2002). A slight modification in these sites downregulates the transcription of *psbA*, but this is the light-responsive regulation in nature. DNA microarray studies of Hihara et al. (2001) stated that in high light conditions, *psbA* gets strongly upregulated. Riediger et al. (2018) reported that in *Synechococcus elongatus* PCC 7942 and *Synechocystis* 6803 and 6714, a specific *psbA* regulation is studied and addressed.

However, *RpaB* is another essential light-responsive gene in cyanobacteria (Fig. 3.1). Hihara et al. (2001) in *Synechocystis* sp. PCC 6803 identified more than 160 HL stress-responsive genes during transcriptomic analysis. The HL regulatory 1 (HLR1) sequence motif is found to be present in the upstream part of such genes which is highly conserved (Kappell and van Waasbergen 2007; Seino et al. 2009; Takahashi et al. 2010; Kadowaki et al. 2016). In *Synechocystis* PCC 6803 and *S. elongatus* PCC 7942, it has been observed that RpaB binds with the HLR1 element present on high light-responsive promoters (Seki et al. 2007; López-Redondo et al. 2010; Kato et al. 2011; Riediger et al. 2019; Yasuda et al. 2020). Kadowaki et al., in 2016, performed an in vivo study under HL condition using ChIP and ChAP analyses to study the binding pattern of *RpaB* with the targeted promoters. These observations suggested that under high light-induced stress, the *RpaB* is deployed as a transcription factor (Hanaoka and Tanaka 2008). *RpaB* and *RpaA* are the OmpR-type response regulator DNA-binding proteins. The roles of these regulators are known for its energy transfer from the PS I and PS II reaction center (Ashby and Mullineaux 1999).

In addition, *Synechocystis* sp. PCC 6803 contains several *ndh* genes which encode different parts of NADPH dehydrogenase. The *ndh* can be present singly or in multiple copies. Regulated transcription of *ndhC*, *ndhK*, and *ndhJ* constitutes an operon, i.e., slr1279 to slr1281. The expression of this operon increases by threefold within 15 min of HL exposure. Contrary to this, the induction of another operon sll0519 to sll0522 consisting of *ndhA*, *ndhG*, *ndhE*, and *ndhI* was found to be low. Studies suggested that in *Synechocystis* PCC 6803, *ndhD* is present in six copies and *ndhF* in three copies (Price et al. 2011). Interestingly, *ndhD2*, *ndhD3*, and *ndhF3* genes showed evident and short-term induction within 15 min of HL exposure by more than eightfold, whereas transcription of other genes was not observed in response to the HL condition. *ndhF3*, *ndhD3*, and an open reading frame are encoded by sll1734 operon and are thought to be involved in the uptake of carbon dioxide in *Synechocystis* PCC 6803 (Ohkawa et al. 2001). In HL condition, cyanobacteria induce sll1734 which encodes putative protein. The genes of *ndhD2* and *ndhF3* operon are also induced by low carbon dioxide (Ohkawa et al. 1998).

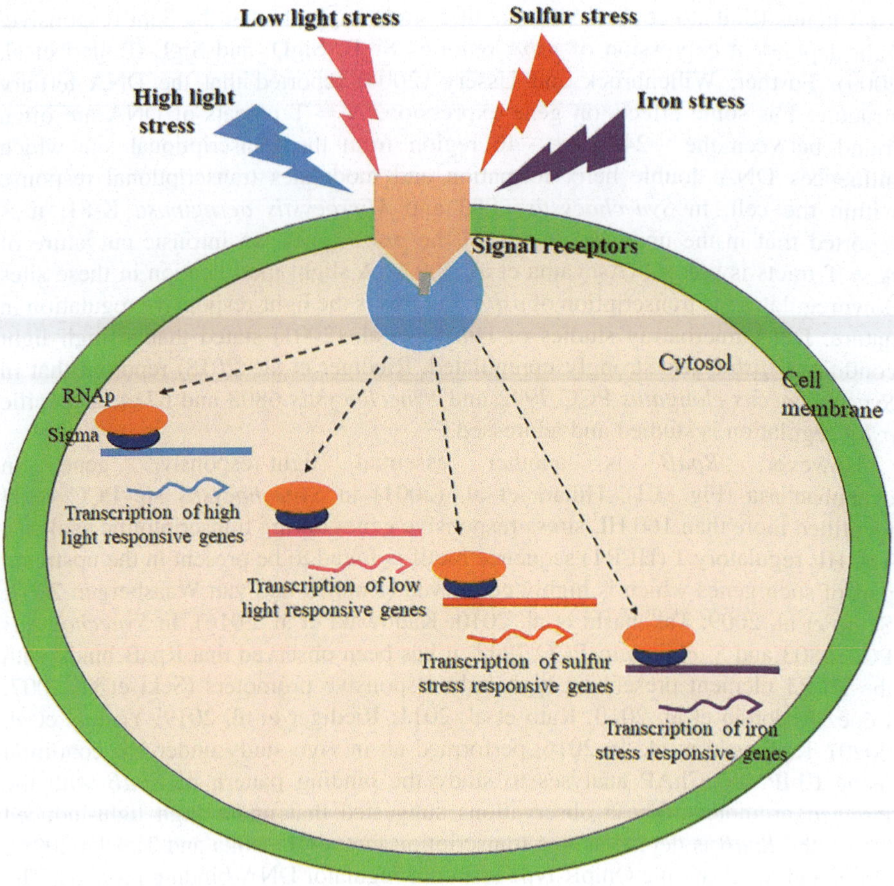

Fig. 3.1 The schematic diagram shows how the cyanobacterial cells modulate their metabolism under different stress conditions

Kaplan and Reinhold (1999) observed that genes *ndh* might also be induced by common signals released by low carbon dioxide and high light.

3.4 Nutrient Stress

Mineral nutrients are elements that are mostly obtained in the form of inorganic ions. Among various environmental difficulties, nutrient scarcity typically serves as a limiting factor for cyanobacterial growth. In response to unfavorable stress conditions, some of the stressed cyanobacterial cells manage to survive and continue their metabolic activity, while some of them undergo programmed cell death. In this part of the chapter, we will discuss specifically the two important elements such as iron and sulfur.

3.4.1 Iron

3.4.1.1 Physiological Functions/Role of Iron

Iron (Fe) plays an important role in the growth of the microorganisms (Hantke 2001). It is involved in the metabolic processes like respiration, photosynthesis, nitrogen fixation, DNA synthesis, biosynthesis of photosynthetic pigments, redox reaction, and many more (Kurisu et al. 2003).

3.4.1.2 Environmental Scenario of Iron

Even though Fe is the most common element on Earth, its availability is constrained due to its low solubility in the oxygen-rich atmosphere (Braun et al. 1990). Concentrations such as 10^{-8} M (minimum effective concentration) and 10^{-7} to 10^{-5} M (optimal concentration) were reported for the growth of microbes (Andrews et al. 2003). Iron is the limiting nutrient affecting various metabolic pathways in cyanobacteria which hampers primary productivity in the ocean (Coale et al. 1996). Reduced and oxidized forms represent two oxidation states of iron in the aqueous solution. Changes from anaerobic to aerobic environment culminate iron oxidation, i.e., from ferrous (reduced) to ferric (oxidized) form. This oxidized form of iron is unavailable for the microbes and causes shortage of Fe (Allen and Vermaas 2010). Free inorganic forms of iron such as Fe^{2+}, Fe^{3+}, and Fe (III)L are biologically available forms. Among them, cyanobacteria use Fe^{2+} and Fe^{3+} (Qiu et al. 2022). Cyanobacteria use Fe for various biochemical processes (Stroebel et al. 2003; Bellenger et al. 2011). Cyanobacteria have high demand for iron, and due to Earth's Great Oxygenation Event, they face iron stress condition (Behrenfeld et al. 1996).

3.4.1.3 Cyanobacterial Responses Under Iron-Limiting Condition

To cope with iron deprivation, cyanobacteria developed highly efficient systems such as activation of iron uptake system and modifications at the structural and molecular levels. Strauss classified three physiological reactions as (i) retrenchment (cell size reduction, phycobilisome degradation, ultrastructural modulations, and pigment reduction), (ii) compensation (induction in the expression of *isiA*), and (iii) acquisition of iron (Ferreira and Straus 1994; Straus 1994).

Blue-green algae showed physiological and biochemical responses such as degradation of phycobilisome; modulations in the photosynthetic process; respiration, nitrogen fixation, and alteration in the utilization of stored iron in cells; and many more (Fraser et al. 2013; Schrader et al. 2011). Low Fe-containing proteins were synthesized by the organisms in the iron-deficient cells (Straus 1994; Hutber et al. 1977; Sandmann and Malkin 1983). Nevertheless, reduction in the content of photopigments and changes in the structure of cyanobacterial cell membrane were observed in iron-limiting condition (Odom et al. 1993; Sandmann 1985; Straus 1994; Riethman and Sherman 1988). Decrease in the size of cells, reduction in the carboxysomes, and/or other morphological changes were reported in the cyanobacteria grown in iron-limiting conditions (Li et al. 2016; Walworth et al. 2016). The ability of the iron-deficient cyanobacterial cells to absorb nutrients is increased by smaller cells because surface-to-volume ratios increase (Leynaert et al. 2004).

As a result, blue-green algae have highly controlled iron equilibrium within their cells. Many of the researchers observed the accumulation of *isiA* in the cyanobacteria that is a homolog of CP43 protein of photosystem II (PS II) (Dühring et al. 2006; Latifi et al. 2005). IsrR is another iron stress-responsive protein present in cyanobacteria (Fig. 3.1). Activation of *isiA* gene is mediated by the Fur protein (Desai et al. 2012) in the iron-limiting cells which causes a shift in the absorbance of chlorophyll *a* (Falk et al. 1995). From time to time several researchers have studied and reported that under prolonged iron limitation, IsiA forms a ring structure around photosystem I monomers, while large aggregation without PS I was also shown in cyanobacteria (Chauhan et al. 2011; Kouřil et al. 2005). Later, Wang et al. (2010) discovered that intense light exposure caused the cyanobacterium *Synechocystis* PCC 6803 to produce unique IsiA ring structures. Daddy et al. (2015) described the role of IsiA which protects the membrane (thylakoid) from oxidative stress. Besides, IsiA also serves as an alternative antenna complex for photosystems (Melkozernov et al. 2003; Andrizhiyevskaya et al. 2002). Additionally, IsiA functions as a chl *a* storage protein and plays a role in the protection of photosystems from photoinhibition (Sandström et al. 2001; Latifi et al. 2005).

Expression of gene *isiA* is transcriptionally and post-transcriptionally regulated. In cyanobacteria, Fur regulates the transcription of *isiA* by binding at the upstream region of the *isiA* (Gonzalez et al. 2011, 2012, 2014). Moreover, *isiA* is also transcriptionally regulated by the other proteins, viz., Pkn22 and PfsR (Cheng and He 2014). IsrR (antisense RNA) controls *isiA* expression post-transcriptionally (Dühring et al. 2006).

The Fur belongs to Crp/Fnr class of protein, plays a role in the regulation of the genes responsible for the incorporation and storage of iron, as well as regulates iron homeostasis (Andrews et al. 2003; Gonzalez et al. 2016). Fur serves as a transcription repressor only when it finds its corepressor (ferrous iron). In addition to *isiA*, the promoter regions of genes such as *isiB*, *irpA*, and *mapA* also have fur binding sites known as Fur boxes (Straus 1994). Gonzalez et al. ascribed that *isiAB* was repressed by Fur at the transcriptional level (Gonzalez et al. 2010). The findings of Hernandez et al. (2010) confirmed the role of Fur in the maintenance of thylakoid membrane configuration and photosynthetic performance using *fur* mutant strains following mutational analysis.

The iron storage protein such as ferritin is present in the marine (*Prochlorococcus*, *Synechococcus*, *Crocosphaera*, and *Trichodesmium*) and some freshwater cyanobacterial strains (Morrissey and Bowler 2012). These proteins consist of 21 monomers and bind to 2000–3000 iron atoms to scavenge the iron from the external environment (Keren et al. 2004). Cyanobacteria experience oxidative stress under iron-limiting condition. Production of ferritin is an adaptive strategy of cyanobacteria to mitigate oxidative damage (Toulzal et al. 2012).

Additionally, DpsA (DNA-binding protein containing heme) is important for cyanobacteria (Dwivedi et al. 1997) to protect the DNA molecules from the oxidative stress under the iron stress condition (Rivers 2009). It is widely present in all the species of *Synechococcus* and few strains of *Prochlorococcus* (Palenik et al. 2006). Sen et al. (2000) stated that the mutant strain of *dpsA* in the freshwater cyanobacterium *Synechococcus elongatus* is lethal under iron-limiting conditions.

Another iron stress-responsive protein of cyanobacteria is PfsR, which resembles with TetR (transcriptional regulator) which plays a role in the gene regulation (Cheng and He 2014). The author also reported that PfsR regulates genes of iron metabolism, namely, *isiA*, *furA*, *fut*, *feoB*, *bfr*, and *ho*, and its own transcription.

IdiA is another iron deficiency-induced protein expressed under iron-deficient condition in the case of cyanobacteria (Michel et al. 1996). CP43 and D1 interact with the IdiA protein and is found toward the outer side of the thylakoid membranes of the cyanobacteria (Exss-Sonne et al. 2000; Lax et al. 2007). IdiA is a homolog of ABC transporter system such as FutA suggesting its role in the iron transportation (Tolle et al. 2002). FutA involved in the prevention of phycobilisome degradation under the Fe-deficient condition (Shcolnick et al. 2009). IdiA is also regulated by another cyanobacterial protein, i.e., IdiB, belonging to the Crp/Fnr class of regulator protein having HTH (helix-turn-helix) motif (Michel et al. 2001; Michel and Pistorius 2004). Expression of *idiB* increases during iron-deficient condition, and in response, it regulates several other genes responsible for iron homeostasis (Yousef et al. 2003).

Under the iron starvation, cyanobacteria increase the uptake of iron by synthesizing siderophores, which chelate the iron from the environment (Kranzler et al. 2014). Siderophore chelates iron present in cyanobacteria in the form of ferric (Kranzler et al. 2013) and forms a complex known as ferri-siderophore complex. The formed ferri-siderophore complex subsequently exported to the periplasmic space through a transport system known as TonB (Nicolaisen et al. 2010; Rudolf et al. 2015).

Under iron-limiting condition, ATP-binding cassette (ABC) is also formed in the cyanobacteria to transport iron (Shcolnick et al. 2009). An ABC-type transport system such as FutABC consists of four neighbor proteins encoded by *futA*, *futA2*, *futB*, and *futC* genes which were reported in *Synechocystis* (Brandt et al. 2009; Badarau et al. 2008; Katoh et al. 2001a). Moreover, FeoB is another transport system found in cyanobacteria which also helps in the transport of iron (Katoh et al. 2001b).

3.4.2 Sulfur

3.4.2.1 Physiological Functions/Role of Sulfur

Sulfur (S) is a vital element for biological organisms, as it takes part in a variety of structural, metabolic, and catalytic processes. S is a crucial component of amino acids as well as numerous metabolites, making it one of the most important macronutrients for every organism due to its versatile nature. Cysteine and methionine play a role in protein formation. Moreover, cysteine also acts as a precursor for the biosynthesis of S-containing compounds such as glutathione, cofactors like Fe-S clusters, heme, siroheme, molybdenum centers, and lipoic acid. A range of biomolecules including biotin, sulfolipid, thiamine, sulfur esters (Co-A, a constituent of several enzymes), thiouridine, and thioredoxin require sulfur for their synthesis and modification (Kharwar et al. 2021a, 2021b, 2021c; Wang et al. 2022).

3.4.2.2 Environmental Scenario of Sulfur

Sulfur is essential for many of Earth's biogeochemical cycles. It is the primary component of sulfide minerals. The concentration of sulfur has shifted throughout time, notably in the different ecosystems (Habicht et al. 2002; Kah et al. 2004; Petrychenko et al. 2005; Gill et al. 2011; Gill et al. 2007; Lowenstein et al. 2003; Takahashi et al. 2011; Giordano et al. 2008; Holmer and Storkholm 2001). In the recent years, the concentration of sulfate in the marine ecosystem has been found nearly 28 mM, indicating that it is the primary sulfate reservoir (Ksionzek et al. 2016), while very low (approximately 10 to 50 μM) concentration of sulfate was observed by Bochenek et al. (2013) in the freshwater systems. Hence, the organism thriving in the freshwater ecosystem face sulfate stress condition. In order to survive in the stressful environmental niches, microbes modulate their cellular metabolism.

3.4.2.3 Cyanobacterial Regulation of Sulfur Metabolism

Cyanobacteria most commonly prefer to uptake sulfate, but they can also uptake cysteine, cystine, methionine, glutathione, thiosulfate, thiocyanate, sulfonate, sulfate ester, ethane sulfonate, taurine, and many other sulfur molecules (Schmidt et al. 1982; Kharwar et al. 2021a, 2021b, 2021c). Thus, metabolism of these compounds begins with intracellular assimilation, i.e., sulfate uptake, and reduction to form cysteine, which is further incorporated into the various downstream sulfur compounds like glutathione, methionine, thioredoxin, S-adenosyl methionine, glutaredoxin, and so on (Kharwar et al. 2021a, 2021b, 2021c). Sulfur not only is involved in nitrogen metabolism but also acts as an intermediate to assimilate carbon. For example, serine and acetyl-CoA are the intermediate products of nitrogen and carbon metabolic pathways, respectively, demarcating the interconnection between nitrogen, sulfur, and carbon metabolisms (Kharwar et al. 2021a, 2021b, 2021c). Cyanobacteria have developed certain strategies to cope with sulfur starvation. This adaptive response is dependent on the ability to detect starvation and adjust gene expression patterns to use alternative sulfate sources (Giordano et al. 2005, 2008). It has been reported that cyanobacteria uptake sulfur from the surrounding environment using two different transporters. These transporter proteins such as H^+/SO_4^{2-} and ABC-type are present on the cyanobacterial membrane (Kaneko et al. 1996). A sulfate permease, i.e., ABC-type transporters, was identified by Green et al. (1989) in *Synechococcus* 6301 which are activated in the sulfur-starved condition. *Synechococcus* PCC 7942 contains sulfate permease genes which are clustered as *cys*W, *cys*T, *cys*A, and *sbp*A on the pANL plasmid of the cyanobacterium (Laudenbach et al. 1991). CysA, i.e., a **n**ucleotide-**b**inding **p**rotein (nbp), forms the periplasmic system, whereas *cysT* encodes the hydrophobic protein, CysT, which spans over the cytoplasmic membrane. CysW is encoded by gene *cysW* which forms a pore in the transportation system. The order of genes involved in sulfur metabolism was identified by Jager in 1992. Further characterization and function of sulfate permease encoding genes were speculated by mutational analysis. The cyanobacterial mutants of *cysA*, *cysT*, and *cysW* were unable to uptake sulfate from the surrounding environment in the sulfate-limiting condition. Based on the mutational analysis, Green and Grossman (1988) discovered the function of *sbpA* as a sulfate-binding protein, while *cysR* as a LysR family of transcriptional regulator. Later, Aguilar-Barajas et al. (2011) discovered that different cyanobacterial species have distinct configurations of gene expression of sulfate permease.

CysR induces the expression of sulfur metabolism genes. This gene is found to be reported in the cyanobacterium *Synechococcus* which also regulates the transcription of *cys* operon genes. *Cys* operon consisted of numerous genes that encode the enzymes involved in the sulfate uptake as well as in the reduction process. Activation of these is noticed in the cyanobacterium, under sulfate-starved conditions (Nicholson and Laudenbach 1995). CysR, the functional homolog of NtcA and BifA, and the N-terminal **SPXX** motif in CysR is involved in the DNA binding and regulates the sulfur metabolism (Nicholson and Laudenbach 1995; Churchill and Suzuki 1989; Suzuki 1989). Laudenbach and Grossman (1991) identified and characterized the *cysR* mutant following insertional mutagenesis and observed lower V_{max} value of sulfate permease under sulfate stress condition (Laudenbach and Grossman 1991). The author, therefore, concluded that the *cysR* gene encoded a protein that is necessary for cyanobacterial metabolism and the regulation of sulfur metabolism. Further, Kharwar et al. (2021a, 2021b, 2021c) expanded the knowledge of *cysR* regulation in cyanobacteria. In addition, a 33 kDa protein known as rhodanese is also accumulated under sulfur-deprived conditions (Laudenbach et al. 1991). Gene *rdhA* encodes a protein which is known for cyanobacterial survival upon sulfur limitation (Laudenbach et al. 1991).

O-acetylserine (thiol) lyase (OAS-TL) and serine acetyltransferase (SAT) are the two enzymes involved in cysteine biosynthesis (Nicholson et al. 1995). These enzymes form cysteine synthase complex (CSC) depending on the concentration of sulfur inside the cell. Sulfate deprivation leads to a drop in the sulfide concentration and OAS accumulation. This result halts the production of OAS from acetyl-coenzyme A (CoA), SAT inactivation, and dissociation of the CS complex. Accumulation of OAS triggers the transcription of genes involved in sulfate transporter to acquire sulfate from the environment, which is further reduced in a multistep process and consumed for the biosynthesis of cysteine via activating OAS-TL until OAS levels drop down. As a result, the CS complex contributes to cysteine cellular homeostasis by sensing the level of sulfide from reductive assimilation. This mechanism of stabilization and dissociation of the complex is well described in plants and bacteria (Yamaguchi et al. 2000; Ikegami et al. 1993). Progress in the knowledge of understanding the feedback regulation of CSC was reported by Wirtz and Hell (2006). Using bioinformatic tools, Kharwar et al. (2021b) confirmed the existence of CSC in cyanobacteria where the electrostatics energy is −26 kcaL/moL.

3.4.2.4 Response of Cyanobacterial Metabolism Under Sulfur Stress Condition

Sulfur is essential for several key tasks, and the deficit of sulfur in cyanobacteria produces a variety of problems including photosynthetic, morphological, biochemical, and physiological processes which have been explained in the present text. Cyanobacteria endure prolonged periods of the sulfur deficiency by slowing down their metabolism and degrading their photosynthetic apparatus, which leads to the bleaching of cyanobacterial cells, a condition known as chlorosis (Collier and Grossman 1994; Davies and Grossman 1998; Ariño et al. 1995). Biosynthesis of NblA protein, i.e., a Clp protease adaptor protein, leads to the onset of chlorosis. When the cyanobacterial cells experience a limited amount of sulfur, they undergo a complex set of morphological changes (Wanner et al. 1986) including ultrastructural changes. Sulfur deprivation includes the attenuation of thylakoid membrane, thickening of cell wall, and accumulation of glycogen granules, carboxysomes (polyhedral bodies), and polyphosphate granules in the cytoplasm of cyanobacteria and allows for the stockpiling of nutrient reserves during the period of abundance (Schmidt et al. 1982; Collier and Grossman 1992, 1994). The most dramatic responses are light-harvesting complex degradation, phycobilisome degradation, and reduction in phycobilin protein contents which serve as nutrient and cysteine reserves (Muller et al. 1997; Gutu et al. 2011; Baier et al. 2014; Karradt et al. 2008). Reduced photosynthetic performance of cyanobacteria is because of the rearrangement of photosynthetic apparatus and reduced oxygen evolution under the sulfur-limiting condition (Collier and Grossman 1994). Decreased activity of PS II upon sulfate deprivation lowers energy transfer and generates toxic oxygen species. Additionally, sulfate stress cause reduction in ATP synthesis, protein synthesis, and nutrient uptake as a consequence of impaired cellular ion homeostasis shown by researchers (Kumaresan et al. 2017; Kharwar and Mishra 2020). In addition, prolonged sulfur limitation also leads to the conversion of cyanobacterial vegetative cells of *Nostoc* into the dormant spores called as akinetes (Kyndiah and Rai 2007). Moreover, accumulation of carbohydrates (Davies and Grossman 1998; Zhang et al. 2020), reduction in the photosynthetic process particularly in the photosynthetic apparatus (Gutu et al. 2011), reduction in the protein content, and increased uptake of the limiting nutrient (Green and Grossman 1988; Bochenek et al. 2013) have also been documented. Reduced nitrogenase enzyme activity in the cyanobacterial strains of *Gloeothece* and *Nostoc* (Ortega-Calvo and Stal 1994; Hifney and Abdel-Basset 2014), and ammonia excretion (Krämer and Schmidt 1989) from the cyanobacterial cells has been reported under sulfate limitation. Alongside, the effect of sulfate limitation on the amino acid (Kiyota et al. 2014, 2012; Kharwar and Mishra 2020) and fatty acid metabolism were also investigated by many workers (Zhang et al. 2020; Kharwar and Mishra 2020; Wang et al. 2021). Decrease in the sulfolipid content of microalgae under the sulfur stress condition indicated acceleration in the cellular sulfur turnover for the mitigation of stress condition (Sugimoto et al. 2007, 2008, 2010; Sato et al. 2017). Sulfur limitation results in hydrogen production in the cyanobacterial strains such as *Gloeocapsa alpicola* (Antal and Lindblad 2005); *Calothrix elenkinii*, *Fischerella muscicola*, *Nostoc calcicola*, *Synechocystis* PCC 6803, and *Scytonema bohneri* (Yodsang et al. 2018); and *Nostoc* sp. (Hifney and Abdel-Basset 2014). Hydrogen production in cyanobacteria can be explained as the inhibition of D1 protein in PS II repairment upon sulfate stress. In addition, sulfur-containing hepatotoxin such as cylindrospermopsin has been affected by the availability of sulfate (Bácsi et al. 2006). It has been reported that microcystin synthesis is also affected by sulfate supply (Long 2010). Similarly, changes in the composition and structure of UV protectant, i.e., mycosporine-like amino acids, in the cyanobacteria were shown by Singh et al. (2010). Upregulation of an encapsulating shell gene, i.e., Synpcc7942_B2662 (Srpl, **S**ulfur **r**egulated **p**lasmid-encoded gene-**l**) (Fig. 3.1) and Synpcc7942_B2661 (cysteine desulfurase cargo), in the freshwater cyanobacterium *Synechococcus elongatus* was discovered (Nichols et al. 2021). The author and his colleague also describe the role of these genes as storage cage for sulfur.

Increased sulfate assimilation capacity is the key response of photosynthetic organisms upon sulfate limitation (Green and Grossman 1988; Laudenbach et al. 1991; Bochenek et al. 2013; Prioretti and Giordano 2016). Moreover, Prioretti et al. (2016) demonstrated higher ATPS activity in *Synechococcus* sp., *Thalassiosira pseudonana*, and *Tertaselmis suecica*, while lower activity in *Amphidinium klebsii* upon sulfur stress. Higher sulfite reductase activity has been noticed in *A. klebsii* upon low sulfate concentration. In addition, enzymes of cysteine biosynthesis were least affected in these organisms, but the activities of these proteins were stimulated under sulfate stress in plants (Ravina et al. 2002). Besides, cysteine and glutathione contents were reduced upon sulfate limitation in cyanobacteria. These S-containing metabolites play a role in the protection of cyanobacterial cell under the oxidative stress condition (Zechmann et al. 2010). The reason for the explanation is upon sulfur starvation, the turnover rate of cysteine and glutathione degradation exceeded that of its biosynthesis pathway due to the reduced ability of cyanobacterial cells to perform sulfur assimilation into cysteine production (Zechmann et al. 2010). Thus, an optimum amount of sulfur is essentially required by the organisms for the maintenance of all these metabolisms.

Several studies reported that how cyanobacterial cells modulated their cellular metabolism under sulfate stress conditions (Schmidt 1988; Schmidt and Jäger 1992; Kharwar et al. 2021a, 2021b, 2021c). Modulation of cellular responses is a dynamic process which occurred at various levels such as transcriptional and translational (Fig. 3.1).

Transcriptomic and proteomic data showed the downregulation of genes involved in the biosynthesis of amino acid, folding of protein, Fe-S cluster, translation, and assembly of ribosomes, while phosphorelay sensor kinase, DNA repair, carbohydrate metabolism, and proteins belongs to membrane were upregulated upon sulfate stress in cyanobacteria. In addition, downregulation of chaperons, adenylyl-sulfate kinase, phosphate acetyltransferase, aspartate aminotransferase, tyrosine phenol-lyase, and adenosylhomocysteinase was observed upon sulfur limitation (Zhang et al. 2008; Hughes et al. 2018; Kumaresan et al. 2017). Although a number of literatures pertaining to the effect of sulfate stress on cyanobacteria have been reported, nevertheless, more research is needed to have a better insight underlying the regulatory processes.

3.5 Conclusions

Light and nutrient play important roles in the cyanobacterial growth and development. Despite the importance of light, the presence of high and low light will lead to stress and causes toxic effect on the healthy growth of cyanobacteria. Besides, nutrients such as iron and sulfur are involved in various processes such as photosynthesis, respiration, nitrogen metabolism, antioxidant systems, etc.; thus, deficiency of these macronutrients leads to alteration in the cellular metabolism. Using omics approaches, such as genomics, transcriptomics, proteomics, and metabolomics, will further help in better understanding the stress signaling and adaptative molecular mechanisms in cyanobacteria.

Acknowledgments We would gratefully acknowledge the University of Lucknow, Lucknow; Pandit Prithi Nath College, Kanpur; and Sunbeam College for Women, Bhagwanpur, Varanasi, for the encouragement and support.

Author Contribution ES, SK, and VK conceptualized the study and retrieved pertaining literature. SK wrote the manuscript with contribution and thorough revision from SK, VK, AM, and ES. All authors have read and agreed to the published version of the manuscript.

Conflicts of Interest The authors declare no conflicts of interest.

References

Adir N, Zer H, Shochat S, Ohad I (2003) Photoinhibition–a historical perspective. Photosynth Res 76:343–370

Aguilar-Barajas E, Díaz-Pérez C, Ramírez-Díaz MI, Riveros-Rosas H, Cervantes C (2011) Bacterial transport of sulfate, molybdate, and related oxyanions. Biometals 24:687–707

Allen JF, Vermaas WFJ (2010) Evolution of photosynthesis. In: eLS. Wiley, Chichester

Anderson JM, Chow WS, Park Y-I (1995) The grand design of photosynthesis: acclimation of the photosynthetic apparatus to environmental cues. Photosynth Res 46:129–139

Anderson JM, Park YI, Chow WS (1997) Photoinactivation and photoprotection of photosystem II in nature. Physiol Plant 100(2):214–223

Anderson LK, Eiserling FA (1986) Asymmetrical core structure in phycobilisomes of the cyanobacterium *Synechocystis* 6701. J Mol Biol 191(3):441–451

Andrews SC, Robinson AK, Rodríguez-Quiñones F (2003) Bacterial iron homeostasis. FEMS Microbiol Rev 27:215–237

Andrizhiyevskaya EG, Schwabe TM, Germano M et al (2002) Spectroscopic properties of PSI-IsiA supercomplexes from the cyanobacterium *Synechococcus* PCC 7942. BBA-Bioenergetics 1556: 265–272

Antal TK, Lindblad P (2005) Production of H_2 by sulphur-deprived cells of the unicellular cyanobacteria *Gloeocapsa alpicola* and *Synechocystis* sp. PCC 6803 during dark incubation with methane or at various extracellular pH. J Appl Microbiol 98(1):114–120

Ariño X, Ortega-Calvo JJ, Hernandez-Marine M et al (1995) Effect of sulfur starvation on the morphology and ultrastructure of the cyanobacterium *Gloeothece* sp. PCC 6909. Arch Microbiol 163(6):447–453

Asayama M, Kato H, Shibato J, Shirai M, Ohyama T (2002) The curved DNA structure in the 5′-upstream region of the light responsive genes: its universality, binding factor and function for cyanobacterial psbA transcription. Nucleic Acids Res 30:4658–4666

Ashby MK, Mullineaux CW (1999) The role of ApcD and ApcF in energy transfer from phycobilisomes to PS I and PS II in a cyanobacterium. Photosynth Res 61(2):169–179
Bácsi I, Vasas G, Surányi G et al (2006) Alteration of cylindrospermopsin production in sulfate-or phosphate-starved cyanobacterium *Aphanizomenon ovalisporum*. FEMS Microbiol Lett 259(2): 303–310
Badarau A, Firbank SJ, Waldron KJ et al (2008) FutA2 is a ferric binding protein from *Synechocystis* PCC 6803. J Biol Chem 283:12520–12527
Baier A, Winkler W, Korte T et al (2014) Degradation of phycobilisomes in *Synechocystis* sp. PCC6803: evidence for essential formation of an NblA1/NblA2 heterodimer and its codegradation by A Clp protease complex. J Biol Chem 289(17):11755–11766
Behrenfeld MJ, Bale AJ, Kolber ZS et al (1996) Confirmation of iron limitation of phytoplankton photosynthesis in the equatorial Pacific Ocean. Nature 383:508–511
Bellenger JP, Wichard T, Xu Y et al (2011) Essential metals for nitrogen fixation in a free-living N2-fixing bacterium: chelation, homeostasis and high use efficiency. Environ Microbiol 13: 1395–1411
Bissati K, Delphin E, Murata N, Etienne A, Kirilovsky D (2000) Photosystem II fluorescence quenching in the cyanobacterium *Synechocystis* PCC 6803: involvement of two different mechanisms. Biochim Biophys Acta 1457:229–242
Biswas S, Niedzwiedzki DM, Pakrasi HB (2022) Introduction of cysteine-mediated quenching in the CP43 protein of photosystem II builds resilience to high-light stress in a cyanobacterium. Biochim Biophys Acta Bioenerg 1863:148580
Bochenek M, Etherington GJ, Koprivova A et al (2013) Transcriptome analysis of the sulfate deficiency response in the marine microalga *Emiliania huxleyi*. New Phytol 199(3):650–662
Brandt AM, Raksajit W, Mulo P et al (2009) Transcriptional regulation and structural modelling of the FutC subunit of an ABC-type iron transporter in *Synechocystis* sp. strain PCC 6803. Arch Microbiol 191:561–570
Braun V, Schaffer S, Hantke K et al (1990) Regulation of gene expression by iron. In: Hauska G, Thauer R (eds) The molecular basis of bacterial metabolism. VCH-Verlagsgesellschaft, Berlin, pp 35–51
Campbell D, Hurry V, Clarke AK, Gustafsson P, Öquist G (1998) Chlorophyll fluorescence analysis of cyanobacterial photosynthesis and acclimation. Microbiol Mol Biol Rev 62(3): 667–683
Chauhan D, Folea IM, Jolley CC et al (2011) A novel photosynthetic strategy for adaptation to low-iron aquatic environments. Biochemistry 50:686–692
Cheng D, He Q (2014) PfsR is a key regulator of iron homeostasis in *Synechocystis* PCC 6803. PLoS One 9(7):e101743
Churchill ME, Suzuki M (1989) 'SPKK' motifs prefer to bind to DNA at A/T-rich sites. EMBO J 8(13):4189–4195
Coale KH, Johnson KS, Fitzwater SE, Gordon RM, Tanner S, Chavez FP, Ferioli L, Sakamoto C, Rogers P, Millero F, Steinberg P, Nightingale P, Cooper D, Cochlan WP, Landry MR, Constantinou J, Rollwagen G, Trasvina A, Kudela R (1996) A massive phytoplankton bloom induced by an ecosystem-scale iron fertilization experiment in the equatorial Pacific Ocean. Nature 383:495–501
Collier JL, Grossman AR (1992) Chlorosis induced by nutrient deprivation in *Synechococcus* sp. strain PCC 7942: not all bleaching is the same. J Bacteriol 174(14):4718–4726
Collier JL, Grossman AR (1994) A small polypeptide triggers complete degradation of light-harvesting phycobiliproteins in nutrient-deprived cyanobacteria. EMBO J 13(5):1039–1047
Daddy S, Zhan J, Jantaro S et al (2015) A novel high light-inducible carotenoid-binding protein complex in the thylakoid membranes of *Synechocystis* PCC 6803. Sci Rep 5:9480
Davies JP, Grossman AR (1998) Responses to deficiencies in macronutrients. In: The molecular biology of chloroplasts and mitochondria in chlamydomonas. Springer, Dordrecht, pp 613–635
Desai DK, Desai FD, Laroche J (2012) Factors influencing the diversity of iron uptake systems in aquatic microorganisms. Front Microbiol 3:362

Dufresne A, Salanoubat M, Partensky F, Artiguenave F, Axmann IM, Barbe V, Hess WR (2003) Genome sequence of the cyanobacterium *Prochlorococcus marinus* SS120, a nearly minimal oxyphototrophic genome. Proc Natl Acad Sci 100(17):10020–10025

Dühring U, Axmann IM, Hess WR et al (2006) An internal antisense RNA regulates expression of the photosynthesis gene isiA. Proc Natl Acad Sci U S A 103:7054–7058

Dwivedi K, Sen A, Bullerjahn GS (1997) Expression and mutagenesis of the dpsA gene of *Synechococcus* sp. PCC 7942, encoding a DNA-binding protein involved in oxidative stress protection. FEMS Microbiol Lett 155:85–91

Ehira S, Ohmori M (2006) NrrA directly regulates expression of hetR during heterocyst differentiation in the cyanobacterium *Anabaena* sp. strain PCC 7120. J Bacteriol 188(24):8520–8525

Exss-Sonne PJ, Toelle KP, Bader EK et al (2000) The IdiA protein of *Synechococcus* sp. PCC 7942 functions in protecting photosystem II under oxidative stress. Photosynth Res 63:145–157

Falk S, Samson G, Bruce D et al (1995) Functional analysis of the iron-stress induced CP43P polypeptide of PSII in the cyanobacterium *Synechococcus* sp. PCC 7942. Photosynth Res 45: 51–60

Ferreira F, Straus NA (1994) Iron deprivation in cyanobacteria. J Appl Phycol 6:199–210

Fraser JM, Tulk SE, Jeans JA et al (2013) Photophysiological and photosynthetic complex changes during iron starvation in *Synechocystis* sp. PCC 6803 and *Synechococcus elongatus* PCC 7942. PLoS One 8(3):e59861

Gill BC, Lyons TW, Saltzman MR (2007) Parallel, high-resolution carbon and sulfur isotope records of the evolving Paleozoic marine sulfur reservoir. Palaeogeogr Palaeoclimatol Palaeoecol 256(3–4):156–173

Gill BC, Lyons TW, Young SA et al (2011) Geochemical evidence for widespread euxinia in the Later Cambrian ocean. Nature 469(7328):80–83

Giordano M, Norici A, Hell R (2005) Sulfur and phytoplankton: acquisition, metabolism and impact on the environment. New Phytol 166(2):371–382

Giordano M, Norici A, Ratti S et al (2008) Role of sulfur for algae: acquisition, metabolism, ecology and evolution. In: Sulfur metabolism in phototrophic organisms. Springer, Dordrecht, pp 397–415

Gonzalez A, Angarica VE, Sancho J et al (2014) The FurA regulon in *Anabaena* sp. PCC 7120: in silico prediction and experimental validation of novel target genes. Nucleic Acids Res 42:4833–4846

Gonzalez A, Bes MT, Barja F et al (2010) Overexpression of FurA in *Anabaena* sp. PCC 7120 reveals new targets for this regulator involved in photosynthesis, iron uptake and cellular morphology. Plant Cell Physiol 51:1900–1914

Gonzalez A, Bes MT, Peleato ML et al (2011) Unraveling the regulatory function of FurA in *Anabaena* sp. PCC 7120 through 2-D DIGE proteomic analysis. J Proteome 74:660–671

Gonzalez A, Bes MT, Peleato ML et al (2016) Expanding the role of FurA as essential global regulator in cyanobacteria. PLoS One 11(3):e0151384

Gonzalez A, Bes MT, Valladares A et al (2012) FurA is the master regulator of iron homeostasis and modulates the expression of tetrapyrrole biosynthesis genes in *Anabaena* sp. PCC 7120. Environ Microbiol 14(12):3175–3187

Green LS, Grossman AR (1988) Changes in sulfate transport characteristics and protein composition of *Anacystis nidulans* R2 during sulfur deprivation. J Bacteriol 170(2):583–587

Green LS, Laudenbach DE, Grossman AR (1989) A region of a cyanobacterial genome required for sulfate transport. Proc Natl Acad Sci 86(6):1949–1953

Gutu A, Alvey RM, Bashour S et al (2011) Sulfate-driven elemental sparing is regulated at the transcriptional and posttranscriptional levels in a filamentous cyanobacterium. J Bacteriol 193(6):1449–1460

Habicht KS, Gade M, Thamdrup B et al (2002) Calibration of sulfate levels in the Archean Ocean. Science 298(5602):2372–2374

Hanaoka M, Tanaka K (2008) Dynamics of RpaB-promoter interaction during high light stress, revealed by chromatin immunoprecipitation (ChIP) analysis in *Synechococcus elongatus* PCC 7942. Plant J Cell Mol Biol 56:327–335

Hantke K (2001) Iron and metal regulation in bacteria. Curr Opin Microbiol 4:172–177

Hassidim M, Keren N, Ohad I, Reinhold L, Kaplan A (1997) Acclimation of *Synechococcus* strain WH7803 to ambient CO_2 concentration and to elevated light intensity. J Phycol 33(5):811–817

Hernandez JA, Alonso I, Pellicer S, Luisa Peleato M, Cases R, Strasser RJ, Barja F, Fillat MF (2010) Mutants of *Anabaena* sp. PCC 7120 lacking alr1690 and α-furA antisense RNA show a pleiotropic phenotype and altered photosynthetic machinery. J Plant Physiol 167:430–437

Hifney AF, Abdel-Basset R (2014) Induction of hydrogen evolution in *Nostoc* spp. by nitrogen and/or sulphur deficiency. Int J Curr Microbiol Appl Sci 3(9):1028–1044

Hihara Y, Kamei A, Kanehisa M, Kaplan A, Ikeuchi M (2001) DNA microarray analysis of cyanobacterial gene expression during acclimation to high light. Plant Cell 13(4):793–806

Holmer M, Storkholm P (2001) Sulfate reduction and sulfur cycling in lake sediments: a review. Freshw Biol 46:431–451

Hughes AR, Sulesky A, Andersson B et al (2018) Sulfate amendment improves the growth and bioremediation capacity of a cyanobacteria cultured on municipal wastewater centrate. Algal Res 32:30–37

Hutber GN, Hutson KG, Rogers LJ (1977) Effect of iron deficiency on levels of two ferredoxins and flavodoxin in a cyanobacterium. FEMS Microbiol Lett 1:193–196

Ikegami F, Ongena G, Sakai R, Itagaki S, Kobori M, Ishikawa T, Kuo YH, Lambein F, Murakoshi I (1993) Biosynthesis of β-(isoxazolin-5-on-2-yl)-l-alanine by cysteine synthase in *Lathyrus sativus*. Phytochemistry 33(1):93–98

Kadowaki T, Nagayama R, Georg J, Nishiyama Y, Wilde A, Hess WR, Hihara Y (2016) A feed-forward loop consisting of the response regulator RpaB and the small RNA PsrR1 controls light acclimation of photosystem I gene expression in the cyanobacterium *Synechocystis* sp. PCC 6803. Plant Cell Physiol 57(4):813–823

Kah LC, Lyons TW, Frank TD (2004) Low marine sulphate and protracted oxygenation of the Proterozoic biosphere. Nature 431(7010):834–838

Kaneko T, Nakamura Y, Sasamoto S, Watanabe A, Kohara M, Matsumoto M, Tabata S (2003) Structural analysis of four large plasmids harboring in a unicellular cyanobacterium, *Synechocystis* sp. PCC 6803. DNA Res 10(5):221–228

Kaneko T, Nakamura Y, Wolk CP, Kuritz T, Sasamoto S, Watanabe A, Tabata S (2001) Complete genomic sequence of the filamentous nitrogen-fixing cyanobacterium *Anabaena* sp. strain PCC 7120 (supplement). DNA Res 8(5):227–253

Kaneko T, Sato S, Kotani H, Tanaka A, Asamizu E, Nakamura Y, Tabata S (1996) Sequence analysis of the genome of the unicellular cyanobacterium *Synechocystis* sp. strain PCC 6803. II. Sequence determination of the entire genome and assignment of potential protein-coding regions. DNA Res 3(3):109–136

Kaplan A, Reinhold L (1999) CO2 concentrating mechanisms in photosynthetic microorganisms. Annu Rev Plant Biol 50(1):539–570

Kappell AD, van Waasbergen LG (2007) The response regulator RpaB binds the high light regulatory 1 sequence upstream of the high-light-inducible hliB gene from the cyanobacterium *Synechocystis* PCC 6803. Arch Microbiol 187(4):337–342

Karradt A, Sobanski J, Mattow J et al (2008) NblA, a key protein of phycobilisome degradation, interacts with ClpC, a HSP100 chaperone partner of a cyanobacterial Clp protease. J Biol Chem 283(47):32394–32403

Kato H, Kubo T, Hayashi M, Kobayashi I, Yagasaki T, Chibazakura T, Watanabe S, Yoshikawa H (2011) Interactions between histidine kinase NblS and the response regulators RpaB and SrrA are involved in the bleaching process of the cyanobacterium *Synechococcus elongatus* PCC 7942. Plant Cell Physiol 52(12):2115–2122

Katoh H, Hagino N, Grossman AR (2001a) Genes essential to iron transport in the cyanobacterium *Synechocystis* sp. strain PCC 6803. J Bacteriol 183:2779–2784

Katoh H, Hagino N, Ogawa T (2001b) Iron-binding of FutA1 subunit of an ABC-type iron transporter in the cyanobacterium *Synechocystis* sp. strain PCC 6803. Plant Cell Physiol 42: 823–827

Keren N, Aurora R, Pakrasi HB (2004) Critical roles of bacterioferritins in iron storage and proliferation of cyanobacteria. Plant Physiol 135:1666–1673

Kharwar S, Bhattacharjee S, Chakraborty S et al (2021c) Regulation of sulfur metabolism, homeostasis and adaptive responses to sulfur limitation in cyanobacteria. Biologia 76(10): 2811–2835

Kharwar S, Bhattacharjee S, Mishra AK (2021a) Disentangling the impact of sulfur limitation on exopolysaccharide and functionality of Alr2882 by in silico approaches in *Anabaena* sp. PCC 7120. Appl Biochem Biotechnol 193(5):1447–1468

Kharwar S, Bhattacharjee S, Mishra AK (2021b) Bioinformatics analysis of enzymes involved in cysteine biosynthesis: first evidence for the formation of cysteine synthase complex in cyanobacteria. 3 Biotech 11(7):1–15

Kharwar S, Mishra AK (2020) Unraveling the complexities underlying sulfur deficiency and starvation in the cyanobacterium *Anabaena* sp. PCC 7120. Environ Exp Bot 172:103966

Kirilovsky D (2007) Photoprotection in cyanobacteria: the orange carotenoid protein (OCP)-related non-photochemical-quenching mechanism. Photosynth Res 93:7–16

Kiyota H, Hirai MY, Ikeuchi M (2014) NblA1/A2-dependent homeostasis of amino acid pools during nitrogen starvation in *Synechocystis* sp. PCC 6803. Metabolites 4(3):517–531

Kiyota H, Ikeuchi M, Hirai MY (2012) Response of amino acid metabolism to sulfur starvation in *Synechocystis* sp. PCC 6803. In: Sulfur metabolism in plants. Springer, Dordrecht, pp 53–59

Kouřil R, Arteni AA, Lax J et al (2005) Structure and functional role of supercomplexes of IsiA and photosystem I in cyanobacterial photosynthesis. FEBS Lett 579:3253–3257

Krämer E, Schmidt A (1989) Nitrite accumulation by *Synechococcus* 6301 as a consequence of carbon-or sulfur-deficiency. FEMS Microbiol Lett 59(1–2):191–195

Kranzler C, Lis H, Finkel OM et al (2014) Coordinated transporter activity shapes high-affinity iron acquisition in cyanobacteria. ISMEJ 8:409–417

Kranzler C, Rudolf M, Keren N et al (2013) Iron in cyanobacteria. Adv Bot Res 65:57–105

Ksionzek KB, Lechtenfeld OJ, McCallister SL et al (2016) Dissolved organic sulfur in the ocean: biogeochemistry of a petagram inventory. Science 354(6311):456–459

Kumaresan V, Nizam F, Ravichandran G et al (2017) Transcriptome changes of blue-green algae, *Arthrospira* sp. in response to sulfate stress. Algal Res 23:96–103

Kurisu G, Zhang HM, Smith JL, Cramer WA (2003) Structure of the cytochrome b6f complex of oxygenic photosynthesis: tuning the cavity. Science 302:1009–1114

Kyndiah O, Rai AN (2007) Induction of sporulation by sulphate limitation in *Nostoc* ANTH, a symbiotic strain capable of colonizing roots of rice plants. NISCAIR 9(14):105

Latifi A, Jeanjean R, Lemeille S et al (2005) Iron starvation leads to oxidative stress in *Anabaena* sp. strain PCC 7120. J Bacteriol 187:6596–6598

Laudenbach DE, Ehrhardt D, Green L et al (1991) Isolation and characterization of a sulfur-regulated gene encoding a periplasmically localized protein with sequence similarity to rhodanese. J Bacteriol 173(9):2751–2760

Laudenbach DE, Grossman AR (1991) Characterization and mutagenesis of sulfur-regulated genes in a cyanobacterium: evidence for function in sulfate transport. J Bacteriol 173(9):2739–2750

Lax JEM, Arteni AA, Boekema E et al (2007) Structural response of photosystem II to iron deficiency: characterization of a new photosystem II-IdiA complex from the cyanobacterium *Thermosynechococcus elongatus* BP-1. Biochim Biophys Acta 1767:528–534

Leynaert A, Bucciarelli E, Claquin P et al (2004) Effect of iron deficiency on diatom cell size and silicic acid uptake kinetics. Limnol Oceanogr 49:1134–1143

Li ZK, Dai GZ, Juneau P et al (2016) Capsular polysaccharides facilitate enhanced iron acquisition by the colonial cyanobacterium, *Microcystis*, sp. isolated from a freshwater lake. J Phycol 52(1): 105–115

Long BM (2010) Evidence that sulfur metabolism plays a role in microcystin production by *Microcystis aeruginosa*. Harmful Algae 9(1):74–81

López-Redondo ML, Moronta F, Salinas P, Espinosa J, Cantos R, Dixon R, Marina A, Contreras A (2010) Environmental control of phosphorylation pathways in a branched two component system: NblS, a branched two component system. Mol Microbiol 78:475–489

Lowenstein TK, Hardie LA, Timofeeff MN et al (2003) Secular variation in seawater chemistry and the origin of calcium chloride basinal brines. Geology 31(10):857–860

Melkozernov AN, Bibby TS, Lin S et al (2003) Time-resolved absorption and emission show that the CP43' antenna ring of iron-stressed *Synechocystis* sp. PCC6803 is efficiently coupled to the photosystem I reaction center core. Biochemistry 42:3893–3903

Michel KP, Pistorius EK (2004) Adaptation of the photosynthetic electron transport chain in cyanobacteria to iron deficiency: the function of IdiA and IsiA. Physiol Plant 119:1–15

Michel KP, Pistorius EK, Golden SS (2001) Unusual regulatory elements for iron deficiency induction of the idiA gene of *Synechococcus elongatus* PCC 7942. J Bacteriol 183:5015–5024

Michel KP, Thole HH, Pistorius EK (1996) IdiA, a 34 kDa protein in the cyanobacteria *Synechococcus* sp. strains PCC 6301 and PCC 7942, is required for growth under iron and manganese limitations. Microbiology 142(Pt9):1635–1645

Morrissey J, Bowler C (2012) Iron utilization in marine cyanobacteria and eukaryotic algae. Front Microbiol 3:43

Muller N, Hauler O, Schenk H (1997) The phycobiliproteins within the cyanoplast of *Cyanophora paradoxa* store carbon, nitrogen, and sulfur for the whole cell. In: Eukaryotism and symbiosis. Springer-Verlag, Berlin, pp 252–260

Mulo P, Sicora C, Aro EM (2009) Cyanobacterial *psbA* gene family: optimization of oxygenic photosynthesis. Cell Mol Life Sci 66(23):3697–3710

Nakamura Y, Kaneko T, Sato S, Ikeuchi M, Katoh H, Sasamoto S, Tabata S (2002) Complete genome structure of the thermophilic cyanobacterium *Thermosynechococcus elongatus* BP-1. DNA Res 9(4):123–130

Nakamura Y, Kaneko T, Sato S, Mimuro M, Miyashita H, Tsuchiya T, Tabata S (2003) Complete genome structure of *Gloeobacter violaceus* PCC 7421, a cyanobacterium that lacks thylakoids. DNA Res 10(4):137–145

Neale PJ, Melis A (1989) Salinity stress enhances photoinhibition of photosynthesis in *Chlamydomonas reinhardtii*. J Plant Physiol 134(5):619–622

Nichols RJ, LaFrance B, Phillips NR et al (2021) Discovery and characterization of a novel family of prokaryotic nanocompartments involved in sulfur metabolism. elife 10:e59288

Nicholson ML, Gaasenbeek M, Laudenbach DE (1995) Two enzymes together capable of cysteine biosynthesis are encoded on a cyanobacterial plasmid. Mol Gen Genet 247(5):623–632

Nicholson ML, Laudenbach DE (1995) Genes encoded on a cyanobacterial plasmid are transcriptionally regulated by sulfur availability and CysR. J Bacteriol 177(8):2143–2150

Nicolaisen K, Hahn A, Valdebenito M et al (2010) The interplay between siderophore secretion and coupled iron and copper transport in the heterocyst-forming cyanobacterium *anabaena* sp. PCC 7120. Biochim Biophys Acta 1798:2131–2140

Niyogi KK (1999) Photoprotection revisited: genetic and molecular approaches. Annu Rev Plant Biol 50(1):333–359

Odom WR, Hodges R, Chitnis PR et al (1993) Characterization of *Synechocystis* sp. PCC 6803 in iron-supplied and iron-deficient media. Plant Mol Biol 23:1255–1264

Ohkawa H, Sonoda M, Katoh H, Ogawa T (1998) The use of mutants in the analysis of the CO_2-concentrating mechanism in cyanobacteria. Can J Bot 76(6):1035–1042

Ohkawa H, Sonoda M, Shibata M, Ogawa T (2001) Localization of NAD(P)H dehydrogenase in the cyanobacterium *Synechocystis* sp. strain PCC 6803. J Bacteriol 183(16):4938–4939

Ortega-Calvo JJ, Stal LJ (1994) Sulphate-limited growth in the N_2-fixing unicellular cyanobacterium *Gloeothece* (Nägeli) sp. PCC 6909. New Phytol 128(2):273–281

Palenik B, Brahamsha B, Larimer FW, Land M, Hauser L, Chain P, Waterbury J (2003) The genome of a motile marine *Synechococcus*. Nature 424(6952):1037–1042

Palenik B, Ren Q, Dupont CL et al (2006) Genome sequence of *Synechococcus* PCC 9311: insights into adaptation to a coastal environment. Proc Natl Acad Sci U S A 103:13555–13559
Petrychenko Y, Peryt TM, Chechel EI (2005) Early Cambrian seawater chemistry from fluid inclusions in halite from Siberian evaporites. Chem Geol 219(1–4):149–161
Pollari M, Ruotsalainen V, Rantama¨ki, S., Tyystja¨rvi, E., Tyystja¨rvi, T. (2009) Simultaneous inactivation of sigma factors B, D interferes with light acclimation of the cyanobacterium *Synechocystis* sp. strain PCC 6803. J Bacteriol 191:3992–4001
Price GD, Badger MR, von Caemmerer S (2011) The prospect of using cyanobacterial bicarbonate transporters to improve leaf photosynthesis in C3 crop plants. Plant Physiol 155(1):20–26
Prioretti L, Giordano M (2016) Direct and indirect influence of sulfur availability on phytoplankton evolutionary trajectories. J Phycol 52(6):1094–1102
Prioretti L, Lebrun R, Gontero B et al (2016) Redox regulation of ATP sulfurylase in microalgae. Biochem Biophys Res Commun 478(4):1555–1562
Qiu GW, Koedooder C, Qiu BS, Shaked Y, Keren N (2022) Iron transport in cyanobacteria–from molecules to communities. Trends Microbiol 30(3):229–240
Rachedi R, Foglino M, Latifi A (2020) Stress signaling in cyanobacteria: A mechanistic overview. Life 10:312
Ravina CG, Chang CI, Tsakraklides GP et al (2002) The sac mutants of *Chlamydomonas reinhardtii* reveal transcriptional and posttranscriptional control of cysteine biosynthesis. Plant Physiol 130(4):2076–2084
Riediger M, Hihara Y, Hess WR (2018) From cyanobacteria and algae to land plants: the RpaB/Ycf27 regulatory network in transition. Perspect Phycol 5:13–25
Riediger M, Kadowaki T, Nagayama R, Georg J, Hihara Y, Hess WR (2019) Biocomputational analyses and experimental validation identify the regulon controlled by the redox-responsive transcription factor RpaB. iScience 15:316–331
Riethman HC, Sherman LA (1988) Purification and characterization of an iron stress-induced chlorophyll-protein from the cyanobacterium *Anacystis nidulans* R2. Biochim Biophys Acta 935:141–151
Rivers AR (2009) Iron limitation and the role of siderophores in marine *Synechococcus*. New College of Florida. Paper for the degree of Doctor. p 3
Rudolf M, Kranzler C, Lis H et al (2015) Multiple modes of iron uptake by the filamentous, siderophore-producing cyanobacterium, *Anabaena* sp. PCC 7120. Mol Microbiol 97:577–588
Sandmann G (1985) Consequences of iron deficiency on photosynthetic and respiratory electron transport in blue-green algae. Photosynth Res 6:261–271
Sandmann G, Malkin R (1983) Iron-sulfur centers and activities of the photosynthetic electron transport chain in iron-deficient cultures of the blue-green alga *Aphanocapsa*. Plant Physiol 73:724–728
Sandström S, Park YI, Oquist G et al (2001) CP43', the isiA gene product, functions as an excitation energy dissipator in the cyanobacterium *Synechococcus* sp. PCC 7942. Photochem Photobiol 74:431–437
Sato N, Kamimura R, Kaneta K et al (2017) Species-specific roles of sulfolipid metabolism in acclimation of photosynthetic microbes to sulfur-starvation stress. PLoS One 12(10):e0186154
Sato N, Ohmori M, Ikeuchi M, Tashiro K, Wolk CP, Kaneko T, Kuhara S (2004) Use of segment-based microarray in the analysis of global gene expression in response to various environmental stresses in the cyanobacterium *Anabaena* sp. PCC 7120. J Gen Appl Microbiol 50(1):1–8
Schena M, Shalon D, Davis RW, Brown PO (1995) Quantitative monitoring of gene expression patterns with a complementary DNA microarray. Science 270(5235):467–470
Schmidt A (1988) [62] sulfur metabolism in cyanobacteria. In: Methods in enzymology. Elsevier, Oxford, pp 572–583
Schmidt A, Erdle I, Köst HP (1982) Changes of C-phycocyanin in *Synechococcus* 6301 in relation to growth on various sulfur compounds materials and methods. Z Naturforsch C 37(10):870–876

Schmidt A, Jäger K (1992) Open questions about sulfur metabolism in plants. Annu Rev Plant Biol 43(1):325–349

Schrader PS, Milligan AJ, Behrenfeld MJ (2011) Surplus photosynthetic antennae complexes underlie diagnostics of iron limitation in a cyanobacterium. PLoS One 6:e18753

Seino Y, Takahashi T, Hihara Y (2009) The response regulator RpaB binds to the upstream element of photosystem I genes to work for positive regulation under low-light conditions in *Synechocystis* sp. strain PCC 6803. J Bacteriol 191:1581–1586

Seki A, Hanaoka M, Akimoto Y, Masuda S, Iwasaki H, Tanaka K (2007) Induction of a group 2 s factor, RPOD3, by high light and the underlying mechanism in *Synechococcus elongatus* PCC 7942. J Biol Chem 282:36887–36894

Sen A, Dwivedi K, Rice KA et al (2000) Growth phase and metal-dependent regulation of the *dpsA* gene in *Synechococcus* sp. strain PCC 7942. Arch Microbiol 173:352–357

Shcolnick S, Summerfield TC, Reytman L et al (2009) The mechanism of iron homeostasis in the unicellular cyanobacterium *Synechocystis* sp. PCC 6803 and its relationship to oxidative stress. Plant Physiol 150:2045–2056

Shibato J, Asayama M, Shirai M (1998) Specific recognition of the cyanobacterial psbA promoter by RNA polymerases containing principal sigma factors. Biochim Biophys Acta 1442:296–303

Sicora CI, Ho FM, Salminen T, Styring S, Aro EM (2009) Transcription of a "silent" cyanobacterial psbA gene is induced by microaerobic conditions. Biochim Biophys Acta 1787:105–112

Singh SP, Klisch M, Sinha RP et al (2010) Sulfur deficiency changes mycosporine-like amino acid (MAA) composition of *Anabaena variabilis* PCC 7937: a possible role of sulfur in MAA bioconversion. Photochem Photobiol 86(4):862–870

Straus NA (1994) Iron deprivation: physiology and gene regulation. In: Bryant DA (ed) The molecular biology of cyanobacteria, Advances in photosynthesis, vol 1. Springer, Dordrecht, pp 731–750

Stroebel D, Choquet Y, Popot JL et al (2003) An atypical haem in the cytochrome b6f complex. Nature 426:413–418

Sugimoto K, Midorikawa T, Tsuzuki M et al (2008) Upregulation of PG synthesis on sulfur-starvation for PS I in *Chlamydomonas*. Biochem Biophys Res Commun 369(2):660–665

Sugimoto K, Sato N, Tsuzuki M (2007) Utilization of a chloroplast membrane sulfolipid as a major internal sulfur source for protein synthesis in the early phase of sulfur starvation in *Chlamydomonas reinhardtii*. FEBS Lett 581(23):4519–4522

Sugimoto K, Tsuzuki M, Sato N (2010) Regulation of synthesis and degradation of a sulfolipid under sulfur-starved conditions and its physiological significance in *Chlamydomonas reinhardtii*. New Phytol 185(3):676–686

Suzuki M (1989) SPXX, a frequent sequence motif in gene regulatory proteins. J Mol Biol 207(1): 61–84

Takahashi H, Kopriva S, Giordano M et al (2011) Sulfur assimilation in photosynthetic organisms: molecular functions and regulations of transporters and assimilatory enzymes. Annu Rev Plant Biol 62:157–184

Takahashi S, Murata N (2008) How do environmental stresses accelerate photoinhibition? Trends Plant Sci 13:178–182

Takahashi T, Nakai N, Muramatsu M, Hihara Y (2010) Role of multiple HLR1 sequences in the regulation of the dual promoters of the *psaAB* genes in *Synechocystis* sp. PCC 6803. J Bacteriol 192:4031–4036

Tolle J, Michel KP, Kruip J et al (2002) Localization and function of the IdiA homologue Slr1295 in the cyanobacterium *Synechocystis* sp. strain PCC 6803. Microbiology 148:3293–3305

Toulzal E, Tagliabue A, Blain S et al (2012) Analysis of the Global Ocean sampling (GOS) project for trends in iron uptake by surface ocean microbes. PLoS One 7(2):e30931

Walworth NG, Fu FX, Webb EA et al (2016) Mechanisms of increased *Trichodesmium* fitness under iron and phosphorus co-limitation in the present and future ocean. Nat Commun 7:12081

Wang Q, Hall CL, Al-Adami MZ et al (2010) IsiA is required for the formation of photosystem I super complexes and for efficient state transition in *Synechocystis* PCC 6803. PLoS One 5: e10432

Wang Q, Zhang Y, Wu H et al (2021) Effects of sulfur limitation on nitrogen and sulfur uptake and lipid accumulation in *Scenedesmus acuminatus*. J Appl Phycol 33(1):301–311

Wang RY, Liu LH, Zhao FJ, Huang XY (2022) Local and systemic response to heterogeneous sulfate resupply after sulfur deficiency in Rice. Int J Mol Sci 23(11):6203

Wanner G, Henkelmann G, Schmidt A et al (1986) Nitrogen and sulfur starvation of the cyanobacterium *Synechococcus* 6301 an ultrastructural, morphometrical, and biochemical comparison. Z Naturforsch C 41(7–8):741–750

Wase N, Ow S, Wright P (2012) A global understanding of light stress in cyanobacteria: environmental and bioproducts perspectives. In: Srivastava AK, Rai AN, Neilan BA (eds) Stress biology of cyanobacteria: molecular mechanisms to cellular responses. CRC Press, Hoboken, p 394

Willenbrock H, Ussery DW (2004) Chromatin architecture and gene expression in *Escherichia coli*. Genome Biol 5:25

Williams SB (2007) A circadian timing mechanism in the cyanobacteria. Adv Microb Physiol 52:229–296

Wirtz M, Hell R (2006) Functional analysis of the cysteine synthase protein complex from plants: structural, biochemical and regulatory properties. J Plant Physiol 163(3):273–286

Yamaguchi Y, Nakamura T, Kusano T, Sano H (2000) Three *Arabidopsis* genes encoding proteins with differential activities for cysteine synthase and β-cyanoalanine synthase. Plant Cell Physiol 41(4):465–476

Yasuda A, Inami D, Hanaoka M (2020) RpaB, an essential response regulator for high-light stress, is extensively involved in transcriptional regulation under light-intensity upshift conditions in *Synechococcus elongatus* PCC 7942. J Gen Appl Microbiol 66(2):73–79

Yodsang P, Raksajit W, Aro EM et al (2018) Factors affecting photobiological hydrogen production in five filamentous cyanobacteria from Thailand. Photosynthetica 56(1):334–341

Yousef N, Pistorius EK, Michel KP (2003) Comparative analysis of idiA and isiA transcription under iron starvation and oxidative stress in *Synechococcus elongatus* PCC 7942 wild-type and selected mutants. Arch Microbiol 180:471–483

Zechmann B, Tomašić A, Horvat L et al (2010) Subcellular distribution of glutathione and cysteine in cyanobacteria. Protoplasma 246(1):65–72

Zhang Y, Huang Z, Zheng H et al (2020) Growth, biochemical composition and photosynthetic performance of *Scenedesmus acuminatus* under different initial sulfur supplies. Algal Res 45: 101728

Zhang Z, Pendse ND, Phillips KN et al (2008) Gene expression patterns of sulfur starvation in *Synechocystis* sp. PCC 6803. BMC Genomics 9(1):1–14

Spirulina: From Ancient Food to Innovative Super Nutrition of the Future and Its Market Scenario as a Source of Nutraceutical

4

Sandeep Kumar Singh, Livleen Shukla, Nisha Yadav, Prashant Kumar Singh, Shiv Mohan Singh, Mukesh Kumar Yadav, Kaushalendra, and Ajay Kumar

Abstract

In the current century, people worldwide are facing hunger due to limitations in the food supply; to combat the scarcity of food, we are looking for an alternate source of food that contains all the necessary nutrients which are present in our normal diet which boost the immune system and provide energy to our body. *Spirulina* belonging to *Cyanophyta* has emerged as food for the future or super nutrition of the future as it contains all the necessary nutrients required in our diet. *Spirulina* is regarded as one of the most studied and commercialized alga having a higher concentration of proteins content. Furthermore, easily digestible nature and various health benefits, the leading world organization such as WHO and FAO consider spirulina as a superfood or future food. In the existing literature, we would focus on the biochemical composition of the alga, properties that make it future food, and prospective related to the algal food and market scenario of the present and future.

S. K. Singh · L. Shukla
Division of Microbiology, Indian Agricultural Research Institute, Pusa, New Delhi, India

N. Yadav
Division of Agricultural Extension, ICAR-Indian Agricultural Research Institute, New Delhi, India

P. K. Singh · M. K. Yadav
Department of Biotechnology, Mizoram University, Aizawl, India

S. M. Singh · A. Kumar (✉)
Department of Botany, Centre of Advanced Studies in Botany, Institute of Science, Banaras Hindu University, Varanasi, India

Kaushalendra
Department of Zoology, Mizoram University, Aizawl, India

B. Neilan et al. (eds.), *Cyanobacterial Biotechnology in the 21st Century*,
https://doi.org/10.1007/978-981-99-0181-4_4

Keywords

Spirulina · Nutraceuticals · Market trend · Toxicological aspect · Nutrition etc.

4.1 Introduction

Cyanobacteria are photoautotrophs found in aquatic habitats (Downing et al. 2001; Miyatake et al. 2013; Charpy et al. 2012). They have a unique ability to fix atmospheric carbon dioxide but also contain photosynthetic bluish-green pigment, referred to as blue-green algae. The characteristic feature abundantly found in cyanobacteria is the presence of heterocysts, whose function is to fix atmospheric nitrogen both free-living and in symbiotic form (Singh et al. 2013). They are found in colonial, coccoid, unicellular, or filamentous forms (Zhang et al. 2011; Sabart et al. 2010; Kazmierczak et al. 2009; Morin et al. 2010).

Spirulina, a photosynthetic, filamentous, and spiral-shaped blue-green algae, discovered from Lake Chad in Mexico during the Aztec period. Blue green algae belonging to kingdom *cyanophyta* are reservoir of phycocyanin (blue colored) and chlorophyll (green) pigment hence referred to as blue green algae. In the recent years, *Spirulina* has been considered a very frequently studied alga. The presence of high protein content and varied amount of other nutrients such as carbohydrates, vitamins, lipids, and minerals make them a popular resource as a future food or superfood for people (Fox 1996; Pelaez 2006).

During the 1990s, most astronauts from NASA used algae as their food in space. Later, they concluded that if the alga being cultured is modified by certain environmental factors, it can be considered as a feed for the space outreach program. Applied Microbiology International in 1977 referred *S. platensis* as a "wonderful future food source" (Sasson 1997). The World Health Organization (WHO) reported that algae could be considered as one of the best food supplements for health. Apart from food supplements they have also found its application in pharmaceutics to combat disease caused by different microbes like bacteria, fungi, mold etc. (Henrikson 1989). In addition, *Spirulina* also plays a crucial role in the production of antibodies, cytokinin, macrophages, and other immunomodulatory biochemicals that significantly inhibit different viral diseases or even cancer (Moorhead et al. 2005; Blinkova et al. 2001).

If we talk about the past, current, and future market scenarios, many countries, for example, the USA, India, China, Japan, and Switzerland, have been working toward the production of *Spirulina* biomass and processing it in different food products as a rich source of nutrition also producing bioactive molecule to treat disease hence numerous industries have been set up across the world who are aiming to increase the production which can serve as the food for future and food for the current population who are dying of hunger and malnutrition.

4.2 Biochemical Composition of *Spirulina* Biomass

As shown in Fig. 4.1, *Spirulina* biomass mainly comprises ~65% protein and 20% carbohydrate; however, other nutrients such as minerals, fats, and moisture constituents are also present.

4.2.1 Proteins

To date, *Spirulina* is the only known algae with the richest source of protein containing nearly 70% protein which is sufficient for the human diet (Phang and Chu 1999). They are building blocks of amino acids leucine, valine, isoleucine, tryptophan, methionine, phenylalanine, threonine, and lysine (Colla et al. 2007). It has also been reported that methionine and cysteine are found in a lower value, whereas albumin and casein are found in a higher value, respectively (Vonshak 1997; Fujisawa et al. 2010). In addition, *Spirulina* contains phycocyanin which is 20% of all protein fractions. The Phycocyanin molecule is special in providing immunity against diseases (Iijima et al. 1982). *Spirulina* is devoid of cellulose, increasing protein digestibility (Cifferi 1983).

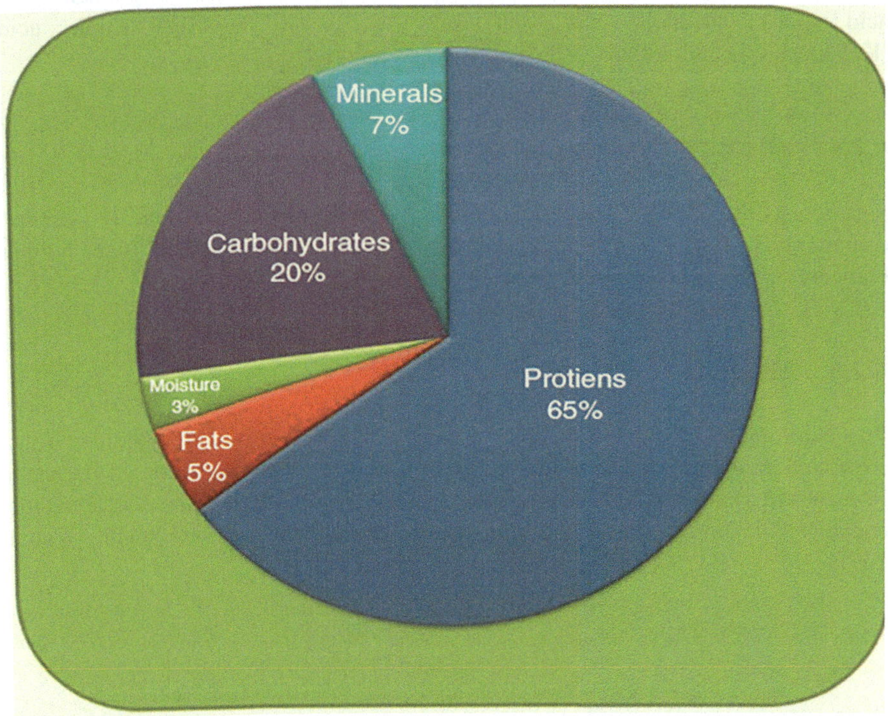

Fig. 4.1 A Pie chart diagram showing the biochemical composition of *Spirulina* biomass

4.2.2 Carbohydrate

Spirulina comprises 15–20% of carbohydrates, which are mainly constituted of sucrose, mannitol, glucose, glycerol, fructose, and sorbitol. *Spirulina* is thought to be eight times higher in mesoinositol than meat and far more than vegetables. According to recent research, calcium spirulan (Ca-SP) is a new sulfated polysaccharide identified in *Spirulina* (Pyne et al. 2017). Certain polysaccharides are also thought to be efficient in providing immune-stimulating and -regulating properties (Baojiang 1994; Evets 1994).

4.2.3 Lipids

Spirulina consist 7% of lipids. Concentration, which mainly contains higher amount of polyunsaturated fatty acids, which are an essential requirement for a healthy life. They are also mostly known as essential fatty acids comprising linoleic, linolenic, and arachidonic. It reduces cholesterol and triglyceride levels in the human body, causing arteriosclerosis and heart disease. The cell membrane of *Spirulina* is mainly composed of lipids. Together with vitamins A and E, they defend cell membranes against antioxidant and free radical assaults. Gamma-linolenic acid (GLA), a fatty acid, has been shown to be an efficient immune stimulant constituting arachidonic acid, eicosapentaenoic acid, stearidonic acid, and docosahexaenoic acid (Kulshreshtha et al. 2008).

4.2.4 Vitamins

Spirulina is the richest source of vitamins, mainly Vitamin A, E, and B complex containing B_1, B_7, B_8, and B_{12}. In addition, Beta-carotene is absorbed by the human body and is also a potent antioxidant (Kapoor and Mehta 1993).

4.2.5 Minerals

The major minerals in *Spirulina* are iron, calcium, phosphorus, and potassium. They are found at high content, which helps protect from anemia, mostly in pregnant women and small children (Puyfoulhoux et al. 2001). In addition, the *Spirulina* contains Calcium and Phosphorus, favoring bone calcification and health (Walter 1997).

4.3 Health Benefits of *Spirulina*

Many researchers around the globe, including countries like Japan, China, India, Europe, the USA, and many more, are studying how this alga is effective as a source of human nutrition. So, after going through the research that has been held on the potential of alga, the following points focus on the possible role of alga on human health:

- Reduce the risk of cancer
- Increase antiviral activity
- Reduce radiation sickness
- Build healthy lactobacillus
- Stimulates the immune system
- Improves wound healing
- Reduces kidney toxicity
- Offers antiaging and neuroprotective benefits
- Overcomes malabsorption and malnutrition

4.3.1 Antioxidant Activity of *Spirulina*

Most research has shown that in vivo and in vitro condition of *Spirulina* has demonstrated significant antioxidant activity. In addition, *Spirulina* alcohol extract inhibits lipid peroxidation more efficiently than chemicals used as antioxidants such as tocopherol. *Spirulina* water extract also shows greater antioxidant activity than chlorogenic and gallic acid.

4.4 Antibacterial and Antiviral Activity of *Spirulina*

Spirulina has antibacterial properties against disease-causing pathogenic bacteria. Incorporating 0.1%, *Spirulina* results in a distinct zone of bacteria, for ex *E. coli* and *S. aureus,* 30 min after injection with nearly zero bacterial count in the blood; this result was acquired due to injection of *Spirulina* (Hayashi et al. 1993). Compared to volatile antibacterial components, petroleum ether, dichloromethane, ethyl acetate extract, and *Spirulina* methanol extract shows more potential antibacterial activity (Ozdemir et al. 2004). *Spirulina* inhibits viral replication at lower doses by inhibiting virus replication at higher concentrations. Furthermore, a water-soluble extract of *Spirulina* suppressed viral cell penetration and replication of the Herpes Simplex Virus Type 1 (HSV-1) in HeLa cell culture in a dose-dependent manner. In addition, the water-soluble extract inhibits viral protein production at 1 mg/mL without impairing host cell functioning. The antiviral activity has also led to the development of a sulfated polysaccharide known as "Calcium Spirulan" (Ca-Sp), which inhibits the reproduction of several enveloped viruses by preventing viral entry into target cells without causing host damage. Hayashi et al. 1993 reported that Ca-Spirulan exhibits antiviral activity against human HIV-1, cytomegalovirus, influenza A virus, measles virus, HSV-1, and mumps virus. Because of its minimal anticoagulant action, extended half-life in the blood, and dose-dependent bioactivity, active Ca-Spirulan can be an excellent promising option for therapeutic potency against HIV-1 and other viruses (Hayakawa et al. 1997).

4.5 Anticancer Activity of *Spirulina*

Phycocyanin extracted from the algae carries numerous antioxidant properties, anti-inflammatory, anticancer activities, etc. (Subbashini et al. 2004). *Spirulina's* synergistic antioxidant and immunological modulating properties help explore possible mechanisms for disrupting tumor-forming cell machinery and play an essential role in preventing cancer. The study by Kumari et al. 2011 concluded *Spirulina's* impact on cancer prevention among tobacco chewers in Kerala, India. It was found that supplementation of *Spirulina* at 1 g/day for 1 year resulted in complete inhibition of cancer cells; hence, regarded as a therapeutic agent (Belay 2002).

4.5.1 *Spirulina's* Role in Immunity Boosting

Spirulina plays a significant role in boosting immunity against viruses, bacteria, etc. by providing resistance against disease by activating macrophages, T and B cells (Schwartz and Shklar 1987). Sulfolipids extracted from *Spirulina* are potent against HIV, herpes virus, cytomegalovirus, influenza virus, etc. They can even inhibit cancer-causing agents (Blinkova et al. 2001). *Spirulina* stimulates macrophages as well as NK and T cell activity. This mechanism causes the release of interferon-gamma (IFN-γ), which finally leads to virus inactivation (Hirahashi et al. 2002; Borchers et al. 2007).

4.5.2 Use of *Spirulina*, Its Perspective as a Source of Functional Feed, and Caution Related to Its Use on Health

Spirulina is considered a functional feed or superfood due to following reasons:

1. Due to its high protein content, it is a great vegan diet that assists in keeping muscles healthy and provides energy by boosting the metabolic rate.
2. Having antioxidant properties helps protect against cardiovascular disorders, viral infection, and cancer and produces antiaging effects.
3. Researchers have demonstrated that 5 g of *Spirulina* contains about 180% more calcium than whole milk.
4. There are also β-carotene and other crucial essential macro- and micronutrients along with vitamins which help in boosting the immunity of human beings.

Many products are produced by amending dried powdered biomass of *Spirulina*, the most known edible products used as a source of food, e.g., bakeries, desserts, juices, snacks, and food. Many recipe books have been published, and products have been commercialized on a large scale. The taste of the products amended with biomass of the alga has been observed that the amount should be no more than 15 g for a pleasant taste of food. Since it is evident from the definition of functional food that it is a food which is fortified with all the nutritional additives and a related word which is being attached to this concept is known as nutraceuticals which are dietary supplements. *Spirulina* is a potential organism for both nutraceuticals and functional foods. Dried biomass of *Spirulina* has been commercialized as functional feed and food. For example, the fatty acid gamma-linolenic acid extracted from *Spirulina* is considered beneficial for treating numerous human deficiency diseases. The photosynthetic pigment phycocyanin is a powerful antioxidant obtained from *Spirulina*, commonly employed in the food industry as a food colorant, emulsifier, thickener, and gelling agent. One of the remarkable properties of the alga is that it lacks cellulose, making it easily digestible as it is the richest source of protein, accounting for nearly 60%. Its cultivation is increased due to its potential to be food for tomorrow but still needs to be considered by an extensive world population. But has the potential to eradicate hunger which many people around the globe still face. Though it is not good in taste, certain techniques can improve taste and other parameters which could make the algae food for tomorrow and remove malnutrition from the world. Even the functional food derived from algae is used as feed for livestock. Technological aspects of the alga need to be explored to achieve economic growth and possess the advantage of growing in a natural environment, giving it the potential to be an alternate source of conventional crops. Due to changing climatic conditions as the land is depleted due to urbanization which eventually leads to the destruction of agricultural land; hence *Spirulina* food becomes an essential source of nutrition. Products marketed with alga biomass should be technologically accepted as per safety standards for food use. Most species of kingdom cyanophyte have been proven to be toxic, but *Spirulina* has proved safe, not secreting any poisonous substance. Still, care needs to be undertaken regarding eliminating any harmful material or metal from the culture system of the algae. Hence, it is significantly important to collect the alga from authentic sources. Its safer side is proved by performing clinical studies using laboratory animals to check alga's toxicity and potential impact. However, it has been confirmed that it is safe for human and livestock use. Moreover, the US health department has also recognized the algae as GRAS (Generally Recognized as Safe). But, safety measures one should keep in mind before consuming it in our diet, sampling needs to be done in case it should not negatively impact human health (Small 2011).

4.5.3 Case Study of *Spirulina* as a Source of Food in India

Mani (2007) conducted a case study in which *Spirulina* biomass was amended in different food at 1 g, 2.5 g, and 5 g and then ranked as per the acceptance level. The following recipes included different types of food and snacks. It was observed that among all the food recipes concerning appearance, texture, and taste, 1 g and 2.5 g were considered acceptable. Hence by this observation, we can prepare food supplements with this amount which would prove sufficient for the normal human diet and helps in combating disease and overcoming the problem of malnutrition.

4.6 Safety Assurance and Toxicological Aspect of *Spirulina*

Another aspect of *Spirulina* is that it produces toxins; some of its supplements have also been detected with microcystins, and its increased levels limit causes liver damage and many other gastrointestinal problems (De Figueiredo et al. 2004). These toxins are produced during cultivation in open ponds as these ponds are subjected to contamination of different blue-green algae, producing toxins with heavy metals that are not found when cultivated in closed photobioreactors. Hence, it becomes important to produce products from *Spirulina* biomass under optimized conditions, daily dose ranging between 10 and 19 g is considered to be safe. But in Africa, the dose reaches up to 40 g daily, which causes negative consequences. Moreover, the negative effect of its excessive consumption leads to nausea, diarrhea, fatigue, or headache (Roy-Lachapelle et al. 2017).

4.6.1 Recent Development and Future of *Spirulina* as a Source of Food

With rising demand for *Spirulina* as a source of nutraceuticals, the transgenic technique has been adopted to improve the quality of feed supplements worldwide (Gaoge et al. 2004). Conjugation is a widely used technique to transfer genes in microalgae, but very little information about *Spirulina* is available; electroporation is also used. The production of nutritional supplements from alga is restricted to the USA, China, and other developing countries. Also, certain studies have reported it as food for women, which helps improve women's well-being by decreasing the chance of menstrual disorders and even reducing neuro-inflammation in lactating women (Patil et al. 2018). Nephrotoxicity is yet another human disease controlled through *Spirulina* intake with camel milk (Hamad et al. 2018).

4.6.2 A Market Scenario of *Spirulina* as a Functional Food

The *Spirulina* market is expected to surge worldwide at a CAGR (Compound Annual Growth Rate) of 8.02% from 2018 to 2023, per MRFR analysis, as *Spirulina* is rich in biochemical compositions and hence used in the majority of food industries as a nutritional supplement. Most food industries have opted to ban the utilization of artificial colors due to their toxic effect, raising safety concerns and demand for natural dyes. *Spirulina* has become an effective natural food color because most food and beverage manufacturers use the blue-colored compound known as phycocyanin. Also, phycocyanin pigment can be mixed with other natural pigments to give it an appealing appearance to a single product or to produce a different product. *Spirulina* has shown its potential in the feed supplement field and has found its way across various sectors like cosmetic and health care personnel. *Spirulina* intake of 1 gram per day is sufficient to bring out children from malnutrition. The alga has also shown its impact in improving physical development and protecting people against disease. Not only for humans, but it has also gained importance as a source of feed for animals, fish, and poultry nutrition. Its demand is increasing due to consumer awareness regarding the negative impact of chemical adulterants on human health. Its demand has increased to minimize its negative impact and have healthy food. Its market is spread across the entire globe. Still, it is mainly concentrated in Asia-Pacific to achieve fast economic growth. Among different types, products from algal powder form are expected to grow at 8.23%. Apart from its application in the food industry, it is widely used in the cosmetic and pharmaceutical industries.

The leading industries impacting the global *Spirulina* market are Cyanotech Corporation (Hawaii), DIC Corporation (Japan), and E.I.D. Parry (India) Limited (India), Tianjin Norland Biotech Co. Ltd. (China), Hydrolina Biotech Private Limited (India), Jiangxi Alga Biotech (China), and Synergy Natural Products Pty Ltd. (Australia).

4.7 Conclusion

Spirulina and its species *Arthrospira maxima* and *Arthrospira platensis* are reservoir of phycocyanin hence referred to as blue green algae due to the presence of phycocyanin. Still, they are considered a superfood or functional feed due to their high protein content and a variable amount of other nutrients like carbohydrates, vitamins, minerals, and lipid, and due to the lack of cellulose, it is easily digestible hence can be considered as a good source of food increasing the metabolic activity of our body. Moreover, the presence of phycocyanin, which gives the algae antioxidant property, helps protect the human body from various disease caused due to virus, bacteria, and other molds by boosting the human body's immune response and also play a crucial role in the regression of tumor growth which causes cancer.

References

Baojiang G (1994) Study on effect and mechanism of polysaccharides of Spirulina on body immune function improvement. In Second Asia-Pacific conference on algal biotechnology, Singapore, p 24

Belay A (2002) The potential application of spirulina (Arthrospira) as a nutritional and therapeutic supplement in health management. J Med Nutr Nutraceut 5(2):27–45

Blinkova LP, Gorobets OB, Baturo AP (2001) Biological activity of Spirulina. Zh Mikrobiol Epidemiol Immunobiol 2:114–118

Borchers AT, Belay A, Keen CL (2007) Spirulina and immunity. In: Spirulina in human nutrition and health. CRC Press, Boca Raton, pp 191–208

Charpy L, Casareto BE, Langlade MJ, Suzuki Y (2012) Cyanobacteria in coral reef ecosystems: a review. J Mar Biol 2012:1

Cifferi O (1983) Spirulina. The edible microorganism. Microbiol Rev 47(4):551–578

Colla LM, Reinehr CO, Reichert C, Costa JAV (2007) Production of biomass and nutraceutical compounds by Spirulina platensis under different temperature and nitrogen regimes. Bioresour Technol 98(7):1489–1493

De Figueiredo DR, Azeiteiro UM, Esteves SM, Gonçalves FJ, Pereira MJ (2004) Microcystin-producing blooms—a serious global public health issue. Ecotoxicol Environ Saf 59(2):151–163

Downing JA, Watson SB, McCauley E (2001) Predicting cyanobacteria dominance in lakes. Can J Fish Aquat Sci 58(10):1905–1908

Evets L (1994) Means to normalise the levels of immunoglobulin E, using the food supplement Spirulina. Grodenski State Medical University Russian Federation Committee of Patents and Trade. Patent (19) RU (11) 2005486. Jan 15, 1994. Russia

Fox DR (1996) Spirulina: production & potential. Aix-en-Province, Edisud, p 232

Fujisawa T, Narikawa R, Okamoto S, Ehira S, Yoshimura H, Suzuki I, Masuda T, Mochimaru M, Takaichi S, Awai K, Sekine M (2010) Genomic structure of an economically important cyanobacterium, Arthrospira (Spirulina) platensis NIES-39. DNA Res 17(2):85–103

Gaoge W, Xuecheng Z, Delin D, Chengkui T (2004) Study on recipient system for transgenic manipulation in Spirulina platensis (Arthrospira). Jpn J Phycol 52:243–245

Hamad EM, Mousa HM, Ashoush IS, Abdel-Salam AM (2018) Nephroprotective effect of camel milk and Spirulina platensis in gentamicin-induced nephrotoxicity in rats. Int J Pharmacol 14(4): 559–565

Hayakawa Y, Hayashi T, Hayashi K, Ozawa T, Niiya K, Sakuragawa N (1997) Calcium spirulan as an inducer of tissue-type plasminogen activator in human fetal lung fibroblasts. Biochim Biophys Acta 1355(3):241–247

Hayashi K, Hayashi T, Morita N, Kojima I (1993) An extract from Spirulina platensis is a selective inhibitor of herpes simplex virus type 1 penetration into HeLa cells. Phytother Res 7(1):76–80

Henrikson R (1989) Earth food spirulina. Ronore Enterprises, Laguna Beach, p 187

Hirahashi T, Matsumoto M, Hazeki K, Saeki Y, Ui M, Seya T (2002) Activation of the human innate immune system by spirulina: augmentation of interferon production and NK cytotoxicity by oral administration of hot water extract of Spirulina platensis. Int Immunopharmacol 2(4): 423–434

Iijima N, Fugii I, Shimamatsu H, Katoh S (1982) Anti-tumor agent and method of treatment therewith. US Patent Pending, Ref (P1150-726), p A82679

Kapoor R, Mehta U (1993) Utilization of β-carotene from Spirulina platensis by rats. Plant Foods Hum Nutr 43(1):1–7

Kazmierczak J, Altermann W, Kremer B, Kempe S, Eriksson PG (2009) Mass occurrence of benthic coccoid cyanobacteria and their role in the production of Neoarchean carbonates of South Africa. Precambrian Res 173(1–4):79–92

Kulshreshtha A, Jarouliya U, Bhadauriya P, Prasad GBKS, Bisen PS (2008) Spirulina in health care management. Curr Pharm Biotechnol 9(5):400–405

Kumari DJ, Babitha B, Jaffar S, Prasad MG, Ibrahim MD, Khan MS (2011) Potential health benefits of Spirulina platensis. Int J Adv Pharm Sci 2:417–422

Mani IU (2007) Spirulina and its therapeutic implications as a food product. In: Spirulina in human nutrition and health. CRC Press, Boca Raton, pp 65–84

Miyatake T, MacGregor BJ, Boschker HT (2013) Depth-related differences in organic substrate utilization by major microbial groups in intertidal marine sediment. Appl Environ Microbiol 79(1):389–392

Moorhead K, Capelli B, Cysewski G (2005) Nature's superfood: spirulina. ISBN #0-9637511-3-1

Morin N, Vallaeys T, Hendrickx L, Natalie L, Wilmotte A (2010) An efficient DNA isolation protocol for filamentous cyanobacteria of the genus Arthrospira. J Microbiol Methods 80(2): 148–154

Ozdemir G, Ulku Karabay N, Dalay MC, Pazarbasi B (2004) Phytotherapy Research: An International Journal Devoted to Pharmacological and Toxicological Evaluation of Natural Product Derivatives 18(9):754–757

Patil J, Matte A, Mallard C, Sandberg M (2018) Spirulina diet to lactating mothers protects the antioxidant system and reduces inflammation in post-natal brain after systemic inflammation. Nutr Neurosci 21(1):59–69

Pelaez F (2006) The historical delivery of antibiotics from microbial natural products—can history repeat? Biochem Pharmacol 71(7):981

Phang SM, Chu WL (1999) University of Malaya Algae Culture Collection (UMACC). Catalogue of strains. Institute of Postgraduate Studies and Research. University of Malaya, Kuala Lumpur

Puyfoulhoux G, Rouanet JM, Besançon P, Baroux B, Baccou JC, Caporiccio B (2001) Iron availability from iron-fortified spirulina by an in vitro digestion/Caco-2 cell culture model. J Agric Food Chem 49(3):1625–1629

Pyne PK, Bhattacharjee P, Srivastav PP (2017) Microalgae (Spirulina platensis) and its bioactive molecules: review. Indian J Nutr 4(2):160

Roy-Lachapelle A, Solliec M, Bouchard MF, Sauvé S (2017) Detection of cyanotoxins in algae dietary supplements. Toxins 9(3):76

Sabart M, Pobel D, Briand E, Combourieu B, Salençon MJ, Humbert JF, Latour D (2010) Spatiotemporal variations in microcystin concentrations and in the proportions of microcystin-producing cells in several Microcystis aeruginosa populations. Appl Environ Microbiol 76(14):4750–4759

Sasson A (1997) Micro biotechnologies: recent developments and prospects for developing countries. Place de Fontenoy, Paris, France: United Nations Educational, Scientific and Cultural Organization (UNESCO), BIOTEC Publication 1/2542; p 11–31

Schwartz J, Shklar G (1987) Regression of experimental hamster cancer by beta carotene and algae extracts. J Oral Maxillofac Surg 45(6):510–515

Singh M, Sharma NK, Prasad SB, Yadav SS, Narayan G, Rai AK (2013) The freshwater cyanobacterium Anabaena doliolum transformed with ApGSMT-DMT exhibited enhanced salt tolerance and protection to nitrogenase activity, but became halophilic. Microbiology 159(Pt_3): 641–648

Small E (2011) 37. Spirulina–food for the universe. Biodiversity 12(4):255–265

Subbashini J, Mahipal SV, Reddy MC, Mallikarjuna Reddy M, Rachamallu A, Reddanna P (2004) Molecular mechanisms in C-Phycocyanin induced apoptosis in human chronic myeloid leukemia cell line K-562. Biochem Pharmacol 68(3):453–462

Vonshak A (ed) (1997) Spirulina platensis arthrospira: physiology, cell-biology and biotechnology. CRC Press, Boca Raton

Walter P (1997) Effects of vegetarian diets on aging and longevity. Nutr Rev 55(1):61–68

Zhang M, Shi X, Yu Y, Kong F (2011) The acclimative changes in photochemistry after colony formation of the cyanobacteria Microcystis aeruginosa 1. J Phycol 47(3):524–532

Response of Cyanobacteria During Abiotic Stress with Special Reference to Membrane Biology: An Overview

5

Pratika Singh, Amrita Srivastava, and Ekta Shukla

Abstract

Cyanobacteria are considered as the first autotroph organisms which are present in almost all possible environments. However, abiotic stresses like temperature, heavy metals, and salt influence the normal metabolic functions and adversely affect its photosynthetic apparatus by increasing reactive oxygen species. Cyanobacteria have evolved themselves in order to respond against harsh conditions, in the form of either change in membrane structure, response of stress-responsive genes, regulatory pathways, or regulating secondary metabolite secretion. Such physiological responses lead to adaptation and acclimatization toward changing environments and provide survival advantage to these ancient organisms. Moreover, the presence of various bioactive compounds like phenolics, flavonoids, alkaloids, phytoprotective metabolites, phytohormones, and mycosporine like amino acids favors cyanobacterial defense strategy, of which the majority is present in plant system too. The purpose of this chapter is to deliver recent knowledge regarding the adaptive response of cyanobacteria including molecular chaperons and two-component system protein. Effects of abiotic stress in membrane structure and defense strategy along with the role of bioactive compounds are also included.

P. Singh · A. Srivastava
Department of Life Science, School of Earth, Biological and Environmental Sciences, Central University of South Bihar, Gaya, Bihar, India
e-mail: amritasrivastava@cub.ac.in

E. Shukla (✉)
Sunbeam College for Women Bhagwanpur, Varanasi, Uttar Pradesh, India

B. Neilan et al. (eds.), *Cyanobacterial Biotechnology in the 21st Century*,
https://doi.org/10.1007/978-981-99-0181-4_5

Keywords

Cyanobacteria · Abiotic stress · Salt stress · Temperature · Secondary metabolite · Molecular chaperons · Membrane

5.1 Introduction

Cyanobacteria are regarded as highly diverse, prokaryotic photoautotrophs inhabiting different niches ranging from extreme hot springs to the cryosphere. They are monophyletic microbes that show the process of oxygen-driven photosynthesis and are considered as key players in the nitrogen and carbon cycle. Owing to these properties, they play pivotal roles in the field of agriculture, aquaculture, and bioremediation. The main reason behind residing in all ecological environments is their capability to adapt, adjust, and cope with various abiotic and biotic stresses. Cyanobacteria can adapt under harsh conditions by undergoing several structural modifications through signaling cascades, secreting secondary metabolites, by stress-responsive gene regulation pathways, metabolic activities, and other regulatory systems (Singh 2014). The high range of habitats is a result of its morphologically diverse structures that include unicellular to complex filamentous structures possessing branches as well-specialized structures like heterocysts, akinetes, and vegetative structures (Singh et al. 2020a). Cyanobacteria possess photosynthetic apparatus similar to that of angiosperms with similarity in cytoplasm and thylakoid membranes (Los et al. 2010). Hence, they are proved to be excellent models to study adaptation and their responses during abiotic stress due to their similarity in lipid profiles and protein assembly.

Stress refers to a condition which disturbs the normal state of organisms. Abiotic stress occurs due to change in environmental parameters like change in temperature, pH, heavy metals, salinity, light, UV, drought, and oxidative stress that alter homeostasis. It majorly targets cellular components like photosystems, respiratory chains, enzymes involved in metabolic activities, and components of the membrane (Srivastava et al. 2013). These lead to the generation of oxidative radicals like superoxide anion (O_2^-), hydroxyl ion (OH°), and hydrogen peroxide (H_2O_2). These reactive oxygen species (ROS) get neutralized or ineffective by upregulating genes responsible for encoding metalloenzymes. The regulation of abiotic stress is tightly regulated by sensors and transcription factors (TFs). Similarly, to respond to stresses, cyanobacteria also make changes in its structures like cell membrane's protein or lipid profile so that it can adjust according to the environment that alters cell physiology and development. The various changes include upregulation of DNA repair mechanism, molecular chaperons like heat shock proteins (HSPs), antioxidants, two-component system proteins, and secondary metabolite secretion like phenols, flavonoids, siderophores, etc.

The current book chapter includes changes occurring in cyanobacterial species during abiotic stresses and how they respond toward them. Adaptive mechanisms are crucial for the growth and development of species. This chapter encompasses mechanisms adopted by cyanobacteria during stress tolerance, role of defense proteins, and tools dedicated to study such responses.

5.2 Abiotic Stress and Adaptive Mechanism

Molecular Chaperons The homeostasis of proteins and its biogenesis are maintained by chaperons since they play a fundamental role in folding, refolding, misfolding, and aggregation. The protein quality control is maintained by major chaperon families like small HSP (cyanobacterial homolog is HspA), Hsp60 (GroEL1, GroEL2), and Hsp70 (DnaK1, DnaK2, DnaK3). These constitutively expressed proteins are majorly responsible for assisting protein folding and inhibit protein aggregation by binding to the hydrophobic surface of non-native proteins (Ellis 2001; Hartl et al. 2011). The molecular weight of small HSPs is in between 12 and 14 KDa with no ATPase activity. It consists of α-crystallin that mediates aggregation along with a N and a C terminal arms. Molecular Chaperons get upregulated during temperature variation, salinity, UV, and osmotic stress and rearrange themselves to form quaternary organization (Srivastava et al. 2013).

Small HSPs are homologous to plant chloroplastic and cytoplasmic HSPs, and the function is directly dependent on its oligomeric stability (Giese and Vierling 2002). GroEL shows affinity for unfolded polypeptides that trigger protein folding with the help of peptidyl-prolyl isomerase and ribosomes. STRING (Search Tool for the Retrieval of Interacting Genes/Proteins) analysis showed the interaction of GroEL with other chaperons like GroS, groL, ClpB, dnaK, thioredoxin, and grpE in *Anabaena* sp. (Fig. 5.1). Thylakoid membrane-situated protein DnaK is responsible for increasing membrane fluidity during temperature upshifts and plays a crucial part in protein transcriptional machinery (Katano et al. 2006).

5.3 Two-Component System Protein

The Environmental signals are perceived by a sensor and transducer called two-component system protein which is found in almost all prokaryotes and smaller animals; however, it is absent in higher animals. These include 45 response regulators (Rres) and 47 histidine kinases (Hiks) candidates identified in *Synechocystis* chromosome (Mizuno et al. 1996). The positive regulation of Hik/Rre system follows the inactivation of Hik under non-stressed condition. Consequently, Rre gets inactivated, and thus, genes activated by two-component system remain silent. Upon phosphorylation, Hik gets activated and further activates Rre through signaling cascades. This leads to an increase in the regulation of gene expression induced by abiotic stress. The positive regulation is well recognized as cold sensor in *Synechocystis* which regulates the cold-inducible genes (Los et al. 2010). Such system provides a means of adaptation by monitoring nutrient uptake, cellular division, and secondary metabolite secretion. The component system consists of membrane-bound sensor kinase which is histidine kinase (called as first component).

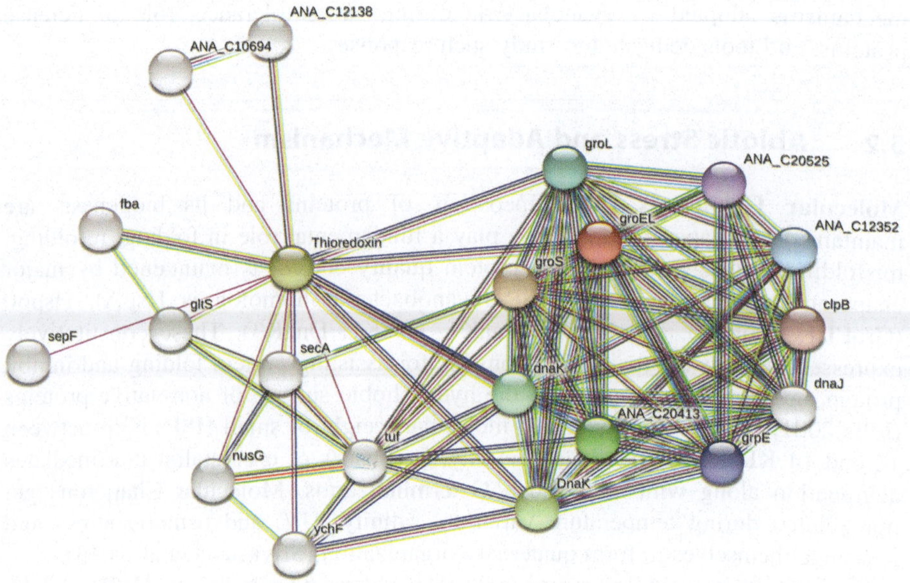

Fig. 5.1 STRING analysis showing molecular chaperon interactions in *Anabaena* sp. Number of nodes, 21; number of edges, 85

It possesses N and C terminal for ligand binding and kinase domain, respectively. ATP-based autophosphorylation occurs at histidine residues that transfer phosphoryl to aspartate residue. The system is also responsible for regulating scytonemin biosynthesis in *Nostoc punctiforme* during oxidative stress mediated by UV rays (Janssen and Soule 2016). Recent studies showed that Hik36-Hik43 and Rre6 are responsible for causing pili-based signal transduction cascade of biofilm formation from planktonic state in *Synechocystis* sp. PCC 6803 under salinity (Kera et al. 2020).

5.4 Effect of Salinity and Membrane Response During Salt Stress

Salt stress is a crucial abiotic stress that has made major parts unproductive globally. Salinization affects organisms severely, thus damaging its structure or inhibiting metabolism. It has been estimated that almost 952 million hectares are affected due to salt stress (Arora et al. 2016). However, the recent trend is alarming as about 1128 million hectares have been affected worldwide (Mandal et al. 2018). Around 15% of area is categorized as severe salinization zone, while the remaining 85% under minimum to fairly affected zone. Since the stress contains an impact of both ionic and osmotic stress, the severity in damage caused in the organism is more. The inhibition in metabolic processes and alteration in morphology, physiology, and biochemical aspects along with the generation of ROS are consequences of salt stress in cyanobacteria. In order to adapt, microbes follow either "salt-in strategy" (like halophiles) or "salt-out strategy" (like cyanobacteria, eukaryotes; Hagemann 2011). The former maintains the turgor pressure and water uptake capacity of cytoplasm and ensures the accumulation of inorganic ions primarily in the form of KCl (Pade and Hagemann 2014). The latter maintains salt balance which is based on two principles: (1) accumulation of compatible solutes for osmotic acclimation by cells and (2) exportation of inorganic ions in order to keep cytoplasmic ions low.

Compatible solutes are known as small molecular weight organic compounds responsible for enhancing cellular osmolarity without interfering with other macromolecules. These do not contain net charge, and varieties include sugar, amino acids, and their derivatives, polyol heterosides (Galinski and Trüper 1994). Freshwater strains, marine strains, moderately halotolerant species, halophilic species, and hypersaline species contain compatible solutes in the form of sucrose/trehalose, glucosylglycerate, glucosylglycerol, glycine betaine, and glutamate betaine, respectively (Welsh 2000; Klähn et al. 2021; Table 5.1). During salt stress, the activity of enzymes like sucrose phosphate phosphatase (Spp) and sucrose phosphate synthase (SpsA) involved in sucrose biosynthesis gets increased while SpsA gets repressed under normal salt condition by the response regulator, Rre39, although the mechanism is unclear (Kirsch et al. 2019). Synthesis of GG involves glucosylglycerol-phosphate synthase (GgpS) and glucosylglycerol-phosphate phosphatase (GgpP) enzymes. LexA and GgpR negatively regulate the *ggpS* gene, while sigma factor SigF induces its transcription (Hagemann et al. 1997). Till date, how LexA gets triggered to target its gene is still in the experimental stage.

Salinity decreases cell volume that restricts the electron transport of photosynthesis (Zhao et al. 2019). It affects PSII and PSI by damaging K/Na ratio and Q_B-non-reduction center and inhibiting D1 protein (Allakhverdiev et al. 2000). Na^+ and Cl^- ion imbalance accumulates ROS, thereby destroying membrane lipid peroxidation and photosynthesis-related apparatus, inhibiting TFs and translation elongation factors EF-G and EF-Tu, and downregulating *psbA* gene regulation responsible for PSII repair (Jimbo et al. 2018; Yang et al. 2020). 0.5 M NaCl is sufficient enough to inhibit D1 protein synthesis due to osmotic stress (Murata et al. 2007).

Recent studies confirm the effect of salt stress on protein synthesis which may be due to three factors: disturbance of polysome and ribosome due to NaCl influx, inactivation of RubisCo and ATP synthesis required for protein synthesis, and inhibition of CO_2 fixation (Murata et al. 2007; Nishiyama and Murata 2014). Salinity affects lipid composition by increasing the amount of unsaturated fatty acid, phosphatidylglycerol, sulfoquinovosyl diacylglycerol, and ratio of monogalactosyldiacylglycerol and digalactosyldiacylglycerol (Sui and Han 2014; Sui et al. 2017).

In order to respond to this abiotic stress, transcriptomic, genomic, and metabolomics data analyses were carried out by Klähn et al. (2021). According to their research, the initial time period of salt stress leads to the reorganization of transcriptomes, upregulation of proteins, and coordinated induction of regulatory

Table 5.1 List of cyanobacterial species with their compatible solutes and stress tolerance capacity (GG, glucosylglycerol; GB, glycine betaine)

S. no.	Species	Compatible solute	Salt tolerance	Reference
1	*Nostoc muscorum*	Sucrose	Stenohaline	Blumwald and Tel-Or (1982)
2	*Anabaena variabilis*	Sucrose	Stenohaline	Erdmann (1983)
3	*Prochlorococcus*		Stenohaline	Klähn et al. (2010)
4	*Synechococcus* sp. PCC 6301	Sucrose	Stenohaline	Blumwald et al. (1983)
5	*Rivularia atra*	Trehalose	Stenohaline	Reed and Stewart (1983)
6	*Synechocystis*	GG	Moderately salt tolerant	Kirsch et al. (2019)
7	*Prochlorococcus* strains	GG	Moderately salt tolerant	
8	*Synechococcus* sp. PCC 7002	GGA	Euryhaline	Klähn et al. (2010)
9	*Synechococcus* strains WH7803 and WH8102	GGA	Salt-treated cells	Klähn et al. (2010)
10	*Aphanothece halophytica* (*Synechococcus* sp. PCC 7418)	GB showed lag phase. GG and proline accumulation	Hypersaline	Fulda et al. (1999)
11	*Prochlorococcus* AS9601	Sucrose	Moderate saline tolerant	Al-Hosani et al. (2015)
12	*Calothrix* strains	GB, ammonia compound glutamate betaine	Hypersaline	Mackay et al. (1984)

RNA involved in stress response. Moreover, Ktr/Kdp system may help during salt shock. Ktr/Kdp system helps in overcoming hyperosmotic shock. This might be due to the fact that Ktr is under control of histidine kinases that play an important role in salt signal. However, this hypothesis requires extensive research.

5.5 Effect of Temperature Stress and Its Response

Cyanobacteria can thrive in a broad range of temperatures; thus, the optimal temperature varies. There are several species that inhabit extreme environments. However, mesophilic cyanobacteria are quite sensitive to temperature stress. Various metabolic activities like photosynthesis, nitrogen fixation, and secondary metabolite production get affected due to temperature fluctuations, and thus, molecular response is a prerequisite for their survival.

Heat shock leads to the induction of several heat shock genes (*hsp*) that in turn activate molecular chaperons (Hsps). These act as heat shock sensors. HrcA binds with controlling inverted repeat of chaperone expression (CIRCE) and represses the transcription of *hsp* genes under optimum temperature (Rachedi et al. 2020). Upon temperature upshifts, HrcA becomes inactive; thereby, induction of *hsp* transcription begins. Formation of thermometer structure occurs by *hsp* mRNAs that under optimum temperature sequester ribosome binding site. These thermometer structures dissociate during heat stress, and thus, the translation process moves further.

Various genes induced during heat stress like *hspA*, *groES*, *dnaI*, *dnaK2*, *sigB*, *sigD*, etc. are responsible for protein folding, protein turnover, and controlling oxidative stress (Suzuki et al. 2005). The increase in membrane fluidity occurs due to heat stress; however, temperature upshifts cannot be perceived through increased membrane fluidity. Other alterations include the inactivation of photosystems, nitrogen fixation, and changes in the composition of polyunsaturated fatty acid in membrane. It is observed that under heat/light stress, monoglucosyldiacylglycerol (MGlcDG) level gets increased in the case of *Synechocystis* and remained stable even at extreme temperature (Balogi et al. 2005). This is possibly because MGlcDG interacts with small HSPs, Hsp17; thus, MGlcDG acts as a heat shock lipid, thereby playing a major role in thermotolerance.

Similarly, presence of cold stress affects mesophilic strains of cyanobacteria by decreasing membrane fluidity, impaired protein folds, and structural alteration in thylakoid membranes. This activates cold shock-induced genes that improve cellular function that are functionally categorized as percepting signal and its transduction, maintaining cell wall and membrane and transcription and translation, and monitoring cellular function. Hik33 regulates cold-induced genes along with genes controlling oxidative and salt stress. Thus, it can be concluded that a common sensor can trigger more than one stress-responsive gene in cyanobacteria.

5.6 Effect of Heavy Metal Stress and Its Response

Cyanobacteria are often challenged by heavy metals and metalloids like aluminum (Al), cadmium (Cd), cobalt (Co), arsenic (As), uranium (U), mercury (Hg), and lead (Pb). These heavy metals do not have any role as nutrient and are mixed into the soil due to natural sources and anthropogenic activities. The first line of defense includes exopolysaccharides (EPS), as these have negative charge due to the presence of uronic acid and can generate stromatolite that helps in heavy metal sequestration (De Philippis et al. 2011). Double mutation of genes *sll1581* and *slr1875* leads to loss of negative charge in EPS (Jittawuttipoka et al. 2013). EPS takes up important metals like iron and provides it to the cell during Cd and Co stress (Kranzler et al. 2013). They also possess cysteine-rich protein called metallothionein that binds with heavy metal to the thiol group of Cys residue. Genes that encode such binding include *smtA* found in *Synechococcus* PCC 7942, and its transcripts get increased upon heavy metal stress (Morby et al. 1993). As resistance mechanism is well studied in cyanobacteria, they possess *arsBHC* tricistronic operon encoded by *arsR* gene. An arsenate reductase, ArsC, responsible for As reduction uses the glutathione/glutaredoxin system; however, the mechanism is not known. Hg (II) affects the photosynthesis ability of cyanobacteria. In response to Hg toxicity, it transforms toxic Hg(II) either to lesser toxic Hg(0) through *slr1849* encoded reductase enzyme, MerA, or to metacinnabar (Lefebvre et al. 2007; Grégoire and Poulain 2014).

Ferredoxins and ferric uptake regulator (FUR)-like proteins have also been identified in cyanobacteria that play an important role in metal homeostasis and protection against oxidative stress due to heavy metals. Ferredoxins utilize iron-sulfur clusters and help in electron transfer. Ferredoxin-induced genes that are widely present in all cyanobacteria are *fed1-fed9* regulated by light and presence of different metals (Cassier-Chauvat and Chauvat 2014). There are three crucial FUR-like regulators present in cyanobacteria important for iron homeostasis: FurA (Sll0567), Slr1738, and Sll1937. In order to survive in heavy metal stress, cyanobacteria synthesize, secrete, and utilize small molecular weight, non-ribosomal secondary metabolite called siderophores (Khan et al. 2017). These chelate ferric ions by forming hexadentate octahedral complexes and transport inside the cells. Recent studies confirm its binding affinity to other metals too thus regarded as metallophores (Singh et al. 2020b). Cyanobacteria produce hydroxamate (e.g., schizokinen and synechobactins A–C) as well as catecholate (e.g., anachelin), kinds of siderophores with ecological significance of chelating and eliminating heavy metals from the environment. Studies confirmed that production of schizokinen in *Anabaena* sp. helps in recovery from cadmium stress (Singh et al. 2016). It was further reported that Cd had negligible effect on siderophore production and possesses thermodynamically stable complex than siderophore-Fe complex. Hence, binding of siderophore with different metals is based on affinity toward siderophore and presence of active pockets.

5.7 Role of Secondary Metabolites as a Signaling Molecule

Cyanobacteria can adapt themselves under various extreme conditions. They secrete secondary metabolites in response to the various stresses, viz., light, UV, osmotic, temperature, etc., which help them to thrive under such conditions. These photosynthetic organisms produce suites of extracellular products known as secondary metabolites. These are natural products that are indirectly required in an organism's primary functions and hence are produced generally after log phase. The metabolites are secreted under specific conditions and are usually unique. Under diverse stress conditions, they produce varied groups of secondary metabolites like polyphenols, alkaloids, terpenoids, and UV-absorbing compounds. These metabolites perform varieties of functions, like photoprotection and antioxidant, and protect against grazers and predators for the defense and survival of the cyanobacterial cells. In the current section, we have summarized important metabolites with their functions like phenolics, terpenoids, photoprotective compounds (MAAs, scytonemin), and antioxidants in particular carotenoids, vitamins, phytohormones, and toxins obtained from cyanobacteria.

5.8 Bioactive Metabolites for Cyanobacterial Adaptations

5.8.1 Phenolics

Phenolic acids are among the key groups of secondary metabolites found to be widely present and distributed in plants and cyanobacteria. This group consists of polyphenols, flavonoids, tannins, and lignins (Thomas and Kim 2011). Cyanobacteria produce phenolic acids to mitigate the photooxidative damage caused by ROS and the hydroxyl radical (OH). Production and deposition of phenolic acids procure better adaptability and tolerance to cyanobacteria under diverse environmental stresses; otherwise, it may lead to free radical deposition and deoxyribose or DNA damage. More recently, Singh et al. (2014) reported a large amount of phenolic acids such as gallic acid, chlorogenic acid, ferulic acid, vanillic acid, and caffeic acid against the salt stress in laboratory-grown cyanobacterial genera, viz., *Plectonema boryanum*, *Anabaena doliolum*, *Oscillatoria acuta*, and *Hapalosiphon intricatus*. The result depicts that phenolic acids can scavenge free radicals under salt stress. Patipong et al. (2019) reported a halotolerant *Halothece* sp. PCC 7418 produces suites of phenolic compounds to stimulate the antioxidant activity of the cells in response to temperature stress. The phlorotannins include phlorethols, fucols, fucophlorethols, fuhalols, and halogenated and sulfated phlorotannins. All these compounds possess promising abilities to repair the damage caused by free radicals under oxidative stress (Kumar et al. 2008).

The other important functions of phenolics include safeguarding against biotic stresses like bacterial settlement and grazing and abiotic stresses like UV radiation and metal toxicity (Lau and Qian 2000; Coleman et al. 2007; Coba et al. 2009; Connan and Stengel 2011). The other metabolites like lignin, flavonoids, phytoalexins, tannins, and anthocyanins are also produced in cyanobacteria and algae under adverse conditions to play a vital role in the defense mechanism of the cell (Stengel et al. 2011). *Microcystis aeruginosa* limits the growth of other microorganisms because of ellagic acids, gallic acids, and catechin (Nakai et al. 2005).

5.8.2 Flavonoids

These are also important antioxidant compounds in addition to phenols present in cyanobacteria and help in their survival. The antioxidative molecule, such as quercetin and rutin, facilitates salt acclimatization of the bacterial cell during stress; similarly, luteolin-7-glucoside and naringenin provide protection against the ROS produced due to high temperature (Singh et al. 2014; Trabelsi et al. 2016;

Jerez-Martel et al. 2017; Patipong et al. 2019; Mallick and Mohn 2000). Cyanobacterial flavonoids having strong antioxidative properties possess pharmacological properties such as protecting kidneys and nerves from damage, inhibiting the proliferation of cancerous cells, and shielding arteries from atherosclerosis (Mulvihill et al. 2010; Assini et al. 2013; Hernández-Aquino and Muriel 2018; Sugumar et al. 2019; Choi et al. 2020).

5.8.3 Alkaloids

Alkaloids are nitrogen-containing metabolites induced under diverse stress conditions by varied organisms including cyanobacteria. Most of the cyanobacterial alkaloids are toxins.

Indole alkaloids are produced by the group of cyanobacteria thriving under extreme conditions for their survival. Till now, more than 80 different indole alkaloids have been reported from different genera of cyanobacteria belonging to subsection V. These alkaloids have been categorized into nine groups (Groups I to IX) based on the carbon skeleton. Group I, the largest group of indole alkaloids, includes hapalindoles produced by cyanobacteria (Mo et al. 2010; Micallef et al. 2014). The presence of vast range of indole alkaloids suggests essential role in ensuring the survival of cyanobacterial cells in extreme conditions. In recent times, most of the terrestrial- and freshwater-inhabiting species such as *Westiella*, *Westiellopsis*, *Fischerella*, and *Hapalosiphon* have been known to produce these secondary metabolites along with some cyanobacterial genera present in extreme environments (Walton and Berry 2016; Demay et al. 2019).

5.8.4 Toxic Metabolites

Cyanobacteria produce a variety of toxins known as cyanotoxins (Gupta et al. 2013). These toxins are mainly produced in blooms on stagnant water by different cyanobacterial species such as *Microcystis* and some of the filamentous strains *Anabaena* and *Nostoc* (Vasas et al. 2004). The cyanotoxins comprise varied chemical structures, viz., ribosomal peptides and NRPs, polyketide alkaloids, and lipopolysaccharides. Some of the cyanobacterial genera like *Microcystis*, *Anabaena*, *Planktothrix*, and *Nostoc* secrete hepatotoxins (which are cyclic peptides produced via NRPs) like microcystin. Another set of cyanobacterial genera release neurotoxin which is found to be β-N-methylamino-L-alanine, a non-protein amino acid (Esterhuizen and Downing 2008; Rastogi and Sinha 2009). Based on their biological properties cyanobacterial toxins have been categorized into cytotoxin targeting cells, neurotoxin targeting nervous system, and hepatotoxin attacking liver cells , whereas dermatoxins target skin, and some are irritants known as endotoxins (Rastogi and Sinha 2009). Other types of toxins include saxitoxin (paralytic shellfish poisons) and anatoxins (Rastogi and Sinha 2009). In spite of their harmful nature, the abovementioned toxins possess important pharmacological properties *viz.* anticancer and anti-rejection drug, to treat viral and microbial infections and also can be used as a biocide for preventing the growth of algae, fungi weeds, and unwanted pests. (Rastogi and Sinha 2009; Vijayakumar and Menakha 2015).

5.9 Photoprotective Compounds

5.9.1 Scytonemin

The extracellular polysaccharide layer of cyanobacteria secretes a photoprotective metabolite termed as scytonemin (Rastogi et al. 2015). Chemically, scytonemin is an indole alkaloid and a UV-filtering agent so it not only acts as an oxidizing agent but also protects cells from UV and short wavelength radiation (Balskus and Walsh 2008; Castenholz and Garcia-Pichel 2012; Rastogi and Incharoensakdi 2014). Scytonemin mainly occurs in green oxidized form, whereas it also exists in red reduced (fuscorhodin) and yellow oxidized form (fuscochlorin) (Garcia-Pichel and Castenholz 1991; Wada et al. 2013). In the recent past, dimethoxy- and tetramethoxyscytonemin derivatives of reduced scytonemin and scytonine pigments have been isolated from *Scytonema* sp. (Bultel-Poncé et al. 2004; Grant and Louda 2013; Rastogi et al. 2014). Experimental studies have put forward that scytonemin reduces the risk from UV-C radiation and can also effectively reduce the inhibition of photosynthesis by UV-A, thus reducing the discoloration of chlorophyll-a when illuminated (Dillon and Castenholz 1999; Gao and Garcia-Pichel 2011; Rastogi et al. 2013).

5.9.2 Mycosporine-Like Amino Acids (MAAs)

Cyanobacteria also produce another photoprotective metabolite known as mycosporine-like acids (MAAs). Chemically, it is made up of cyclohexenone or cyclohexenimine units conjugated with the structural analog of an amino acid (serine; Tarasuntisuk et al. 2018). The variation in absorption potential is owing to the presence of different nitrogen-containing structures and side groups (Sinha et al. 2007; Derikvand et al. 2017). MAAs are low molecular weight molecules categorized in 30 different groups present in the cytosol or the outer cell membrane (Nazifi et al. 2014). The exact mechanism responsible for the production of MAAs is still unclear, but abiotic stress such as UV and osmotic shock plays a vital role in the production of MAAs (Pathak et al. 2019). MAAs provide photoprotection to the cell without producing ROS and thus act as an antioxidative agent (Sinha and Häder 2008; Geraldes et al. 2020). Production of MAA has been observed in the members of the orders *Synechococcales*, *Chroococcales*, *Oscillatoriales*, and *Nostocales* (Jain et al. 2017). The cyanobacterium *Sphaerospermopsis torques-reginae* ITEP-024 dwelling in freshwater under low salinity conditions, when exposed to UV radiation, produces imino-mycosporines, shinorine, and porphyra-334 for adaptation (Geraldes et al. 2020).

Aphanothece (halotolerant species) produce mycosprorine-2-glycine (M2G) which has numerous biological roles. Some noteworthy roles include scavenging free radicals (Nishida et al. 2020), oxidative stress protection (Cheewinthamrongrod et al. 2016), and osmoregulation (Patipong et al. 2017).

5.9.3 Terpenoids

Cyanobacteria produce diverse range of terpenoid/isoterpenoid compouds (having 55,000 and more metabolites) (Breitmaier 2006). The group includes monoterpenoids, hemiterpenoids, sesquiterpenoids, diterpenoids, triterpenoids (steroids), sesterterpenoids, and tetraterpenoids (carotenoids) on the basis of isoprene units (Kandi et al. 2015). The plethora of cyanobacterial terpenoids play a significant rolef in ensuring cyanobacterial survival under extreme conditions, in cellular processes (like respiration and photosynthesis) and in the biogenesis and stability of cell membrane (Belin et al. 2018; Devi et al. 2020). Recently Devi et al. 2020 reported two cyanobacterial genera from Drang salt mine India i.e. *Cylindrospermum muscicola* HPUSD12 (halophilic cyanobacteria) and *Phormidium* sp. produce terpenoids which functions as an antioxidative compound to protect the cells against free radical damage. Production of sesquiterpenoid geosmin has been reported in some of the freshwater, aquatic, and terrestrial cyanobacteria, viz., *Calothrix* PCC 7507 (filamentous) and *Synechocystis* sp. PCC 6803 (unicellular), *Nostoc punctiforme* PCC 73102, and *Nostoc* sp. PCC 7120 (Höckelmann et al. 2009; Dienst et al. 2020). These sesquiterpenoids help in cellular defense mechanism.

A pentacyclic triterpenoid called as 2-Methylhopanoid (2-MeBHP), is found to be present in natural cyanobacterial-dominated mats and laboratory-grown cultures (Summons et al. 1999; Jahnke et al. 2004). It also acts as a biomarker for environmental settings and provides resistance to cyanobacteria against freezing/thawing, osmotic shocks, and pH variations and thus warrants the presence of cyanobacteria in different habitats like on desert soil covering, hydrothermal springs, hypersaline lakes, Antarctic water, and Artic soil (Agger et al. 2008).

5.9.4 Carotenoids

This is the essential group of metabolites belonging to the class terpenes/terpenoids. It has been further categorized into two groups, that is, carotenes, such as β-carotene, and xanthophylls, such as zeaxanthin and lutein, depending upon the presence and absence of oxygen. Additional carotenoids generally found are echinenone, canthaxanthin, and myxoxanthophyll. Carotenoids are the light-harvesting accessory pigments and are also important for energy dissipation process (Hirschberg and Chamovitz 1994; Merhan 2017). Carotenoids are lipophilic secondary metabolites found to be important to facilitate cyanobacterial protection against photooxidative damage caused due to high light and UV radiation during photosynthesis. For the photoprotection of PSII specially against singlet oxygen, echinenone and zeaxanthin play a pivotal role (Kusama et al. 2015). Now and then, many cyanobacterial species inhabiting freshwater, terrestrial, marine, and arid regions have been explored for the presence of high amount of carotenoids and its derivatives (Patias et al. 2017; Lopes et al. 2020; Pagels et al. 2020; Boucar et al. 2021). Carotenoids perform singlet oxygen quenching, releasing excessive light as heat through the xanthophyll cycle and radical scavenging (Latifi et al. 2009).

5.9.5 Antioxidants

The formation of ROS such as hydrogen peroxides, superoxides, and hydroxyl radicals is an unnecessary evil during the process of photosynthesis and respiration. Such ROS causes damage to the biomolecules within the cells (Latifi et al. 2009). To prevent the production of ROS, the first mechanism involved is non-photochemical quenching (NPQ) or dissipation of excess energy as heat via using the carotenoid zeaxanthin. To mitigate the production of the higher amount of ROS which can lead to oxidative damage, cyanobacteria produce various enzymes such as superoxide dismutases (SOD), catalases, and peroxidases as well as non-enzymatic molecules such as carotenoids and phycobiliproteins (PBP; Reuter et al. 1994; Banerjee et al. 2013).

5.9.6 Vitamins

Vitamins have been categorized into water-soluble (vitamins B and C) and fat-soluble (vitamins A, D, E, and K). Vitamin B is known to be synthesized mostly in photosynthetic organisms. It has been observed that vitamins D, K, and B12 are present in some of the cyanobacteria and microalgae, which are found to be absent in higher plants (Del Mondo et al. 2020).

High concentrations of β-carotene (pro-vitamin A) have been reported in the cyanobacterial genera *Arthrospira maxima*, *Anabaena cylindrica*, and *Synechococcus* sp. (Aaronson et al. 1977). This pro-vitamin A carotenoid functions as a protective agent against oxidative damage (Santiago-Morales et al. 2018).

Cyanobacteria are also an important source of vitamin B, a water-soluble vitamin. It is required in several metabolic pathways and serves as a nutrient for other aquatic organisms (Helliwell et al. 2016; Del Mondo et al. 2020). *Nodularia spumigena* accumulates B1 to combat salinity and temperature stresses which indicates the role of vitamin B in acclamation in extreme habitats (Sylvander et al. 2013). Various studies have suggested the presence of a high amount of B2, B3, B5, B6, B9, and B12 in freshwater and dried cells of cyanobacteria (Edelmann et al. 2019). Further, vitamin D (very low) has also been observed in cyanobacteria; this lipid-soluble vitamin might be important to reduce the damage caused by UV radiation (Ljubic et al. 2020). Another vitamin, i.e., E, has also been detected in cyanobacterial strains which protect against photooxidative damage (Krieger-Liszkay and Trebst 2006;

Mudimu et al. 2017; Santiago-Morales et al. 2018). Vitamin K1 is found in most of the cyanobacteria such as *Anabaena cylindrica*, *Anabaena variabilis*, *Mastigocladus laminosus*, *Nostoc muscorum*, *Prochlorococcus* sp., *Anacystis nidulans*, and *Synechocystis* sp. PCC 6803 (Collins and Jones 1981; Sakuragi and Bryant 2006). Vitamin K2 menaquinone has also been reported in *Gloeobacter violaceus* and *Synechococcus* sp. PCC 7002 (Mimuro et al. 2005; Sakuragi et al. 2005).

5.9.7 Phytohormones

It is well known that cyanobacteria accumulate and release diverse phytohormones responsible for growth and development similar to higher plants such as auxins, gibberellins (GA), cytokinins (CKs), and ethylene (ET) (Stirk et al. 2002; Gayathri et al. 2015). The indole-3-acetic acid (IAA) and cytokinin (CK) hormones are secreted by symbiotic cyanobacteria (Hussain and Hasnain 2010). Production of the same amount of IAA and CK in rice and wheat root cells by endophytic *Nostoc* was recently reported by Hussain et al. (2013). In cyanobacteria, ABA (abscisic acid) has been known to function against salt stress conditions unlike other algae, where it is synthesized in other types of stress (Kobayashi et al. 1997; Yoshida et al. 2003, 2004; Lu et al. 2014). Additionally, it can be laid down that cyanobacteria are the prodigious sources of numerous plant hormones such as jasmonic acid and its derivatives.

5.10 Tools Dedicated to Study Omic Analysis

In the recent past along with conventional biology, molecular and computational biology have also contributed immensely in providing better insights into cell and system biology and also in the development of high-throughput methods. These technical advancements have answered key aspects related to the omic approach such as genomics, proteomics, transcriptomics, and metabolomics at the cellular level. In the past two decades, the application of genomics and proteomics approaches has accelerated the investigations of cyanobacterial stress responses at a system level. System biology of cyanobacterial stress response demands mammoth genome-wide analysis for generating microarray data from the samples found in diverse stress conditions. Furthermore, the techniques like NMR, GC, and GEM need to be updated to identify a number of proteins involved in stress regulation as well as metabolites produced. These advanced skills will render in improving our abilities to connect physiological aspects with omic data for the development of advanced tools for a better understanding of complex data sets.

5.11 Conclusion

Cyanobacteria have always been considered as an excellent model to study the effect of abiotic stress and thereby their response toward it. Studies reveal the response of membrane structure, defense protein, and bioactive metabolite as the prime protective effects; however, several questions are still unanswered. The link between all these defense mechanisms is still unclear; the role of membrane lipids requires special attention. The underlying mechanisms behind ecologically successful organisms can be monitored through advanced proteomics and metabolomics approaches.

Author's Contribution ES and AS initiated the preparation of the manuscript. PS, AS, and ES conceptualized the idea and revised the manuscript. PS and ES drafted the manuscript. All authors have read and approved the final manuscript.

Conflict of Interest The authors declare that they have no conflict of interest.

References

Aaronson S, Dhawale SW, Patni NJ, DeAngelis B, Frank O, Baker H (1977) The cell content and secretion of water-soluble vitamins by several freshwater algae. Arch Microbiol 112:57–59

Agger SA, Lopez-Gallego F, Hoye TR, Schmidt-Dannert C (2008) Identification of sesquiterpene synthases from *Nostoc punctiforme* PCC 73102 and *Nostoc* sp. strain PCC 7120. J Bacteriol 190:6084–6096

Al-Hosani S, Oudah MM, Henschel A, Yousef LF (2015) Global transcriptome analysis of salt acclimated *Prochlorococcus* AS9601. Microbiol Res 176:21–28

Allakhverdiev SI, Sakamoto A, Nishiyama Y, Inaba M, Murata N (2000) Ionic and osmotic effects of NaCl-induced inactivation of photosystem I and II in *Synechococcus* sp. Plant Physiol 123: 1047–1056

Arora S, Singh YP, Vanza M, Sahni D (2016) Bioremediation of saline and sodic soils through halophilic bacteria to enhance agricultural production. J Soil Water Conserv 15:302–305

Assini JM, Mulvihill EE, Huff MW (2013) Citrus flavonoids and lipid metabolism. Curr Opin Lipidol 24:34–40

Balogi Z, Torok Z, Balogh G, Josvay K, Shigapova N, Vierling E, Vigh L, Horvath I (2005) "Heat shock lipid" in cyanobacteria during heat/light-acclimation. Arch Biochem Biophys 436:346–354

Balskus EP, Walsh CT (2008) Investigating the initial steps in the biosynthesis of cyanobacterial sunscreen scytonemin. J Am Chem Soc 130:15260–15261

Banerjee M, Raghavan PS, Ballal A, Rajaram H, Apte SK (2013) Oxidative stress management in the filamentous, heterocystous, diazotrophic cyanobacterium, *anabaena* PCC7120. Photosynth Res 118:59–70

Belin BJ, Busset N, Giraud E, Molinaro A, Silipo A, Newman DK (2018) Hopanoid lipids: from membranes to plant–bacteria interactions. Nat Rev Microbiol 16:304

Blumwald E, Mehlhorn RJ, Packer L (1983) Studies of osmoregulation in salt adaptation of cyanobacteria with ESR spin-probe techniques. Proc Natl Acad Sci U S A 80:2599–2602

Blumwald E, Tel-Or E (1982) Osmoregulation and cell composition in salt-adaptation of *Nostoc muscorum*. Arch Microbiol 132:168–172

Boucar MCM, Shen LQ, Wang K, Zhang ZC, Qiu BS (2021) UV-B irradiation enhances the production of unique mycosporine like amino acids and carotenoids in the subaerial cyanobacterium *Pseudanabaena* sp. CCNU1. Eur J Phycol 56:316–323

Breitmaier E (2006) Terpenes: importance, general structure, and biosynthesis. In: Terpenes flavors fragrances, pharmaca, pheromones, vol 1. Wiley, Weinheim, pp 1–3

Bultel-Poncé V, Felix-Theodose F, Sarthou C, Ponge JF, Bodo B (2004) New pigments from the terrestrial cyanobacterium *Scytonema* sp. collected on the Mitaraka inselberg, French Guyana. J Nat Prod 67:678–681

Cassier-Chauvat C, Chauvat F (2014) Function and regulation of ferredoxins in the cyanobacterium, *Synechocystis* PCC6803: recent advances. Life (Basel) 4(4):666–680

Castenholz RW, Garcia-Pichel F (2012) Cyanobacterial responses to UV radiation. In: Whitton BA (ed) Ecology of cyanobacteria II. Springer, Dordrecht, pp 481–499

Cheewinthamrongrod V, Kageyama H, Palaga T, Takabe T, Waditee-Sirisattha R (2016) DNA damage protecting and free radical scavenging properties of mycosporine-2-glycine from the Dead Sea cyanobacterium in A375 human melanoma cell lines. J Photochem Photobiol B Biol 164:289–295

Choi J, Lee DH, Jang H, Park SY, Seol JW (2020) Naringenin exerts anticancer effects by inducing tumor cell death and inhibiting angiogenesis in malignant melanoma. Int J Med Sci 17:3049

Coba FDL, Aguilera J, Figueroa FL, Galvez MVD, Herrera E (2009) Antioxidant activity of mycosporine-like amino acids isolated from three red macroalgae and one marine lichen. J Appl Phycol 21:161–169

Coleman RA, Ramchunder SJ, Moody AJ, Foogo A (2007) An enzyme in snail saliva induces herbivore-resistance in a marine alga. Funct Ecol 21:101–106

Collins MD, Jones D (1981) Distribution of isoprenoid quinone structural types in bacteria and their taxonomic implication. Microbiol Rev 45:316–354

Connan S, Stengel DB (2011) Impacts of ambient salinity and copper on brown algae: 2. Interactive effects on phenolic pool and assessment of metal binding capacity of phlorotannin. Aquat Toxicol 14:1–13

De Philippis R, Colica G, Micheletti E (2011) Exopolysaccharide-producing cyanobacteria in heavy metal removal from water: molecular basis and practical applicability of the biosorption process. Appl Microbiol Biotechnol 92(4):697–708

Del Mondo A, Smerilli A, Sané E, Sansone C, Brunet C (2020) Challenging microalgal vitamins for human health. Microb Cell Factories 19:201

Demay J, Bernard C, Reinhardt A, Marie B (2019) Natural products from cyanobacteria: focus on beneficial activities. Mar Drugs 17:320

Derikvand P, Llewellyn CA, Purton S (2017) Cyanobacterial metabolites as a source of sunscreens and moisturizers: a comparison with current synthetic compounds. Eur J Phycol 52:43–56

Devi S, Rani N, Sagar A (2020) GC-MS analysis and antioxidant activity of two species of cyanobacteria isolated from Drang salt mine of district Mandi, Himachal Pradesh, India. Plant Arch 20:7505–7510

Dienst D, Wichmann J, Mantovani O, Rodrigues JS, Lindberg P (2020) High density cultivation for efficient sesquiterpenoid biosynthesis in *Synechocystis* sp. PCC 6803. Sci Rep 10:5932

Dillon JG, Castenholz RW (1999) Scytonemin, a cyanobacterial sheath pigment, protects against UVC radiation: implications for early photosynthetic life. J Phycol 35:673–681

Edelmann M, Aalto S, Chamlagain B, Kariluoto S, Piironen V (2019) Riboflavin, niacin, folate and vitamin B12 in commercial microalgae powders. J Food Compos Anal 82:103226

Ellis RJ (2001) Macromolecular crowding: an important but neglected aspect of the intracellular environment. Curr Opin Struct Biol 11:114

Erdmann N (1983) Organic osmoregulatory solutes in blue-green algae. Z Pflanzenphysiol 110: 147–155

Esterhuizen M, Downing TG (2008) β-N-methylamino-L-alanine (BMAA) in novel South African cyanobacterial isolates. Ecotoxicol Environ Saf 71:309–313

Fulda S, Huckauf J, Schoor A, Hagemann M (1999) Analysis of stress responses in the cyanobacterial strains *Synechococcus* sp. PCC 7942, *Synechocystis* sp. PCC 6803, and *Synechococcus* sp. PCC 7418: osmolyte accumulation and stress proteins synthesis. J Plant Physiol 154:240–249

Galinski EA, Trüper HG (1994) Microbial behaviour in salt-stressed ecosystems. FEMS Microbiol Rev 15:95–108

Gao Q, Garcia-Pichel F (2011) Microbial ultraviolet sunscreens. Nat Rev Microbiol 9:791–802

Garcia-Pichel F, Castenholz RW (1991) Characterization and biological implications of scytonemin, a cyanobacterial sheath pigment. J Phycol 27:395–409

Gayathri M, Kumar PS, Prabha AML, Muralitharan G (2015) In vitro regeneration of *Arachis hypogaea* L., and *Moringa oleifera* Lam. Using extracellular phytohormones from *Aphanothece* sp. MBDU 515. Algal Res 7:100–105

Geraldes V, de Medeiros LS, Lima ST, Alvarenga DO, Gacesa R, Long PF, Fiore MF, Pinto E (2020) Genetic and biochemical evidence for redundant pathways leading to mycosporine-like amino acid biosynthesis in the cyanobacterium Sphaerospermopsis torques-reginae ITEP-024. Harmful Algae 35:177–187

Giese KC, Vierling E (2002) Changes in oligomerization are essential for the chaperone activity of a small heat shock protein in vivo and in vitro. J Biol Chem 277:46310

Grant CS, Louda JW (2013) Scytonemin-imine, a mahogany-colored UV/Vis sunscreen of cyanobacteria exposed to intense solar radiation. Org Geochem 65:29–36

Grégoire DS, Poulain AJ (2014) A little bit of light goes a long way: the role of phototrophs on mercury cycling. Metallomics 6(3):396–407

Gupta V, Ratha SK, Sood A, Chaudhary V, Prasanna R (2013) New insights into the biodiversity and applications of cyanobacteria (blue-green algae)-prospects and challenges. Algal Res 2:79–97

Hagemann M (2011) Molecular biology of cyanobacterial salt acclimation. FEMS Microbiol Rev 35(1):87–123

Hagemann M, Schoor A, Jeanjean R, Zuther E, Joset F (1997) The stpA gene form *Synechocystis* sp. strain PCC 6803 encodes the glucosyl glycerol-phosphate phosphatase involved in cyanobacterial osmotic response to salt shock. J Bacteriol 179:1727–1733

Hartl FU, Bracher A, Hayer-Hartl M (2011) Molecular chaperones in protein folding and proteostasis. Nature 475:324

Helliwell KE, Lawrence AD, Holzer A, Kudahl UJ, Sasso S, Kräutler B, Scanlan DJ, Warren MJ, Smith AG (2016) Cyanobacteria and eukaryotic algae use different chemical variants of vitamin B12. Curr Biol 26:999–1008

Hernández-Aquino E, Muriel P (2018) Beneficial effects of naringenin in liver diseases: molecular mechanisms. World J Gastroenterol 24:1679

Hirschberg J, Chamovitz D (1994) Carotenoids in cyanobacteria. In: Bryant DA (ed) The molecular biology of cyanobacteria, Advances in photosynthesis, vol 1, 1st edn. Springer, Dordrecht, pp 559–579

Höckelmann C, Becher PG, von Reuss SH, Jüttner F (2009) Sesquiterpenes of the geosmin-producing cyanobacterium *Calothrix* PCC 7507 and their toxicity to invertebrates. Z Naturforsch C 64:49–55

Hussain A, Hamayun M, Shah ST (2013) Root colonization and phytostimulation by phytohormones producing entophytic *Nostoc* sp. AH-12. Curr Microbiol 67:624–630

Hussain A, Hasnain S (2010) Phytostimulation and biofertilization in wheat by cyanobacteria. J Ind Microbiol Biotechnol 38:85–92

Jahnke LL, Embaye T, Hope J, Turk KA, Van Zuilen M, Des Marais DJ, Farmer JD, Summons RE (2004) Lipid biomarker and carbon isotopic signatures for stromatolite-forming, microbial mat communities and Phormidium cultures from Yellowstone National Park. Geobiology 2:31–47

Jain S, Prajapat G, Abrar M, Ledwani L, Singh A, Agrawal A (2017) Cyanobacteria as efficient producers of mycosporine-like amino acids. J Basic Microbiol 57:715–727

Janssen J, Soule T (2016) Gene expression of a two-component regulatory system associated with sunscreen biosynthesis in the cyanobacterium *Nostoc punctiforme* ATCC 29133. FEMS Microbiol Lett 363:fnv235

Jerez-Martel I, García-Poza S, Rodríguez-Martel G, Rico M, Afonso-Olivares C, Gómez-Pinchetti JL (2017) Phenolic profile and antioxidant activity of crude extracts from microalgae and cyanobacteria strains. J Food Qual 2017:2924508

Jimbo H, Yutthanasirikul R, Nagano T, Hisabori T, Hihara T, Nishiyama Y (2018) Oxidation of translation factor EF-Tu inhibits the repair of photosystem II. Plant Physiol 176:2691–2699

Jittawuttipoka T, Planchon M, Spalla O, Benzerara K, Guyot F, Cassier-Chauvat C, Chauvat F (2013) Multidisciplinary evidences that *Synechocystis* PCC6803 exopolysaccharides operate in cell sedimentation and protection against salt and metal stresses. PLoS One 8(2):e55564

Kandi S, Godishala V, Rao P, Ramana KV (2015) Biomedical significance of terpenes: an insight. Biomedicine 3:8–10

Katano Y, Nimura-Matsune K, Yoshikawa H (2006) Involvement of DnaK3, one of the three DnaK proteins of cyanobacterium *Synechococcus* sp. PCC7942, in translational process on the surface of the thylakoid membrane. Biosci Biotechnol Biochem 70:1592–1598

Kera K, Yoshizawa Y, Shigehara T, Nagayama T, Tsujii M, Tochigi S, Uozumi N (2020) Hik36–Hik43 and Rre6 act as a two-component regulatory system to control cell aggregation in *Synechocystis* sp. PCC6803. Sci Rep 10:1940

Khan A, Singh P, Srivastava A (2017) Synthesis, nature and utility of universal iron chelator - siderophore: a review. Microbiol Res 212-213:103–111

Kirsch F, Klähn S, Hagemann M (2019) Salt-regulated accumulation of the compatible solutes sucrose and glucosylglycerol in cyanobacteria and its biotechnological potential. Front Microbiol 10:2139

Klähn S, Mikkat S, Riediger M, Georg J, Hess WR, Hagemann M (2021) Integrative analysis of the salt stress response in cyanobacteria. Biol Direct 16:26

Klähn S, Steglich C, Hess WR, Hagemann M (2010) Glucosylglycerate: a secondary compatible solute common to marine cyanobacteria from nitrogen-poor environments. Environ Microbiol 12:83–94

Kobayashi M, Hirai N, Kurimura Y, Ohigashi H, Tsuji Y (1997) Abscisic acid-dependent algal morphogenesis in the unicellular green alga *Haematococcus pluvialis*. Plant Growth Regul 22: 79–85

Kranzler C, Rudolf M, Keren N, Schleiff E (2013) Iron in cyanobacteria. Adv Bot Res 65:57–105

Krieger-Liszkay A, Trebst A (2006) Tocopherol is the scavenger of singlet oxygen produced by the triplet states of chlorophyll in the PSII reaction centre. J Exp Bot 57:1677–1684

Kumar CS, Ganesan P, Suresh PV, Bhaskar N (2008) Seaweeds as a source of nutritionally beneficial compounds – a review. J Food Sci Technol 45:1–13

Kusama Y, Inoue S, Jimbo H, Takaichi S, Sonoike K, Hihara Y, Nishiyama Y (2015) Zeaxanthin and echinenone protect the repair of photosystem II from inhibition by singlet oxygen in *Synechocystis* sp. PCC 6803. Plant Cell Physiol 56:906–916

Latifi A, Ruiz M, Zhang CC (2009) Oxidative stress in cyanobacteria. FEMS Microbiol Rev 33: 258–278

Lau SCK, Qian PY (2000) Inhibitory effect of phenolic compounds and marine bacteria on larval settlement of the barnacle *Balanus Amphitrite* darwin. Biofouling 16:47

Lefebvre DD, Kelly D, Budd K (2007) Biotransformation of Hg(II) by cyanobacteria. Appl Environ Microbiol 73(1):243–249

Ljubic A, Jacobsen C, Holdt SL, Jakobsen J (2020) Microalgae Nannochloropsis oceanica as a future new natural source of vitamin D3. Food Chem 320:126627

Lopes G, Clarinha D, Vasconcelos V (2020) Carotenoids from cyanobacteria: a biotechnological approach for the topical treatment of psoriasis. Microorganisms 8:302

Los DA, Zorina A, Sinetova M, Kryazhov S, Mironov K, Zinchenko VV (2010) Stress sensors and signal transducers in cyanobacteria. Sensors 10:2386–2415

Lu Y, Tarkowská D, Turecková V, Luo T, Xin Y, Li J, Wang Q, Jiao N, Strnad M, Xu J (2014) Antagonistic roles of abscisic acid and cytokinin in oleaginous microalga Nannochloropsis oceanica upon nitrogen- depletion expand the evolutionary breadth of phytohormone function. Plant J 80:52–68

Mackay MA, Norton RS, Borowitzka LJ (1984) Organic osmoregulatory solutes in cyanobacteria. J Gen Microbiol 130:2177–2191

Mallick N, Mohn FH (2000) Reactive oxygen species: response of algal cells. J Plant Physiol 157: 183–193

Mandal S, Raju R, Kumar A, Kumar P, Sharma PC (2018) Current status of research, technology response and policy needs of salt-affected soils in India – a review. Ind Soc Coastal Agri Res 36: 40–53

Merhan O (2017) The biochemistry and antioxidant properties of carotenoids. In: Carotenoids, vol 5. IntechOpen, London, p 51

Micallef ML, Sharma D, Bunn BM, Gerwick L, Viswanathan R, Moffitt MC (2014) Comparative analysis of hapalindole, ambiguine and welwitindolinone gene clusters and reconstitution of indole-isonitrile biosynthesis from cyanobacteria. BMC Microbiol 14:1–18

Mimuro M, Tsuchiya T, Inoue H, Sakuragi Y, Itoh Y, Gotoh T, Miyashita H, Bryant DA, Kobayashi M (2005) The secondary electron acceptor of photosystem I in *Gloeobacter violaceus* PCC 7421 is menaquinone-4 that is synthesized by a unique but unknown pathway. FEBS Lett 579:3493–3496

Mizuno T, Kaneko T, Tabata S (1996) Compilation of all genes encoding bacterial two-component signal transducers in the genome of the cyanobacterium, *Synechocystis* sp. strain PCC 6803. DNA Res 3(6):407–414

Mo S, Krunic A, Santaerio BD, Franzblau SG, Orjala J (2010) Hapalindole-related alkaloids from the cultured cyanobacterium *Fischerella ambigua*. Phytochemistry 71:2116–2123

Morby AP, Turner JS, Huckle JW, Robinson NJ (1993) SmtB is a metal-dependent repressor of the cyanobacterial metallothionein gene smtA: identification of a Zn inhibited DNA-protein complex. Nucleic Acids Res 21(4):921–925

Mudimu O, Koopmann IK, Rybalka N, Friedl T, Schulz R, Bilger W (2017) Screening of microalgae and cyanobacteria strains for α-tocopherol content at different growth phases and the influence of nitrate reduction on α-tocopherol production. J Appl Phycol 29:2867–2875

Mulvihill EE, Assini JM, Sutherland BG, DiMattia AS, Khami M, Koppes JB, Sawyez CG, Whitman SC, Huff MW (2010) Naringenin decreases progression of atherosclerosis by improving dyslipidemia in high-fat–fed low-density lipoprotein receptor–null mice. Arterioscler Thromb Vasc Biol 30:742–748

Murata N, Takahashi S, Nishiyama Y, Allakhverdiev SI (2007) Photoinhibition of photosystem II under environmental stress. Biochim Biophys Acta Bioenerg 1767:414–421

Nakai S, Yamada S, Hosomi M (2005) Anti-cyanobacterial fatty acids released from *Myriophyllum spicatum*. Hydrobiologia 543:71–78

Nazifi E, Wada N, Asano T, Nishiuchi T, Iwamuro Y, Chinaka S, Matsugo S, Sakamoto T (2014) Characterization of the chemical diversity of glycosylated mycosporine-like amino acids in the terrestrial cyanobacterium *Nostoc* commune. J Photochem Photobiol B 142C:154–168

Nishida Y, Kumagai Y, Michiba S, Yasui H, Kishimura H (2020) Efficient extraction and antioxidant capacity of mycosporine-like amino acids from red alga Dulse *Palmaria palmata* in Japan. Mar Drugs 18:502

Nishiyama Y, Murata N (2014) Revised scheme for the mechanism of photoinhibition and its application to enhance the abiotic stress tolerance of the photosynthetic machinery. Appl Microbiol Biotechnol 98:8777–8796

Pade N, Hagemann M (2014) Salt acclimation of cyanobacteria and their application in biotechnology. Life (Basel, Switzerland) 5(1):25–49

Pagels F, Salvaterra D, Amaro HM, Lopes G, Sousa-Pinto I, Vasconcelos V, Guedes AC (2020) Bioactive potential of *Cyanobium* sp. pigment-rich extracts. J Appl Phycol 32:3031–3040

Pathak J, Ahmed H, Singh SP, Häder DP, Sinha RP (2019) Genetic regulation of scytonemin and mycosporine-like amino acids (MAAs) biosynthesis in cyanobacteria. Plant Gene 17:100172

Patias LD, Fernandes AS, Petry FC, Mercadante AZ, Jacob-Lopes E, Zepka LQ (2017) Carotenoid profile of three microalgae/cyanobacteria species with peroxyl radical scavenger capacity. Food Res Int 100:260–266

Patipong T, Hibino T, Waditee-Sirisattha R, Kageyama H (2017) Efficient bioproduction of mycosporine-2-glycine, which functions as potential osmoprotectant, using *Escherichia coli* cells. Nat Prod Commun 12:12

Patipong T, Hibino T, Waditee-Sirisattha R, Kageyama H (2019) Induction of antioxidative activity and antioxidant molecules in the halotolerant cyanobacterium *Halothece* sp. PCC 7418 by temperature shift. Nat Prod Commun 14:1934578X1986568

Rachedi R, Foglino M, Latifi A (2020) Stress signaling in cyanobacteria: a mechanistic overview. Life 10(12):312

Rastogi RP, Incharoensakdi A (2014) Characterization of UV-screening compounds, mycosporine-like amino acids, and scytonemin in the cyanobacterium *Lyngbya* sp. CU2555. FEMS Microbiol Ecol 87:244–256

Rastogi RP, Sinha RP (2009) Biotechnological and industrial significance of cyanobacterial secondary metabolites. Biotechnol Adv 27:521–539

Rastogi RP, Sinha RP, Incharoensakdi A (2013) Partial characterization, UV-induction and photoprotective function of sunscreen pigment, scytonemin from *Rivularia* sp. HKAR-4. Chemosphere 93:1874–1878

Rastogi RP, Sonani RR, Madamwar D (2014) The high-energy radiation protectant extracellular sheath pigment scytonemin and its reduced counterpart in the cyanobacterium *Scytonema* sp. R77DM. Bioresour Technol 171:396–400

Rastogi RP, Sonani RR, Madamwar D (2015) Cyanobacterial sunscreen scytonemin: role in photoprotection and biomedical research. Appl Biochem Biotechnol 176:1551–1563

Reed JR, Stewart W (1983) Physiological responses of *Rivularia atra* to salinity: osmotic adjustment in hyposaline media. New Phytol 95:595–603

Reuter W, Westermann M, Brass S, Ernst A, Böger P, Wehrmeyer W (1994) Structure, composition, and assembly of paracrystalline phycobiliproteins in *Synechocystis* sp. strain BO 8402 and of phycobilisomes in the derivative strain BO 9201. J Bacteriol 176:896–904

Sakuragi Y, Bryant DA (2006) Genetic manipulation of quinone biosynthesis in cyanobacteria. In: Golbeck JH (ed) Photosystem I: the light-driven plastocyanin: ferredoxin oxidoreductase. Springer, Dordrecht, pp 205–222

Sakuragi Y, Zybailov B, Shen G, Bryant DA, Golbeck JH, Diner BA, Karygina I, Pushkar Y, Stehlik D (2005) Recruitment of a foreign quinone into the A1 site of photosystem I: characterization of a menB rubA double deletion mutant in *Synechococcus* sp. PCC 7002 devoid of FX, FA, and FB and containing plastoquinone or exchanged 9, 10-anthraquinone. J Biol Chem 280: 12371–12381

Santiago-Morales IS, Trujillo-Valle L, Márquez-Rocha FJ, Hernández JFL (2018) Tocopherols, phycocyanin and superoxide dismutase from microalgae: as potential food antioxidants. Appl Food Biotechnol 5:19–27

Singh A, Kaushik MS, Srivastava MT, Nath D, Mishra AK (2016) Siderophore mediated attenuation of cadmium toxicity by paddy field cyanobacterium *Anabaena oryzae*. Algal Res 16:63–68

Singh DP, Prabha R, Meena KK, Sharma L, Sharma AK (2014) Induced accumulation of polyphenolics and flavonoids in cyanobacteria under salt stress protects organisms through enhanced antioxidant activity. Am J Plant Sci 2014:43916

Singh P, Khan A, Srivastava A (2020a) Heterocyst and akinete differentiation in cyanobacteria: a view toward cyanobacterial symbiosis. In: Singh PK, Kumar A, Shrivastav AK (eds) Advances in cyanobacterial biology. Elsevier, London, pp 235–248

Singh P, Khan A, Kumar R, Kumar R, Singh VK, Srivastava A (2020b) Recent developments in siderotyping: procedure and application. World J Microbiol Biotechnol 36(12):178

Singh S (2014) A review on possible elicitor molecules of cyanobacteria: their role in improving plant growth and providing tolerance against biotic or abiotic stress. J Appl Microbiol 117(5): 1221–1244

Sinha RP, Häder DP (2008) UV-protectants in cyanobacteria. Plant Sci 174:278–289

Sinha RP, Singh SP, Häder DP (2007) Database on mycosporines and mycosporine-like amino acids (MAAs) in fungi, cyanobacteria, macroalgae, phytoplankton and animals. J Photochem Photobiol B Biol 89:29–35

Srivastava A, Rai A, Neilan B (2013) Understanding the mechanisms of abiotic stress management in cyanobacteria with special reference to proteomics. In: Srivastava A, Rai A, Neilan B (eds) Stress biology of cyanobacteria (molecular mechanisms to cellular responses). Taylor and Francis, Boca Raton, pp 93–112

Stengel DB, Connan S, Popper ZA (2011) Algal chemodiversity and bioactivity: sources of natural variability and implications for commercial application. Biotechnol Adv 29:483–501

Stirk WA, Ordog V, Staden JV, Jager K (2002) Cytokinins and auxin-like activity in Cyanophyta and microalgae. J Appl Phycol 14:215–221

Sugumar M, Sevanan M, Sekar S (2019) Neuroprotective effect of naringenin against MPTP-induced oxidative stress. Int J Neurosci 129:534–539

Sui N, Han GL (2014) Salt-induced photoinhibition of PSII is alleviated in halophyte *Thellungiella halophila* by increases of unsaturated fatty acids in membrane lipids. Acta Physiol Plant 36: 983–992

Sui N, Tian S, Wang W, Wang M, Fan H (2017) Overexpression of glycerol-3-phosphate acyltransferase from *Suaeda salsa* improves salt tolerance in *Arabidopsis*. Front Plant Sci 8: 1337

Summons RE, Jahnke LL, Hope JM, Logan GA (1999) 2-methylhopanoids as biomarkers for cyanobacterial oxygenic photosynthesis. Nature 400:554–557

Suzuki I, Kanesaki Y, Hayashi H, Hall JJ, Simon WJ, Slabas AR, Murata N (2005) The histidine kinase Hik34 is involved in thermotolerance by regulating the expression of heat shock genes in synechocystis. Plant Physiol 138(3):1409–1421

Sylvander P, Häubner N, Snoeijs P (2013) The thiamine content of phytoplankton cells is affected by abiotic stress and growth rate. Microb Ecol 65:566–577

Tarasuntisuk S, Patipong T, Hibino T, Waditee-Sirisattha R, Kageyama H (2018) Inhibitory effects of mycosporine-2-glycine isolated from a halotolerant cyanobacterium on protein glycation and collagenase activity. Lett Appl Microbiol 67:314–320

Thomas NV, Kim SK (2011) Potential pharmacological applications of polyphenolic derivatives from marine brown algae. Environ Toxicol Pharmacol 32:325–335

Trabelsi L, Mnari A, Abdel-Daim MM, Abid-Essafi S, Aleya L (2016) Therapeutic properties in Tunisian hot springs: first evidence of phenolic compounds in the cyanobacterium *Leptolyngbya* sp. biomass, capsular polysaccharides and releasing polysaccharides. BMC Complement Altern Med 16:515

Vasas G, Gáspár A, Páger C, Surányi G, Máthé C, Hamvas MM, Borbely G (2004) Analysis of cyanobacterial toxins (anatoxin-a, cylindrospermopsin, microcystin-LR) by capillary electrophoresis. Electrophoresis 25(1):108–115

Vijayakumar S, Menakha M (2015) Pharmaceutical applications of cyanobacteria—a review. J Acute Med 5:15–23

Wada N, Sakamoto T, Matsugo S (2013) Multiple roles of photosynthetic and sunscreen pigments in cyanobacteria focusing on the oxidative stress. Metabolites 3:463–483

Walton K, Berry JP (2016) Indole alkaloids of the Stigonematales (Cyanophyta): chemical diversity, biosynthesis and biological activity. Mar Drugs 14:73

Welsh DT (2000) Ecological significance of compatible solute accumulation by micro-organisms: from single cells to global climate. FEMS Microbiol Rev 24(3):263–290

Yang Z, Li JL, Liu LN, Xie Q, Sui N (2020) Photosynthetic regulation under salt stress and salt-tolerance mechanism of sweet sorghum. Front Plant Sci 10:1722
Yoshida K, Igarashi E, Mukai M, Hirata K, Miyamoto K (2003) Induction of tolerance to oxidative stress in the green alga, *Chlamydomonas reinhardtii*, by abscisic acid. Plant Cell Environ 26: 451–457
Yoshida K, Igarashi E, Wakatsuki E, Miyamoto K, Hirata K (2004) Mitigation of osmotic and salt stresses by abscisic acid through reduction of stress-derived oxidative damage in *Chlamydomonas reinhardtii*. Plant Sci 167:1335–1341
Zhao YQ, Ma YC, Duan HM, Liu RR, Song J (2019) Traits of fatty acid accumulation in dimorphic seeds of the euhalophyte *Suaeda salsa* in saline conditions. Plant Biosyst 153:514–520

Microalgal Bio-pigments: Production and Enhancement Strategies to Enrich Microalgae-Derived Pigments

6

Alka Devi, Mohneesh Kalwani, Krutika Patil, Arti Kumari, Aruna Tyagi, Pratyoosh Shukla, and Sunil Pabbi

Abstract

Due to increased demand for naturally produced pigments, microalgae signify a biotechnologically important source for a wide variety of pigments which includes chlorophylls (green in colour; *Chls a*, *b*, *d* and *f*), carotenoids (red, orange and yellow; β-carotene, astaxanthin, lutein, fucoxanthin and lycopene) and phycobiliproteins (pink, phycoerythrin; blue, phycocyanin; and pinkish blue, allophycocyanin). These pigments are marketable in cosmetics, food and feed industries because synthetic dyes have a negative impact on humans and their surroundings. However, under optimal growth conditions, the production rate is slow, which makes them unable to use at the industrial level and economically feasible. Therefore, various strategies, such as nutrimental and physical modulations, were utilised to overpopulate the bio-pigments. In this chapter, the discussion is based on commercially important pigment distribution across the microalgal groups, the application of the pigments with their biosynthesis and the various studies conducted for the enhanced production of bio-pigments.

A. Devi · K. Patil · S. Pabbi (✉)
Centre for Conservation and Utilisation of Blue Green Algae (CCUBGA), Division of Microbiology, ICAR-Indian Agricultural Research Institute, New Delhi, India

M. Kalwani
Centre for Conservation and Utilisation of Blue Green Algae (CCUBGA), Division of Microbiology, ICAR-Indian Agricultural Research Institute, New Delhi, India

School of Biotechnology, Banaras Hindu University, Varanasi, Uttar Pradesh, India

A. Kumari · A. Tyagi
Division of Biochemistry, ICAR-Indian Agricultural Research Institute, New Delhi, India

P. Shukla
School of Biotechnology, Banaras Hindu University, Varanasi, Uttar Pradesh, India

B. Neilan et al. (eds.), *Cyanobacterial Biotechnology in the 21st Century*,
https://doi.org/10.1007/978-981-99-0181-4_6

Keywords

Microalgae · Bio-pigments · Carotenoids · Genetic engineering · Phycobiliproteins

6.1 Introduction

Microalgae and photolithoautotrophs are diversified in different habitats, i.e. freshwater, marine and terrestrial ecosystems. They are considered to be one of the pioneer species inhabitants on earth. Because of their ability to quickly acclimatise under different ecosystems, they are expansive for global applications (Patel et al. 2022; Patel et al. 2021). In addition, they are rich reservoirs of bio-commodities, viz. proteins, pigments, vitamins, minerals, lipids and polysaccharides (Silva et al. 2020; Devi et al. 2022; Kalwani et al. 2022). All these metabolites extracted from microalgae have physiological functions in different industrial and healthcare sectors, improving human health and well-being (Patel et al. 2021; Choi et al. 2019; Hong et al. 2019).

Microalgae store a diverse variety of photoprotective and light-harvesting pigments, i.e. carotenes, xanthophyll, chlorophyll, phycobiliproteins, etc. (Duppeti et al. 2017). Light absorption takes place by the conjugate system, which also forms the backbone of microalgae. Chlorophyll comprises the porphyrin ring, which constitutes four pyrrole rings and is a biosynthetic derivative of protoporphyrin IX; Mg^{2+} which holds a centric position in porphyrin is held by covalent and coordinate bonds. Even carotenoids constitute polyene complex composed of "ionone", which gives them orange or yellow colour; hence, they are referred to as carotenoids. If their chemical structures lack oxygen atoms, they are called xanthophyll. Commonly known carotenoids and xanthophylls are β-carotene, α-carotene, astaxanthin, lutein, lycopene, astaxanthin, cryptoxanthin, etc., produced by *Haematococcus*, *Chlorella*, *Scenedesmus*, *Botryococcus*, etc. (Srivastava et al. 2022; Zahedi Dizaji et al. 2021; Chekanov et al. 2021). Furthermore, three types of phycobiliproteins present in algae are phycoerythrin (red colour; absorbance 540–570 nm), allophycocyanin (blue colour; absorbance 540–570 nm) and phycocyanin (blue colour; absorbance 651–655 nm) which are found in cyanobacteria, rhodophytes, etc. (Sonani et al. 2017). During photosynthesis, phycobilisome-captured light is transferred to chlorophyll. They are comprised of heteromonomers with two subunits, α- and β-subunit (Chakdar and Pabbi 2017). Numerous biotic and abiotic stress factors, viz. light, temperature, pH and salinity, and also various other biotechnological interventions, e.g. genetic and metabolic engineering, along with the application of physical and chemical mutagens, lead to enhancement in the production of the microalgal pigment (Musa et al. 2019; Yun et al. 2020; Mogany et al. 2021).

Microalgae have a great potential application in improving the health of human beings and animals, as they bear antioxidant, neuroprotective, anti-carcinogenic, anti-bacterial, anti-obesity and anti-inflammatory properties. In addition, they are widely used as food colourants due to their non-toxic and eco-friendly nature

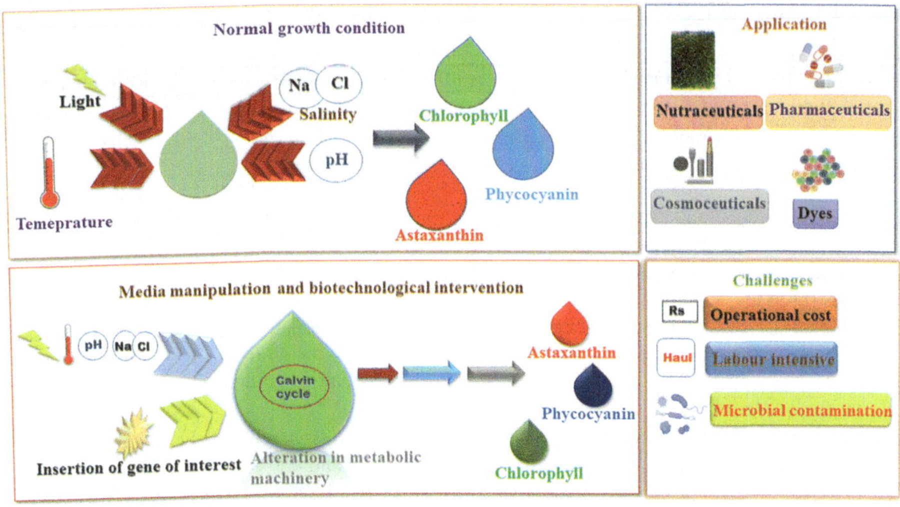

Fig. 6.1 Overview of different applications of microalgal bio-pigments with various enhancement strategies used for pigment production

(Chakdar and Pabbi 2017; Wang et al. 2018). Furthermore, bio-pigments have their application in the nutraceutical industry, helping eliminate various health-related ailments (Ambati et al. 2019). Hence, synthetic chemicals can be replaced with bio-pigments as these chemicals bear carcinogenic properties, lethal to humans and animals.

Although several plant-based bio-pigments are available, microalgae-derived pigments are considered to be highly active. This chapter summarises different types of pigments with their applications and enhancement strategies through media optimisation and biotechnological approach (Fig. 6.1).

6.2 Bio-pigments and Microalgae

Pigments absorb visible light and emit the unabsorbed light that determines the compound's colour. Pigments can be synthetically produced (organic acids, inorganic chemical compounds and petroleum) in addition to biologically (parts of plants, bugs and microalgae). Bio-pigments are utilised in various products, from food, feed and cosmetics to nutraceuticals and research (labels for antibodies and receptors) (Kobylewski and Jacobson 2010). Since synthetic pigments are toxic and harmful, natural pigments are considered sustainable sources (Alam et al. 2018).

Biotechnologically, microalgae are the enthralling resource of pigments and other value-added products. Microalgae produce coloured pigments, such as green-coloured chlorophylls; red-, orange- and yellow-coloured carotenoids; and red- and blue-coloured phycobiliproteins, and other important compounds such as fatty

Table 6.1 Microalgal classification and bio-pigment production

Phylum and common name	Characteristics	Cell size (μm)	Pigments	Reference
Chlorophyta (green microalgae)	Unicellular, multicellular, filamentous, siphonous, thallus and primarily freshwater algae	300–1000	Chlorophylls *a* and *b*, *β*-carotene, prasinoxanthin, siphonaxanthin, astaxanthin	Graham and Wilcox (2000), Van den Hoek et al. (1995), and Lee (2018)
Diatomophyceae/ Diatoms (brown microalgae)	Unicellular: They occur either as solitary cells or in colonies	2–200	Chlorophylls *a* and *c*, *β*-carotene, fucoxanthin, diadinoxanthin	Hasle and Syvertsen (1997) and Canter-Lund and Lund (1995)
Cryptophytes (cryptomonads)	Unicellular and widely distributed in both freshwater and marine environments	Up to 50	Chlorophylls *a* and *c*, carotenoids and phycobiliproteins	Graham and Wilcox (2000) and Van den Hoek et al. (1995)
Cyanobacteria (blue-green microalgae)	Unicellular to filamentous and include colonial species	0.5–40	Chlorophyll *a*, xanthophyll and phycobiliproteins	Graham and Wilcox (2000) and Martins et al. (2018)
Euglenophyta	Flagellates commonly found in freshwater	15–500	Chlorophylls *a* and *b*, *β*-carotene, diadinoxanthin and neoxanthin	Graham and Wilcox (2000) and Barsanti and Gualtieri (2020)
Dinophyta (dinoflagellates)	Marine, flagellated and unicellular	10–400	Chlorophylls *a* and *c*, carotenoid (*β*-carotene), peridinin	Graham and Wilcox (2000) and Carty and Parrow (2015)

acids, vitamins, starch and lipids (Borowitzka 2013; Draaisma et al. 2013). Moreover, microalgal pigments can be synthesised in much higher concentrations compared to higher plants.

Microalgae are classified into 13 phyla with many smaller groups with incomplete details, and microalgae are known as a heterogeneous group of cryptogamic plants (Reynolds 2006). The important phylum with the characteristics and the type of pigment secretion is described in Table 6.1.

Phycobiliproteins also exhibit antioxidant, anti-inflammatory, neuroprotective and hepatoprotective properties (Santiago-Santos et al. 2004; Spolaore et al. 2006). Nowadays, *Spirulina* sp. for phycocyanin, *Dunaliella* sp. for *β*-carotene and astaxanthin from *Haematococcus* sp. are gaining importance (Begum et al. 2016).

6.3 Biosynthesis of the Pigments

The chemical structure of the bio-pigments is categorised into two groups, which are known as tetrapyrroles and isoprenoids. Tetrapyrroles are the linear or cyclic structures of four pyrrole-derived compounds synthesised via a very conserved pathway in both prokaryotes and eukaryotes, with the only exception of synthesis of the first common precursor, i.e. aminolevulinic acid (ALA). Chlorophyll and phycobiliproteins (PBP) are the types of tetrapyrrole. ALA biosynthesis is an essential process in both photosynthetic organisms (Beale pathway or C5 pathway) and heterotroph organisms (Shemin or C4 pathway) (Bradshaw et al. 1993; Wu et al. 2019; Bryant et al. 2020). Succinyl-CoA and glycine condensation by the ALA synthase (ALAS) enzyme in the mitochondrion leads to the production of ALA, followed by transport to the cytosol. ALA is converted to protoporphyrin IX (proto), a metal-free tetrapyrrole ring, through several enzymatic steps and acts as a precursor for phycobilin and chlorophyll synthesis and a substrate for ferrochelatase (FeCH) and magnesium chelatase. FeCH supplements the ferrous iron for heme synthesis, while magnesium chelatase supplements Mg^{2+} to synthesise Chl biosynthetic intermediate (Mg-Proto) (Sobotka et al. 2008). Heme oxygenase enzyme catalyses the conversion of the heme macrocycle ring into biliverdin IXa, isomeric form of phycobiliproteins phycocyanobilin and phycoerythrobilin, a significant chromophore ligated to apoproteins which leads to synthesis of phycobiliproteins (Fig. 6.2) (Saini et al. 2019).

Isoprenoids are the five organic carbon compounds arranged in a specific pattern, e.g. carotenoids. Carotenoid's biosynthesis has a common metabolic pathway across the species in which isopentenyl diphosphate (IPP) acts as a central intermediate for isoprenoid biosynthesis and dimethylallyl diphosphate (DMAPP) as a precursor, and the pathway is referred to as mevalonic acid pathway (MVA). Carotenoid biosynthesis is also occurred by MEP (methylerythritol 4-phosphate) pathway involving glyceraldehyde-3-phosphate and pyruvate. In cyanobacteria and algae, the MEP pathway is the only pathway followed for IPP or DMAPP synthesis (Fig. 6.3) (Zhao et al. 2013).

Phycobiliproteins (PBPs) are light-harvesting, coloured, fluorescent and water-soluble proteins which are abundantly present on the photosynthetic membrane in all the classes of kingdom algae. PBPs are the antennae which efficiently harvest solar energy for the photosynthesis process. In photosynthetic organisms, there are two distinct reaction centres (RC), P700 and P680, which are known as photosystem I (PSI) and photosystem II (PSII), respectively. The light-harvesting antennae are associated with this RC (Bearden and Malkin 1974). In the photosynthetic lamellae, the excitation energy is transferred to photosystems II and I by some protein assemblies known as phycobilisomes (PBSs) (MacColl 1998). They can absorb the visible range (450–670 nm) of light. PBPs have core and peripheral rods and are also the backup source of nutrient's storage for nitrogen, sulphur or carbon (Parmar et al. 2011). The PBSs are supramolecular molecules composed of a core molecule which is attached with two polypeptide chains known as α- and β-chain which are separated by peptide bond.

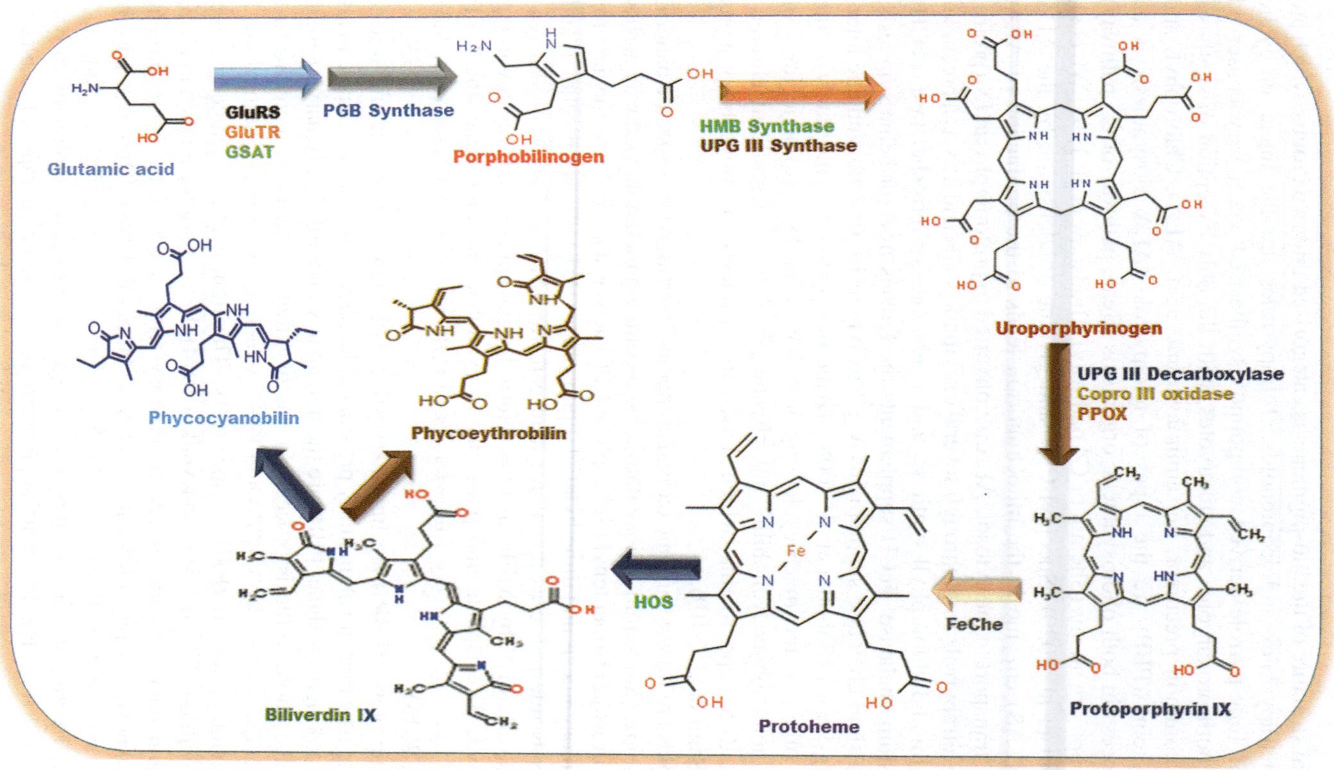

Fig. 6.2 Phycobiliprotein's synthesis pathway. Phycobiliprotein biosynthesis is similar to the biosynthesis of tetrapyrroles. Synthesis begins with glutamic acid, which is converted to biliverdin. Biliverdin is further reduced to phycocyanobilin and phycoerythrobilin, which are subsequently isomerised to generate phycocyanin and phycoerythrin. Major enzymes involved in biosynthesis of phycobiliproteins are: PGB Synthase: UPG III Decarboxylase: Uroporphyrinogen Decarboxylase Porphobilinogen Synthase *GluTR*: Glutamyl t-RNA Reductase *GluRS*: Glutamyl t-RNA Synthase, HMB Synthase: Hydroxymethylbilane Synthase, *PPOX*: Protoporphyrinogen IX Oxidase GSAT: Glutamate –1-Semialdehyde Aminotransferase, Uroporphyrinogen III Synthase Copro III oxidase: Coproporphyrinogen III Oxidase, HOS: Heme Oxygenase, FeChe: Fe Chetalase (Srivastava et al. 2022)

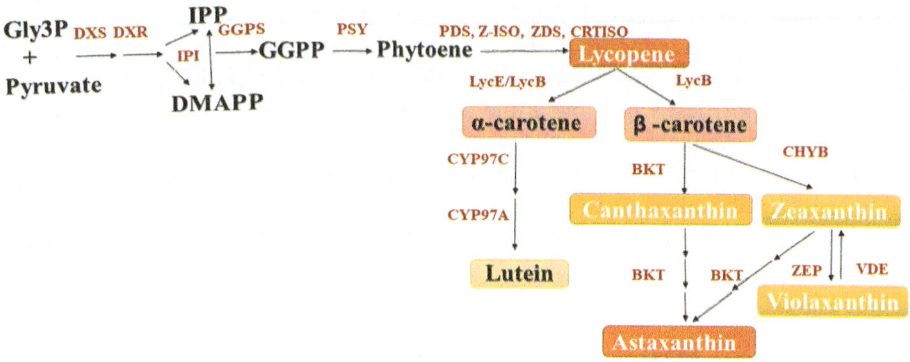

Fig. 6.3 MEP or MVA (mevalonate) pathways for carotenoid biosynthesis. In MEP pathway, pyruvate and glyceraldehyde-3-phosphate (Gly3P) are reduced to isopentenyl pyrophosphate (IPP) and dimethylallyl diphosphate (DMAPP). IPP is converted into various carotenoids through enzyme-catalysed pathways using geranylgeranyl pyrophosphate (GGPP) as a precursor. Major enzymes involved synthesis of microalgal carotenoids are DXS, 1-deoxy-D-xylulose 5-phosphate synthase; DXR, 1-deoxy-D-xylulose 5-phosphate reductoisomerase; GGPS, GGPP synthase; PSY, phytoene synthase; PDS, phytoene desaturase; Z-ISO, ζ-carotene isomerase; ZDS, ζ-carotene desaturase; CRTISO, carotenoid isomerase; LycE, ε-cyclase; LycB, β-cyclase; CYP97, carotene ε-hydroxylase; CHYB, β-carotene hydroxylase; ZEP, zeaxanthin epoxidase; VDE, violaxanthin de-epoxidase; and BKT, β-carotene ketolase Source: Saini et al. (2020) and Srivastava et al. (2022)

PBSs also constitute a high percentage of cyanobacteria's total protein content (Singh et al. 2015). Phycobiliproteins are characterised (on the bases of energy) into three main categories: phycocyanins ($\lambda_{max} \sim 620$ nm), phycoerythrins ($\lambda_{max} \sim 565$ nm) and allophycocyanins ($\lambda_{max} \sim 650$ nm) (Chakdar and Pabbi 2016). The transient change in the ratio of PC (phycocyanin) and PE (phycoerythrin) is regulated by light harvesting and energy transfer mechanism during environmental changes. In the photosynthetic membrane, the periphery and core regions contain about 20% of PC and APC (allophycocyanins), while PE is in the distal region of the PBs, helping in adaptation during drastic environmental changes. Besides PE, various colourless linker polypeptides help stabilise the PB's structure (Kannaujiya et al. 2017a, 2017b).

6.4 Industrial Application of Microalgal Pigments

The most commonly used microalgae which are exploited for pigment production are *Chlorella*, *Nannochloropsis*, *Haematococcus*, *Dunaliella*, *Spirulina*, *Phaeodactylum*, etc., which are recognised as safe and stable by the Food and Drug Administration (FDA); hence, these microalgae are generally referred to as "Generally Recognised as Safe (GRAS)" (Fernandes et al. 2020). Chlorophyll, one of the major bio-pigments, and its derivatives are used as green dyes in the nutraceutical-based industry (Viera et al. 2019). In 2019, Food and Agriculture

Organization (FAO)-approved chlorophyll can be used in food as it is stable to changes in physical conditions. Carotenoids are another important microalgal pigments which have found their application in the food industry due to their antioxidant properties, which block the oxidation of fats. They are widely used as a beverage color enhancer as well, in food products, in confectionery, etc. In the healthcare sector, they treat cardiovascular or mental disorders and different cancers (Honda 2020; Ghosh et al. 2015). Among other groups of cyanobacteria, *Spirulina* is a significant reservoir of phycocyanin with large industrial and healthcare applications.

6.5 Various Methods Used to Enhance Bio-pigment Production

Various strategies have been implemented for the enhanced production of bio-pigments in microalgae. These strategies are divided into different classes (nutrimental factors, physical factors and biotechnical methods). Nutrimental factors involve the manipulation of growth media nutritional composition and physical condition in which they are growing; biotechnological tools also play a vital part in enhanced pigment production.

6.5.1 Effect of Nutrimental Factors on Bio-pigment Synthesis

The impact of cultivation conditions and changes in the media composition affects bio-pigment production. Phytohormones (bioactive compounds) can endorse metabolic shifts, stimulating cell growth. A study by Chen et al. (2020) concludes that auxins were the most efficient phytohormone for *Chromochloris zofingiensis*. Auxins cause higher astaxanthin and lipid accumulation, with the highest content of 13.1 mg g^{-1}. Other phytohormones such as indoleacetic acid and indole propionic acid also show an efficient effect on astaxanthin accumulation, i.e. 7.8 mg L^{-1} and 10 mg L^{-1}, respectively. In *K. aperta* DMGFW_2, *Rhodella* sp. APOT_15 and *B. submarina* APSW_11, use of salicylic acid resulted in an elevation in the protein content of the cell.

Nitrate is considered one of the important components of the cell. Because nitrate helps in the signalling and expression of gene apart from acting as a nitrogen source for photosynthesis, it was observed that increasing nitrate concentration in the medium improves biomass growth and pigment and lipid content by threefold and with the addition of methyl jasmonate, carotenoid and lipid production increased (Mc Gee et al. 2020). *Chlorella vulgaris* also show increased biomass growth and pigments, while low nitrate (<28.4 mg L^{-1}) in media leads to higher lipid accumulation content (42%). Nitrate-rich media (170 mg L^{-1}) result in higher biomass, carotenoids and chlorophyll-*a*, such as 800, 2 and 6 mg L^{-1}, respectively. Another important media component is iron, which acts as the building block for many proteins. It is essential for photosynthetic electron transfer and nitrogen and sulphur assimilation. In *Chlorella vulgaris*, it was observed that lower iron concentration leads to biomass with higher chlorophyll concentration, while glucose results in lower biomass production (Caetano et al. 2020). Magnesium (Mg^{2+}) and copper (Cu^{2+}) play an important role in photosynthetic pigment and biomass production. In *Arthrospira platensis*, it was observed that Cu^{2+} (0.5 μM) and Mg^{2+} (0.8 mM) cause higher biomass with the elevated amount of phycocyanin, allophycocyanin and chlorophyll *a*. High levels of chl*a* were obtained at 0.5 μM Cu^{2+}. In contrast, higher β-carotene concentrations of 2400 ± 80 μg mL^{-1} were obtained Cu^{2+}-deficient media (Urek & Kerimoglu, 2019).

To extract phycobiliproteins from *Spirulina* (*Arthrospira*) *platensis*, a novel approach utilising various protic ionic liquids (PILs), mechanical agitation and thermal heating was suggested. At 35 °C and pH 6.50, a solvent system was created using N-methyl-2-hydroxyethylammonium acetate (2-HEAA)/N-methyl-2-hydroxyethylammonium formate (2-HEAF) protic ionic liquids in a 1:1 v/v ratio. This system extracted phycobiliproteins in a solvent/biomass ratio of 6.59 mL g^{-1}. Phycocyanin, allophycocyanin and phycoerythrin were extracted in the maximum concentration of 1.65 g L^{-1}, 1.70 g L^{-1} and 0.64 g L^{-1}, respectively, with a purity index of 0.50 (cosmetics and foods) (Rodrigues et al. 2019). N_2-fixing algae show different growth patterns in the media's nitrogen (N)- content. *Trichormus* sp. IMU26, a filamentous cyanobacterium, is known for fast growth and high phycobiliprotein (PBP) production during N_2-fixing conditions. The formation of PBP and growth are typically impacted by nitrogenous source concentration. For example, nitrogen promotes higher soluble saccharide and protein content but decreases growth rates and PBP production. However, it was found that in *Trichormus* sp. IMU26, short-term N-P deprivation leads to increased PBP production but no change in growth rate. The cells' other activities are affected by the elevated PBP, including changes in carotenoid-pigment interactions, a quick induction of zeaxanthin and beta-carotene production and a modest decrease in the amounts of echinenone and canthaxanthin (Haddad et al. 2020). In the case of *Anabaena variabilis* IMU8, the PBP ratio was limited when grown on N-availability. PBP concentration decreased by 29% on 12-day incubation in N-replete cells. The PC, APC and PE levels decreased by approximately 59.1%, 58.6% and 30.2% in the N-P-deficient medium. When N-depleted cells and N-replete cells were tested, it was found that N-supply slightly boosted proliferation and total saccharide production while not affecting total protein production. However, PBP production decreased (Haddad et al. 2019).

Various optimisation strategies are also employed to boost PBP production. For example, Saini et al. (2020) optimised *Anabaena variabilis* CCC421 using a design matrix approach known as GA-Fuzzy, an evolutionary process combining genetic algorithm (GA) and fuzzy logic methodology (FLM). As a result, the increased PBP production was 2.13 times higher, or 408.5 mg L^{-1}, at the optimal confluence of variables (FAC 0.153 g L^{-1}, K_2HPO_4 0.2 g L^{-1} and trace metals 0.5 ml L^{-1}).

The creation of the best ultrasound-assisted extraction (UAE) for phycobiliproteins (PBPs) in *Oscillatoria* sp. (BTA 170) followed the Taguchi technique. This methodology tuned four process parameters, including the solid-to-liquid ratio, duty cycle, electrical acoustic intensity and pH. The signal-to-noise (S/N) ratio was used to calculate the optimal circumstances to produce a higher yield of PBPs. Analysis of variance (ANOVA) was used to determine the statistically significant parameters. As a result, improved PC (94.10%), PE (95.20%) and APC (90.54%) production were accomplished using the Taguchi methodology and variables such as solid-to-liquid ratio (0.2 g mL^{-1}), electrical acoustic intensity (16.99 w cm^{-2}), duty cycle (75%) and pH 7 (Sharma et al. 2020).

Salinity also affects bio-pigment production in various microalgae. For example, when *Porphyra haitanensis* was exposed to salinity of 40%, 36% and 30%, there was no effect on chlorophyll content. However, a 40% increase in carotenoid, phycocyanin and phycoerythrin contents was observed compared to the control. The reason behind enhanced salinity production would be oxidative stress, as carotenoid, phycocyanin and phycoerythrin function as antioxidants (Zhang et al. 2020).

6.5.2 Effect of Physical Factors on Bio-pigment Synthesis

Light is the most important factor for maintaining microalgal physiology and photosynthesis because the photoperiod determines the light energy required for metabolic activities. Numerous investigations have been conducted to ascertain how photoperiod affects the development of microalgae and the synthesis of secondary metabolites. The photoperiod directly emphasises nitrogen and carbon metabolism. During daylight, the C/N ratio increases rapidly, leading to slower protein synthesis than carbohydrate synthesis in high increments in photosynthetically active radiance (PAR). Carbon from polymers of stored carbohydrates can be transferred to proteins at night. Carbon/nitrogen ratio in biomass has a linear relationship with PAR (Table 6.2) (Benavente-Valdés et al. 2016). Industrially important microalga *Tetraselmis* sp. CTP4 is an important source for carotenoid extraction because of its balanced biochemical profile. Schüler et al. (2020) investigated different light intensities and found that lutein levels increased 1.5-fold (3.21 0.18 mg g^{-1} DW) under higher light intensities (170 and 280 mol m^{-2} s^{-1}), while beta-carotene contents increased by 3-fold (3.17 0.18 mg g^{-1} DW) under low light intensities (33 mol m^{-2} s^{-1}).

Fatty acid, total protein, β-carotene, chlorophyll a and growth rate are known to be regulated by photoperiod and irradiance rate in *Chlorella vulgaris*, *Haematococcus pluvialis*, *Dunaliella salina* and *Spirulina platensis*. Maximum growth rates, levels of saturated fatty acids, protein content and β-carotene were seen in *C. vulgaris* at an illumination of 100 mmol photons m^{-2} s^{-1} and a photoperiod of 16:8 h of light/dark. Monounsaturated and polyunsaturated fatty acids and chlorophyll decrease in response to increased light intensity and duration. Total saturated fatty acids (SFA) were measured at 100 mmol photons m^{-2} s^{-1} and 16:8 h of photoperiod, whereas monounsaturated (MUFA) and polyunsaturated (PUFA) fatty acids were estimated at 37.5 mmol photons m^{-2} s^{-1} and 8:16 h of photoperiod, respectively (Khoeyi et al. 2012).

Table 6.2 Effect of physical methods on enhancing bio-pigments in microalgae

Species	Pigment	Methodology	Production	References
Chlamydomonas reinhardtii	Carotenoid	Use of organic dye rhodamine 8G to increase photon flux density	36% increase	Ramanna et al. (2018)
Cyanobium sp.	Carotenoid	Treatment with red light	10% increase	Pagels et al. (2020a)
Arthrospira platensis	Carotenoid	Exposure to blue and red light (30:70)	Threefold increase	Lima et al. (2018)
Pseudanabaena sp.	PE	Greenlight	63% increase	Mishra et al. (2012)
Arthrospira maxima	PBPs	Use of blue light-emitting diodes (LED)	PC: 3.20 mg g^{-1} APC: 0.19 mg g^{-1} PE: 0.97 mg g^{-1}	Park and Dinh (2019)
Nostoc	PE	Growing at pH 5.0	Highest (111 μg mg^{-1}) phycoerythrin	Keithellakpam et al. (2015)
Phormidium and *Nostoc commune*	PC	Growing at pH 6.0	Maximum output of 2.04 mg L^{-1} d^{-1}	Pagels et al. (2020b)
Arthrospira platensis	Carotenoids	Using 0.1 g L^{-1} of sodium nitrate	45.5 mg g^{-1} of dry weight	El Baky et al. (2020)
Anabaena variabilis	PC	Amending medium with nitrate (2 mM) and glucose (0.5 g L^{-1})	78.4 mg g^{-1}	Cottas et al. (2020)
Prochlorococcus sp.	Lutein	Urea as a nitrogen source	3.34 mg g^{-1}	Erdoğan et al. (2016)
Geitlerinema amphibium	Astaxanthin and lutein	Wright's cryptophyte (WC) artificial media and bold basal media (BBM)	Enhanced astaxanthin: 130% Lutin: 234%	D'Alessandro et al. (2019)

PC, phycocyanin; PE, phycoerythrin; APC, allophycocyanin

Because there was more energy available for astaxanthin production at the high light intensity in the *Haematococcus pluvialis* culture, there was an enhanced rate of nutrient depletion. For example, when using acetate as a carbon source, *H. pluvialis* demonstrated a peak growth rate of 0.243 L d^{-1} at a light intensity of 170 mmol photons m^{-2} s^{-1}. In contrast, an investigation revealed a maximum growth rate (0.195 L d-1) in light intensity of 40 mmol photons m^{-2} s^{-1}. Similarly, β-carotene and phycocyanin in *Dunaliella salina* and *Spirulina platensis* improved with increasing light exposure (Benavente-Valdés et al. 2016).

The concentration of cells, cell size, biomass output, proximate composition, pigment content and fatty acid content are also impacted by the light spectrum (various light colours). In a study on *C. vulgaris* by Sánchez-Saavedra et al. in 2020, it was found that while green and yellow lights were better during exponential growth, blue light was better for increasing biomass productivity and lipid content. White light had a higher protein content (34.42%), while green light had a higher protein content (31.64%).

Pseudanabaena mucicola is used for phycobiliprotein production. It resulted in higher ($P < 0.05$) production of phycobiliprotein 237.01 mg g^{-1} with a purity of 1.14 (food grade blue colour) on white light and wastewater as the medium (Khatoon et al. 2018). Higher phycobiliproteins were attained under blue LED in *Arthrospira maximum*, including 3.20 mg g1 phycocyanin (PC), 0.19 mg g1 allophycocyanin (APC) and 0.97 mg g1 phycoerythrin (PE). The red/white light cultivation strategy resulted in higher chlorophyll production, while low blue light concentration resulted in lower chlorophyll concentrations (Park and Dinh 2019).

Two-phase light cultivation systems seem very useful in stimulating pigment production, which helps in increasing economic turnover. *Cyanobium* sp. was used for carotenoids and phycobiliproteins under white/red light. The optimal cultivation condition was 10:4 days in white/red light, which resulted in biomass of 137.4 ± 0.8 mg L^{-1} d^{-1}, total phycobiliproteins 17.0 ± 0.2 mg L^{-1} d^{-1} and carotenoids 4.5 ± 0.2 mg L^{-1} d^{-1} (Pagels et al. 2020a, 2020b).

Galdieria sulphuraria was grown at 1% CO_2 and a photon fluence rate of 200 mol photons m^{-2} s^{-1}, which produced enhanced biomass productivity at 34 ° C, which was reported to be 10 g m^{-2} d^{-1} dry weight. The formation of phycobilin and improved growth are attributed to adding CO_2 in *G. sulphuraria*. Enhanced phycocyanin and allophycocyanin concentration was also reported to be 0.28 g m^{-2} d^{-1} and 0.23 g m^{-2} d^{-1} (Carbone et al. 2020).

Temperature and pH are two parameters that impact PBP production and are crucial in deactivating PBPs. For example, when the culture of *Oscillatoria* sp. was cultivated at different pH ranges from 2 to 12, it was observed that the K_d value at pH 7 was the lowest for PBPs.

At pH 7, PC deactivates more quickly after the temperature change, followed by APC and PE. For PBPs, it was discovered that entropy and enthalpy decreased at increasing acidic and basic conditions. PBPs are advised for use in various industries due to their capacity to tolerate thermal and pH stress (Tiwari et al. 2019). Agar of exceptional quality is produced by the commercial marine red alga *Gelidium amansii*; cultivation attempts have been unsuccessful. It is unclear how high temperatures, in particular, affect this alga's ability to photosynthesise. Through the experiment, it was discovered that photosynthesis required a temperature range of 22 to 26 °C. Several processes, including carbon assimilation, light-harvesting pigments and PSII, operate at higher temperatures (30 °C).

Reaction centres and acceptor side were damaged. Chl-*a* and carotenoids were decreased at 30 °C, while phycobiliproteins increased at 30 °C when cells were incubated for 12 h. The evidence suggests that phycobiliproteins may be crucial in reducing the oxidative damage brought on by high temperatures (Li et al. 2019).

Furcellaria lumbricalis (red algae), R-phycoerythrin, gave the highest yields at 20 °C and pH 6 when detected using a fluorescence detector (0.13%) and photodiode array detector (0.43%) (Saluri et al. 2019).

6.5.3 Use of Molecular Tools for Bio-pigment Synthesis

In the past two decades, photosynthetic organisms have drawn much attention because they can create goods with a high industrial value while having less impact on the environment's carbon dioxide, water and nutrient levels. In addition, several photoautotrophs' complete genome sequences are now readily available because of advancements in molecular biology, thereby extending the possibility of genetic modification (Table 6.3) (Srivastava et al. 2020). Recently, a variety of biological tools have been employed to improve the bio-pigments in cyanobacteria, such as CRISPR/Cas9, *Cpf1* techniques, CRISPR interference, development and characterisation of endogenous/inducible promoters and expression of heterologous genes (Markley et al. 2015; Gordon et al. 2016; Li et al. 2016; Ungerer and Pakrasi 2016; Liu and Pakrasi 2018).

Carotenoids are major bio-pigments, particularly in the food and pharmaceutical industries, and they are crucial for preventing and treating UV-B radiation-induced cell damage. To increase the production rate, several microalgae and cyanobacteria have undergone genetic modification. The genes involved in the biosynthesis of carotenoids (C40) are, namely, CrtE (GGPP synthase), which catalyses the biosynthesis of geranylgeranyl diphosphate (GGPP), from the precursors, namely, farnesyl diphosphate and isopentenyl phosphate, and CrtB (phytoene synthase), which converts GGPP to the first carotenoid, phytoene; after that, phytoene is converted to lycopene in a series of steps which are catalysed by *CrtP* (phytoene desaturase), CrtQ (β-carotene desaturase), *Z-ISO* (β-carotene isomerase) and *CrtH* (cis-carotene isomerase) (Sugiyama and Takaichi 2020).

Echinenone and/or canthaxanthin (food and industrially important) biosynthesis is catalysed by β-carotene ketolase (*CrtO*), where β-carotene acts as the substrate. Gao et al. (2020) reported the genetic modification of *Nostoc* sp. PCC 7120 for the overexpression of gene by the heterologous expression of *crtO* gene from *Nostoc flagelliforme*. Overexpression of *crtO* increased the pigments' production rate by 16%; similarly in transgenic *Nostoc* sp. PCC 7120, the osmotic stress condition enhanced the expression by 80%. Another successful method for the enhanced production is the protein expression system. In a study by Betterle and Melis (2019), overexpression of the β-phellandrene synthase (*PHLS*) and geranyl diphosphate synthase (*GPPS*) genes was carried out by utilising fusion constructs as vectors for protein overexpression in *Synechocystis* sp. PCC 6803. Overexpression of these genes resulted in the enhanced accumulation of *GPPS* and *PHLS* by 4% and 10%, respectively, of the total cellular protein. Increased accumulation results in delayed catalytic activity and the synthesis of 24 mg of PHL g-1 biomass in transformant *Synechocystis*.

Table 6.3 Enhancement of bio-pigments by biotechnological approaches

Microorganism	Bio-pigments	Methodology	Production	References
Haematococcus pluvialis	Carotenoid and astaxanthin	Use of chemical mutagen N-nitro-N-nitrosoguanidine	23–59% increase	Sandesh Kamath et al. (2008)
Lyngbya	Carotenoid	Irradiation with UV light	125% increase	Kokabi et al. (2019)
Chlorogloeopsis fritschii	Canthaxanthin	Exposure to low-level UV-B light	9-fold increase	Llewellyn et al. (2020)
Chlorella vulgaris	Violaxanthin	Random mutagenesis	3.18 times increase	Kim et al. (2020)
Synechocystis	Zeaxanthin	Overexpression of β-carotene hydroxylase *crtR*	150% increase	Lagarde et al. (2000)
Synechocystis	Astaxanthin	Transfer of *E. coli* β-carotenoid ketolase (*crtW*) and hydroxylase (*crtZ*) gene	500-fold increase	Diao et al. (2020)
Anabaena sp. PCC 7120	Canthaxanthin and echinenone	Introduction of *Nostoc flagelliforme* β-carotene ketolase (*crtO*) gene	80% and 16% increase, respectively	Gao et al. (2020)
Phaeodactylum tricornutum	Fucoxanthin	Overexpression of genes: violaxanthin de-epoxidase (*Vde*), Vde-related (*Vdr*) genes and zeaxanthin epoxidase 3 (*Zep3*)	4-fold increase	Manfellotto et al. (2020)
Haematococcus pluvialis	Astaxanthin	Overexpression of phytoene desaturase (*pds*)	67% increase	Galarza et al. (2018)
Scenedesmus	β-Carotene	Transformation of phytoene synthase (*psy*) gene	2-fold increase	Chen et al. (2017)
Chlamydomonas reinhardtii	Canthaxanthin	microRNA-mediated silencing of *ATG1* and ATG8 genes (involved in the control of carotenoid biosynthesis by autophagy)	2.34-fold rise	Tran et al. (2019)
Chlamydomonas reinhardtii	Zeaxanthin	CRISPR-Cas9 (clustered regularly interspaced short palindromic repeats- CRISPR-associated protein 9)-mediated knockout of zeaxanthin epoxidase (*ZEP1*) gene	56-fold rise	Baek et al. (2018)
Porphyridium sp.	Phycoerythrin	Through CRISPR/Cas9 RNP	63.30% increase	Jeon et al. (2021)
Nannochloropsis sp.	Violaxanthin	Irradiation with γ-ray	43.54% increase	Park et al. (2021)

Orange carotenoid protein (OCP) facilitates a photoprotective mechanism in cyanobacteria as OCP become photoactivated in the presence of blue-green light, which causes them to interact with light-harvesting structures called phycobilisomes and divert extra photosynthetically absorbed energy to dissipate heat. OCP (Or gene) is also well-recognised for promoting carotene accumulation in higher plants. In the study conducted by Kumari et al. 2020, about 1.7-fold increased carotenoid content with increased cell size and appearance of multicellularity (grouping of both orange and green transgenic cells) in a mutant of *Chlamydomonas reinhardtii*—mCrOr (R106H) was obtained by cloning heterologous *mBoOr* (from *Brassica oleracea*). OCP protein structure consists of an *N*-terminal domain (NTD) and a C-terminal domain (CTD). Two other domains, helical carotenoid proteins (HCPs) and CTDHs, are homologous to NTD and CTD, respectively. Both domains can bind the carotenoids. It was observed that amino acid conversion is important for binding carotenoids with these proteins in *Fremyella diplosiphon* and the model algae *Chlamydomonas reinhardtii*. Thus, the system for overexpressing carotenoids will be developed by the in vitro reconstitution of these purified proteins and evaluating their probable carotenoid binding ability (Wei et al. 2019).

Lutein is one of the pharmaceutically essential carotenoids, and its biosynthesis is catalysed by the genes *psy*, *pds* and *lut1*. Ma et al. (2020) showed a low incubation temperature (20 °C) in *Chlamydomonas* sp. JSC4 resulted in the upregulation of these genes and genes involved in fatty acid biosynthesis (*bccp* and *bc*). These results are related to diverting the metabolic flux from carbohydrates towards carotenoid, protein and fatty acid biosynthesis.

Photosynthesis is the primary factor for microalgal growth that involves utilising light.

The quality of light and thermodynamics of photosynthesis directly affect biomass production. In microalgae, the anatomy of the light-absorbing antenna plays a significant role. A giant antenna absorbs high light radiance, leading to light-saturated photosynthesis and photodamage. Therefore, a reduction in chlorophyll antenna could contribute to enhanced biomass production. EMS mutagenesis was done to reduce chlorophyll antenna in *Chlorella saccharophila*, resulting in high light tolerance. Mutagenesis resulted in chlorophyll fluorescence reduction by 21% and PSII and PSI antenna size reduction by 37% and 15% reduction, respectively, and this reduction helped increase biomass productivity by 27% and non-photochemical quenching (Patil et al. 2020).

Photosynthesis thermodynamics also limits biomass productivity. When light energy is transformed into chemical energy, there is an excellent level of thermodynamic inefficiency because the light-harvesting antenna has ten times fast light-capturing ability.

So, reduction in the size of the antenna does not fully solve the problem. Hence, adjusting light-harvesting antenna sizes through the development of a translational control system to enhance photosynthetic performance will help in expressing chlorophyllide a, an oxygenase (CAO) gene with *Nab1* (NGFI-A-binding protein 1, a cytosolic translation repressor gene which helps in light regulation in PSII system) translational repressor binding site at 5' mRNA extension. *Nab1*-mediated repression of *CAO* expression will continuously modify levels of chlorophyll *b* and sizes of light harvesting. This method results in twofold greater biomass productivity (Negi et al. 2020).

The phycobiliproteins (PBPs) are the light-harvesting protein complexes of tetrapyrrole (noncyclic linear or open chain) as prosthetic groups, which are composed of four pyrrole rings (A, B, C and D) incorporated in biliproteins (α, β and γ). The structure of the PBPs commonly consists of heterodimers of αβ-subunits, each carrying the cysteinyl residues linked to bilins via a thioether linkage. In cyanobacteria, phycobilins are classified into three types based on their chromophore structure: phycocyanobilin (PCB), phycoerythrobilin (PEB) and phycourobilin (PUB) (Kannaujiya et al. 2017a, 2017b). In *Nostoc* sp. PCC 7120, *CpcE/F* heterodimer is covalently linked to α-subunits of phycocyanin and phycoerythrin. *CpcE/F* heterodimer belongs to the family of lyases and has chromophore-detaching activity. Between the inner and outer layers of the CpcE and CpcF, a cavity is formed in which the phycocyanobilin chromophore can be carved. *CpcE/F* structure was used to develop the structure lyase *PecE/F*, which catalyses the isomerisation reaction for the conversion of phycocyanobilin into phycoviolobilin. *NblB* was moulded using *CpcE* and catalysed phycobilisome degradation, leading to low chromophore-detaching activity (Zhao et al. 2017). *NblA/B* also promotes the segregation of phycobilisomes, cores and allophycocyanin trimers. *NblA/B* assists in lyase-catalysed and lyase-independent chromophore transfers among core protein, biliproteins, the rod protein, *ApcF*, *CpcA* and sensory biliproteins which balance and complete the metabolic requirements (Hu et al. 2019).

Regulatory promoters are essential for the overexpression of genes. Two promoters, A2520 and A2579, showed strong expression patterns in *Synechococcus* sp. PCC 7002 with a twofold increase in the expression of gene under light conditions when compared to dark. These promoters can genetically modify genes or regulatory systems (Ruffing et al. 2016).

6.6 Conclusion

Nowadays, several industries have gained interest in natural products because of their assorted application and properties. Hence, in that regard cyanobacterial photopigments have been widely explored at an industrial level, such as food and nutraceuticals, mainly due to their high production rate and resistance to changes in pH, temperature and other physical factors. Furthermore, recent research and developments in computational biology, genetic engineering and structural biology have led researchers to understand bio-pigments' nature and molecular structure. Hence, metabolic engineering is considered an intriguing approach for improving bio-pigment production at an industrial scale with economic benefits.

References

Alam T, Najam L, Al-Harrasi A (2018) Extraction of natural pigments from marine algae. JAMS 23:81–91

Ambati RR, Gogisetty D, Aswathanarayana RG, Ravi S, Bikkina PN, Bo L et al (2019) Industrial potential of carotenoid pigments from microalgae: current trends and future prospects. Crit Rev Food Sci Nutr 59(12):1880–1902. https://doi.org/10.1080/10408398.2018.1432561

Baek K, Yu J, Jeong J, Sim SJ, Bae S, Jin E (2018) Photoautotrophic production of macular pigment in a *Chlamydomonas reinhardtii* strain generated by using DNA-free CRISPR-Cas9 RNP-mediated mutagenesis. Biotechnol Bioeng 115(3):719–728. https://doi.org/10.1002/bit.26499

Barsanti L, Gualtieri P (2020) Anatomy of *Euglena gracilis*. In: Handbook of algal science, technology and medicine. Academic Press, London

Bearden AJ, Malkin R (1974) Primary photochemical reactions in chloroplast photosynthesis. Q Rev Biophys 7(2):131–177

Begum H, Yusoff FM, Banerjee S, Khatoon H, Shariff M (2016) Availability and utilisation of pigments from microalgae. Crit Rev Food Sci Nutr 56(13):2209–2222. https://doi.org/10.1080/10408398.2013.764841

Benavente-Valdés JR, Aguilar C, Contreras-Esquivel JC, Méndez-Zavala A, Montañez J (2016) Strategies to enhance the production of photosynthetic pigments and lipids in Chlorophyceae species. Biotechnol Rep (Amst) 10:117–125. https://doi.org/10.1016/j.btre.2016.04.001

Betterle N, Melis A (2019) Photosynthetic generation of heterologous terpenoids in cyanobacteria. Biotechnol Bioeng 116(8):2041–2051. https://doi.org/10.1002/bit.26988

Borowitzka MA (2013) High-value products from microalgae—their development and commercialisation. J Appl Phycol 25(3):743–756. https://doi.org/10.1007/s10811-013-9983-9

Bradshaw RE, Dixon SW, Raitt DC, Pillar TM (1993) Isolation and nucleotide sequence of the 5-aminolevulinate synthase gene from *Aspergillus nidulans*. Curr Genet 23(5–6):501–507. https://doi.org/10.1007/BF00312642

Bryant DA, Hunter CN, Warren MJ (2020) Biosynthesis of the modified tetrapyrroles—the pigments of life. J Biol Chem 295:6888

Caetano N, Melo AR, Gorgich M, Branco-Vieira M, Martins AA, Mata TM (2020) Influence of cultivation conditions on the bioenergy potential and bio-compounds of *Chlorella vulgaris*. Energy Rep 6:378–384. https://doi.org/10.1016/j.egyr.2019.08.076

Canter-Lund H, Lund JWG (1995) Freshwater algae: their microscopic world explored. Biopress Limited, Bristol

Carbone DA, Olivieri G, Pollio A, Melkonian M (2020) Biomass and phycobiliprotein production of *Galdieria sulphuraria*, immobilised on a twin-layer porous substrate photobioreactor. Appl Microbiol Biotechnol 104(7):3109–3119. https://doi.org/10.1007/s00253-020-10383-8

Carty S, Parrow MW (2015) Dinoflagellates. In: Freshwater algae of North America. Academic Press, London, pp 773–807

Chakdar H, Pabbi S (2016) Cyanobacterial phycobilins: production, purification, and regulation. In: Frontier discoveries and innovations in interdisciplinary microbiology. Springer, New Delhi, pp 45–69

Chakdar H, Pabbi S (2017) Algal pigments for human health and cosmeceuticals. In: Rastogi RP, Madamwar D, Pandey A (eds) Algal green chemistry. Elsevier, Amsterdam, pp 171–188

Chekanov K, Litvinov D, Fedorenko T, Chivkunova O, Lobakova E (2021) Combined production of astaxanthin and β-carotene in a new strain of the microalga *Bracteacoccus aggregatus* bm5/15 (Ippas c-2045) cultivated in photobioreactor. Biology 10(7):1–17. https://doi.org/10.3390/biology10070643

Chen CY, Kao AL, Tsai ZC, Shen YM, Kao PH, Ng IS et al (2017) Expression of synthetic phytoene synthase gene to enhance β-carotene production in *Scenedesmus* sp. CPC2. Biotechnol J 12(11):1700204. https://doi.org/10.1002/biot.201700204

Chen JH, Wei D, Lim PE (2020) Enhanced coproduction of astaxanthin and lipids by the green microalga *Chromochloris zofingiensis*: selected phytohormones as positive stimulators. Bioresour Technol 295:122242. https://doi.org/10.1016/j.biortech.2019.122242

Choi YY, Patel AK, Hong ME, Chang WS, Sim SJ (2019) Microalgae bioenergy carbon capture utilisation and storage (BECCS) technology: an emerging sustainable bioprocess for reduced CO2 emission and biofuel production. Bioresour Technol Rep 7:100270. https://doi.org/10.1016/j.biteb.2019.100270

Cottas AG, Cunha WR, Ribeiro EJ, de Souza FJ (2020) Influence of medium composition on the production of phycocyanin from *Anabaena variabilis*. Ind Biotechnol 16(2):45–49. https://doi.org/10.1089/ind.2019.29196.agc

D'Alessandro EB, Soares AT, de Oliveira D'Alessandro NC, Antoniosi Filho NR (2019) Potential use of a thermal water cyanobacterium as raw material to produce biodiesel and pigments. Bioprocess Biosyst Eng 42(12):2015–2022. https://doi.org/10.1007/s00449-019-02196-5

Devi A, Patil K, Kumar A, Gupta N, Malik A, Pabbi S (2022) Production of valuable platform chemicals through microalgal routes utilizing waste streams. Bioresource Technol Rep:101071

Diao J, Song X, Zhang L, Cui J, Chen L, Zhang W (2020) Tailoring cyanobacteria as a new platform for highly efficient synthesis of astaxanthin. Metab Eng 61:275–287. https://doi.org/10.1016/j.ymben.2020.07.003

Draaisma RB, Wijffels RH, Slegers PM, Brentner LB, Roy A, Barbosa MJ (2013) Food commodities from microalgae. Curr Opin Biotechnol 24(2):169–177. https://doi.org/10.1016/j.copbio.2012.09.012

Duppeti H, Chakraborty S, Das BS, Mallick N, Kotamreddy JNR (2017) Rapid assessment of algal biomass and pigment contents using diffuse reflectance spectroscopy and chemometrics. Algal Res 27:274–285. https://doi.org/10.1016/j.algal.2017.09.016

El Baky HHA, El Baroty GS, Mostafa EM (2020) Optimisation growth of *Spirulina* (Arthrospira) platensis in photobioreactor under varied nitrogen concentration for maximised biomass, carotenoids and lipid contents. Recent Pat Food Nutr Agric 11(1):40–48. https://doi.org/10.2174/2212798410666181227125229

Erdoğan A, Demirel Z, Eroğlu AE, Dalay MC (2016) Carotenoid profile in Prochlorococcus sp. and enrichment of lutein using different nitrogen sources. J Appl Phycol 28(6):3251–3257. https://doi.org/10.1007/s10811-016-0861-0

Fernandes AS, Nass PP, Oliveira A, Zepka LQ (2020) Chlorophylls as food additives. In: Pigments from microalgae handbook. Springer, Cham, pp 391–420

Galarza JI, Gimpel JA, Rojas V, Arredondo-Vega BO, Henríquez V (2018) Over-accumulation of astaxanthin in *Haematococcus pluvialis* through chloroplast genetic engineering. Algal Res 31: 291–297. https://doi.org/10.1016/j.algal.2018.02.024

Gao X, Xu H, Zhu Z, She Y, Ye S (2020) Improved production of echinenone and canthaxanthin in transgenic *Nostoc sp.* PCC 7120 overexpressing a heterologous *crtO* gene from *Nostoc flagelliforme*. Microbiol Res 236:126455. https://doi.org/10.1016/j.micres.2020.126455

Ghosh T, Paliwal C, Maurya R, Mishra S (2015) Microalgal rainbow colours for nutraceutical and pharmaceutical applications. In: Plant biology and biotechnology. Springer, New Delhi, pp 777–791

Gordon GC, Korosh TC, Cameron JC, Markley AL, Begemann MB, Pfleger BF (2016) CRISPR interference as a titratable, trans-acting regulatory tool for metabolic engineering in the cyanobacterium *Synechococcus* sp. strain PCC 7002. Metab Eng 38:170–179. https://doi.org/10.1016/j.ymben.2016.07.007

Graham L, Wilcox L (2000) Algae. Prentice Hall, Englewood Cliffs, NJ

Haddad MF, Dayioglu T, Nalbantoğlu B, Cakmak T (2019) Long term diazotrophic cultivation induces phycobiliprotein production in *Anabaena variabilis* IMU8. Biocell 43(4):327–333. https://doi.org/10.32604/biocell.2019.08522

Haddad MF, Dayioglu T, Yaman M, Nalbantoglu B, Cakmak T (2020) Long-term diazotrophic cultivation of *Trichormus* sp. IMU26: evaluation of physiological changes related to elevated phycobiliprotein content. J Appl Phycol 32(2):881–888. https://doi.org/10.1007/s10811-019-02012-3

Hasle GR, Syvertsen EE (1997) Marine diatoms. In: Tomas CR (ed) Identifying marine diatoms and dinoflagellates. Academic Press, San Diego, pp 5–38

Honda M (2020) Nutraceutical and pharmaceutical applications of carotenoids. In: Pigments from microalgae handbook. Springer, Cham, pp 449–469

Hong ME, Chang WS, Patel AK, Oh MS, Lee JJ, Sim SJ (2019) Microalgae based carbon sequestration by converting LNG-fired waste CO2 into red gold astaxanthin: the potential applicability. Energies 12(9):1718. https://doi.org/10.3390/en12091718

Hu PP, Hou JY, Xu YL, Niu NN, Zhao C, Lu L et al (2019) The role of lyases, *NblA* and *NblB* proteins and bilin chromophore transfer in restructuring the cyanobacterial light harvesting complex. Plant J 102:529

Jeon MS, Han SI, Jeon M, Choi YE (2021) Enhancement of phycoerythrin productivity in *Porphyridium purpureum* using the clustered regularly interspaced short palindromic repeats/ CRISPR-associated protein 9 ribonucleoprotein system. Bioresour Technol 330:124974. https:// doi.org/10.1016/j.biortech.2021.124974

Kalwani M, Devi A, Patil K, Kumari A, Dalvi V, Malik A et al (2022) Microalgae-mediated wastewater treatment and enrichment of wastewater-cultivated biomass for biofuel production. In: Expanding horizon of cyanobacterial biology. Academic Press, London, pp 259–281

Kannaujiya VK, Sundaram S, Sinha RP (2017a) Gene manipulation and biosynthesis of phycobiliproteins. In: Phycobiliproteins: recent developments and future applications. Springer, Singapore, pp 45–69

Kannaujiya VK, Sundaram S, Sinha RP (2017b) Structural and functional significance of phycobiliproteins. In: Phycobiliproteins: recent developments and future applications. Springer, Singapore, pp 21–44

Keithellakpam OS, Nath TO, Oinam AS, Thingujam I, Oinam G, Dutt SG (2015) Effect of external pH on cyanobacterial phycobiliproteins production and ammonium excretion. J Appl Biol Biotechnol 3(4):38–42

Khatoon H, Leong LK, Rahman NA, Mian S, Begum H, Banerjee S, Endut A (2018) Effects of different light source and media on growth and production of phycobiliprotein from freshwater cyanobacteria. Bioresour Technol 249:652–658

Khoeyi ZA, Seyfabadi J, Ramezanpour Z (2012) Effect of light intensity and photoperiod on biomass and fatty acid composition of the microalgae, *Chlorella vulgaris*. Aquac Int 20(1): 41–49. https://doi.org/10.1007/s10499-011-9440-1

Kim J, Kim M, Lee S, Jin E (2020) Development of a *Chlorella vulgaris* mutant by chemical mutagenesis as a producer for natural violaxanthin. Algal Res 46:101790. https://doi.org/10. 1016/j.algal.2020.101790

Kobylewski S, Jacobson MF (2010) Food dyes: a rainbow of risks. Center for Science in the Public Interest, Washington

Kokabi M, Yousefzadi M, Soltani M, Arman M (2019) Effects of different UV radiation on photoprotective pigments and antioxidant activity of the hot-spring cyanobacterium *Leptolyngbya* cf. fragilis. Phycol Res 67(3):215–220. https://doi.org/10.1111/pre.12374

Kumari S, Vira C, Lali AM, Prakash G (2020) Heterologous expression of a mutant Orange gene from *Brassica oleracea* increases carotenoids and induces phenotypic changes in the microalga *Chlamydomonas reinhardtii*. Algal Res 47:101871. https://doi.org/10.1016/j.algal.2020. 101871

Lagarde D, Beuf L, Vermaas W (2000) Increased production of zeaxanthin and other pigments by application of genetic engineering techniques to *Synechocystis* sp. strain PCC 6803. Appl Environ Microbiol 66(1):64–72. https://doi.org/10.1128/AEM.66.1.64-72.2000

Lee R (2018) Chlorophyta. In: Phycology. Cambridge University Press, Cambridge, pp 133–230. https://doi.org/10.1017/9781316407219.009

Li H, Shen CR, Huang CH, Sung LY, Wu MY, Hu YC (2016) *CRISPR-Cas9* for the genome engineering of cyanobacteria and succinate production. Metab Eng 38:293–302. https://doi.org/ 10.1016/j.ymben.2016.09.006

Li Y, Liu J, Zhang L, Pang T, Qin R (2019) Effects of temperature on the photosynthetic performance in mature thalli of the red alga *Gelidium amansii* (Gelidiaceae). Aquaculture 512:734320. https://doi.org/10.1016/j.aquaculture.2019.734320

Lima GM, Teixeira PCN, Teixeira CMLL, Filócomo D, Lage CLS (2018) Influence of spectral light quality on the pigment concentrations and biomass productivity of *Arthrospira platensis*. Algal Res 31:157–166. https://doi.org/10.1016/j.algal.2018.02.012

Liu D, Pakrasi HB (2018) Exploring native genetic elements as plug-in tools for synthetic biology in the cyanobacterium *Synechocystis* sp. PCC 6803. Microb Cell Factories 17(1):48. https://doi.org/10.1186/s12934-018-0897-8

Llewellyn CA, Airs RL, Farnham G, Greig C (2020) Synthesis, regulation and degradation of carotenoids under low level UV-B radiation in the filamentous cyanobacterium *Chlorogloeopsis fritschii* PCC 6912. Front Microbiol 11:163. https://doi.org/10.3389/fmicb.2020.00163

Ma R, Zhao X, Ho SH, Shi X, Liu L, Xie Y et al (2020) Co-production of lutein and fatty acid in microalga *Chlamydomonas sp.* JSC4 in response to different temperatures with gene expression profiles. Algal Res 47:101821. https://doi.org/10.1016/j.algal.2020.101821

MacColl R (1998) Cyanobacterial phycobilisomes. J Struct Biol 124:311–334

Manfellotto F, Stella GR, Falciatore A, Brunet C, Ferrante MI (2020) Engineering the unicellular alga *Phaeodactylum tricornutum* for enhancing carotenoid production. Antioxidants 9(8):757. https://doi.org/10.3390/antiox9080757

Markley AL, Begemann MB, Clarke RE, Gordon GC, Pfleger BF (2015) Synthetic biology toolbox for controlling gene expression in the cyanobacterium *Synechococcus* sp. strain PCC 7002. ACS Synth Biol 4(5):595–603. https://doi.org/10.1021/sb500260k

Martins BMC, Tooke AK, Thomas P, Locke JCW (2018) Cell size control driven by the circadian clock and environment in cyanobacteria. Proc Natl Acad Sci U S A 115(48):E11415–E11424. https://doi.org/10.1073/pnas.1811309115

Mc Gee D, Archer L, Fleming GTA, Gillespie E, Touzet N (2020) The effect of nutrient and phytohormone supplementation on the growth, pigment yields and biochemical composition of newly isolated microalgae. Process Biochem 92:61–68. https://doi.org/10.1016/j.procbio.2020.03.001

Mishra SK, Shrivastav A, Maurya RR, Patidar SK, Haldar S, Mishra S (2012) Effect of light quality on the C-phycoerythrin production in marine cyanobacteria *Pseudanabaena* sp. isolated from Gujarat coast, India. Protein Expr Purif 81(1):5–10. https://doi.org/10.1016/j.pep.2011.08.011

Mogany T, Bhola V, Ramanna L, Bux F (2021) Photosynthesis and pigment production: elucidation of the interactive effects of nutrients and light on *Chlamydomonas reinhardtii*. Bioprocess Biosyst Eng 45:187

Musa M, Ayoko GA, Ward A, Rösch C, Brown RJ, Rainey TJ (2019) Factors affecting microalgae production for biofuels and the potentials of chemometric methods in assessing and optimising productivity. Cell 8(8):851. https://doi.org/10.3390/cells8080851

Negi S, Perrine Z, Friedland N, Kumar A, Tokutsu R, Minagawa J et al (2020) Light-regulation of light harvesting antenna size substantially enhances photosynthetic efficiency and biomass yield in green algae. Plant J 103(2):584–603. https://doi.org/10.1111/tpj.14751

Pagels F, Lopes G, Vasconcelos V, Guedes AC (2020a) White and red LEDs as two-phase batch for cyanobacterial pigments production. Bioresour Technol 307:123105. https://doi.org/10.1016/j.biortech.2020.123105

Pagels F, Salvaterra D, Amaro HM, Lopes G, Sousa-Pinto I, Vasconcelos V, Guedes A (2020b) Factorial optimisation of upstream process for *Cyanobium* sp. pigments production. J Appl Phycol 32(6):3861–3872

Park J, Dinh TB (2019) Contrasting effects of monochromatic LED lighting on growth, pigments and photosynthesis in the commercially important cyanobacterium *Arthrospira maxima*. Bioresour Technol 291:121846. https://doi.org/10.1016/j.biortech.2019.121846

Park SB, Yun JH, Ryu AJ, Yun J, Kim JW, Lee S et al (2021) Development of a novel nannochloropsis strain with enhanced violaxanthin yield for large-scale production. Microb Cell Factories 20(1):43. https://doi.org/10.1186/s12934-021-01535-0

Parmar A, Singh NK, Kaushal A, Sonawala S, Madamwar D (2011) Purification, characterization and comparison of phycoerythrins from three different marine cyanobacterial cultures. Bioresour Technol 102:1795–1802

Patel AK, Albarico FPJB, Perumal PK, Vadrale AP, Nian CT, Chau HTB et al (2022) Algae as an emerging source of bioactive pigments. Bioresour Technol 351:126910. https://doi.org/10.1016/j.biortech.2022.126910

Patel AK, Singhania RR, Dong CD, Obulisami PK, Sim SJ (2021) Mixotrophic biorefinery: a promising algal platform for sustainable biofuels and high value coproducts. Renew Sust Energ Rev 152:111669. https://doi.org/10.1016/j.rser.2021.111669

Patil S, Prakash G, Lali AM (2020) Reduced chlorophyll antenna mutants of *Chlorella saccharophila* for higher photosynthetic efficiency and biomass productivity under high light intensities. J Appl Phycol 16:1–9

Ramanna L, Rawat I, Zerrouki D, Bux F (2018) A novel organic dye-based approach to increase photon flux density for enhanced microalgal pigment production. J Clean Prod 198:187–194. https://doi.org/10.1016/j.jclepro.2018.07.016

Reynolds CS (2006) Ecology of phytoplankton. Cambridge University Press, Cambridge

Rodrigues RDP, de Lima PF, de Santiago-Aguiar RS, Rocha MVP (2019) Evaluation of protic ionic liquids as potential solvents for the heating extraction of phycobiliproteins from *Spirulina (Arthrospira) platensis*. Algal Res 38:101391. https://doi.org/10.1016/j.algal.2018.101391

Ruffing AM, Jensen TJ, Strickland LM (2016) Genetic tools for advancement of *Synechococcus* sp. PCC 7002 as a cyanobacterial chassis. Microb Cell Factories 15(1):190. https://doi.org/10.1186/s12934-016-0584-6

Saini DK, Chakdar H, Pabbi S, Shukla P (2019) Enhancing production of microalgal biopigments through metabolic and genetic engineering. CRC. Crit Rev Food Sci Nutr 60:1–5

Saini DK, Yadav D, Pabbi S, Chhabra D, Shukla P (2020) Phycobiliproteins from *Anabaena variabilis* CCC421 and its production enhancement strategies using combinatory evolutionary algorithm approach. Bioresour Technol 309:123347. https://doi.org/10.1016/j.biortech.2020.123347

Saluri M, Kaldmäe M, Tuvikene R (2019) Extraction and quantification of phycobiliproteins from the red alga *Furcellaria lumbricalis*. Algal Res 37:115–123. https://doi.org/10.1016/j.algal.2018.11.013

Sánchez-Saavedra MP, Sauceda-Carvajal D, Castro-Ochoa FY, Molina-Cárdenas CA (2020) The use of light spectra to improve the growth and lipid content of *Chlorella vulgaris* for biofuels production. Bioenergy Res 13(2):487–498. https://doi.org/10.1007/s12155-019-10070-1

Sandesh Kamath BS, Vidhyavathi R, Sarada R, Ravishankar GA (2008) Enhancement of carotenoids by mutation and stress induced carotenogenic genes in *Haematococcus pluvialis* mutants. Bioresour Technol 99(18):8667–8673. https://doi.org/10.1016/j.biortech.2008.04.013

Santiago-Santos MC, Ponce-Noyola T, Olvera-Ramírez R, Ortega-López J, Cañizares-Villanueva RO (2004) Extraction and purification of phycocyanin from *Calothrix* sp. Process Biochem 39(12):2047–2052. https://doi.org/10.1016/j.procbio.2003.10.007

Schüler LM, Santos T, Pereira H, Duarte P, Katkam NG, Florindo C et al (2020) Improved production of lutein and β-carotene by thermal and light intensity upshifts in the marine microalga *Tetraselmis* sp. Algal Res 45:101732. https://doi.org/10.1016/j.algal.2019.101732

Sharma R, Bhunia B, Mondal A, Kanti Bandyopadhyay T, Devi I, Oinam G et al (2020) Statistical optimisation of process parameters for improvement of phycobiliproteins (PBPs) yield using ultrasound-assisted extraction and its kinetic study. Ultrason Sonochem 60:104762. https://doi.org/10.1016/j.ultsonch.2019.104762

Silva SC, Ferreira ICFR, Dias MM, Barreiro MF (2020) Microalgae-derived pigments: a 10-year bibliometric review and industry and market trend analysis. Molecules 25(15):3406. https://doi.org/10.3390/molecules25153406

Singh NK, Sonani RR, Rastogi RP, Madamwar D (2015) The phycobilisomes: an early requisite for efficient photosynthesis in cyanobacteria. EXCLI J 14:268–289. https://doi.org/10.17179/excli2014-723

Sobotka R, McLean S, Zuberova M, Hunter CN, Tichy M (2008) The C-terminal extension of ferrochelatase is critical for enzyme activity and for functioning of the tetrapyrrole pathway in Synechocystis strain PCC 6803. J Bacteriol 190(6):2086–2095

Sonani RR, Rastogi RP, Madamwar D (2017) In: Rastogi RP, Madamwar D, Pandey A (eds) Algal green chemistry. Elsevier, Amsterdam, pp 91–120

Spolaore P, Joannis-Cassan C, Duran E, Isambert A (2006) Commercial applications of microalgae. J Biosci Bioeng 101(2):87–96. https://doi.org/10.1263/jbb.101.87

Srivastava A, Kalwani M, Chakdar H, Pabbi S, Shukla P (2022) Biosynthesis and biotechnological interventions for commercial production of microalgal pigments: a review. Bioresour Technol 352:127071. https://doi.org/10.1016/j.biortech.2022.127071

Srivastava A, Villalobos MB, Singh RK (2020) Engineering photosynthetic microbes for sustainable bioenergy production. In: Contemporary environmental issues and challenges in era of climate change. Springer, Singapore, pp 183–198

Sugiyama K, Takaichi S (2020) Carotenogenesis in cyanobacteria: *CruA*/*CruP*-type and *CrtL*-type lycopene cyclases. J Gen Appl Microbiol 66(2):53–58. https://doi.org/10.2323/jgam.2020.01.005

Tiwari ON, Bhunia B, Chakraborty S, Goswami S, Devi I (2019) Strategies for improved production of phycobiliproteins (PBPs) by *Oscillatoria* sp. BTA170 and evaluation of its thermodynamic and kinetic stability. Biochem Eng J 145:153–161. https://doi.org/10.1016/j.bej.2019.02.016

Tran QG, Cho K, Kim U, Yun JH, Cho DH, Heo J et al (2019) Enhancement of β-carotene production by regulating the autophagy-carotenoid biosynthesis seesaw in *Chlamydomonas reinhardtii*. Bioresour Technol 292:121937

Ungerer J, Pakrasi HB (2016) *Cpf1* is a versatile tool for *CRISPR* genome editing across diverse species of cyanobacteria. Sci Rep 6:39681. https://doi.org/10.1038/srep39681

Urek RO, Kerimoglu Y (2019) Evaluation of effects of Mg^{2+} and Cu^{2+} on pigment-metabolite production and photosystem II activity of *Arthrospira platensis* Gomont 1892. Turk J Fish Aquat Sci 19:873–883

Van den Hoek C, Mann DG, Jahns HM (1995) Algae: an introduction to phycology. Cambridge University Press, Cambridge

Viera I, Pérez-Gálvez A, Roca M (2019) Green natural colorants. Molecules 24(1):154. https://doi.org/10.3390/molecules24010154

Wang F, Huang L, Gao B, Zhang C (2018) Optimum production conditions, purification, identification, and antioxidant activity of violaxanthin from microalga *Eustigmatos* cf. polyphem (Eustigmatophyceae). Mar Drugs 16(6). https://doi.org/10.3390/md16060190

Wei DT, Bao H, Dominguez-Martin MA, Kerfeld CA (2019) Characterisation of novel homologs to the C-terminal domain of the Orange carotenoid protein. FASEB J 33(S1):779–745. https://doi.org/10.1096/fasebj.2019.33.1_supplement.779.45

Wu Y, Liao W, Dawuda MM, Hu L, Yu J (2019) 5-aminolevulinic acid (ALA) biosynthetic and metabolic pathways and its role in higher plants: a review. Plant Growth Regul 87(2):357–374. https://doi.org/10.1007/s10725-018-0463-8

Yun HS, Kim YS, Yoon HS (2020) Characterisation of *Chlorella sorokiniana* and *Chlorella vulgaris* fatty acid components under a wide range of light intensity and growth temperature for their use as biological resources. Heliyon 6(7):e04447. https://doi.org/10.1016/j.heliyon.2020.e04447

Zahedi Dizaji S, Attaran Fariman G, Zahedi MM (2021) Pigment content analysis in two HAB forming dinoflagellate species during the growth period. J Appl Phycol 33(2):807–817. https://doi.org/10.1007/s10811-020-02331-w

Zhang Z, Yang ZW, Zhong CH, Ding L, Zheng YY, Zhang G (2020) Effects of concentrated seawater on photosynthetic pigment contents in *Porphyra haitanensis*. Pol J Environ Stud 29:29

Zhao L, Chang WC, Xiao Y, Liu HW, Liu P (2013) Methylerythritol phosphate pathway of isoprenoid biosynthesis. Annu Rev Biochem 82:497–530

Zhao C, Höppner A, Xu QZ, Gärtner W, Scheer H, Zhou M et al (2017) Structures and enzymatic mechanisms of phycobiliprotein lyases *CpcE/F* and *PecE/F*. Proc Natl Acad Sci U S A 114(50):13170–13175. https://doi.org/10.1073/pnas.1715495114

Bioprospecting and Mechanisms of Cyanobacterial Hydrogen Production and Recent Development for Its Enhancement as a Clean Energy

7

Rahul Prasad Singh, Priya Yadav, Indrajeet Kumar, Ajay Kumar, and Rajan Kumar Gupta

Abstract

Energy security is a major concern for economic strength worldwide because of fossil resource depletion and rising costs. The synthesis of biohydrogen from cyanobacteria is a viable alternative clean and renewable energy source with significant viable potential. Cyanobacteria are highly relevant and valuable as prospective hydrogen producers because they produce hydrogen from water due to solar energy conversion. Furthermore, cyanobacteria have high photosynthetic efficiency and produce a large amount of biomass, which is used as a fourth-generation feedstock to produce biohydrogen. The yield of cyanobacterial biohydrogen has been improved in various ways with limited information in a systematic way. That's why the current state of research in the field of cyanobacterial hydrogen production enhancement is discussed with earlier published reports in this chapter. The major aim of this chapter is to discuss cyanobacterial hydrogen production, characteristics and roles of nitrogenase and hydrogenase enzymes concerned with hydrogen production, the various mechanisms of hydrogen production, recent metabolic pathway developments, modern photobioreactor efficiency, new cyanobacterial molecular genetic engineering and synthetic cyanobacterial biology. Finally, the major limitations to more efficient cyanobacterial hydrogen production and improvements for future commercialisation are also discussed.

R. P. Singh · P. Yadav · I. Kumar · R. K. Gupta (✉)
Laboratory of Algal Research, Centre of Advanced Study in Botany, Institute of Science, Banaras Hindu University, Varanasi, India
e-mail: rajang.bot@bhu.ac.in

A. Kumar
Agriculture Research Organization, Volcani Center, Department of Postharvest Science, Rishon Lezzion, Israel

B. Neilan et al. (eds.), *Cyanobacterial Biotechnology in the 21st Century*,
https://doi.org/10.1007/978-981-99-0181-4_7

Keywords

Bioprospecting · Biohydrogen · Cyanobacteria · Nitrogenase · Hydrogenase · Photobioreactor · Genetic engineering · Synthetic biology

7.1 Introduction

Energy security has recently emerged as a significant impediment to global economic success. Worldwide energy consumption has risen considerably with global population growth and modernisation, resulting in a global energy crisis, rising energy prices and global warming (Nagarajan et al. 2017). During millions of years, solar energy was naturally stored in the form of fossil fuels due to biotic and abiotic processes, and fossil energy continues to dominate global primary energy use (Wang et al. 2020). Unfortunately, fossil fuels are in limited supply and will soon run out, making it important to find alternative energy sources.

Hydrogen production from phototrophic organisms is commercially appealing due to its potential as a profitable, most versatile, high energy output, capable and sustainable renewable energy carrier that can be used to replace traditional fossil fuels (Bolatkhan et al. 2019; Hosseini and Wahid 2016; Khetkorn et al. 2017). Hydrogen will become a primary fuel in the future and potentially resolve global environmental issues. Furthermore, the use of hydrogen for transportation has advanced in a variety of industries, and it is now used in automobiles and aeroplanes (Wang et al. 2020). According to the literature, by 2050, more than 400 million automobiles, 15–20 million trucks and 5 million buses could be powered by hydrogen, making hydrogen a key biofuel in the future (Mostafaeipour et al. 2016; Sekoai et al. 2019). The use of sunlight to produce hydrogen should be a first step towards producing sustainable fuels using renewable solar energy (Chatzitakis et al. 2013). The use of biological hydrogen energy instead of conventional energy reduces CO_2 emissions and global warming (Schipper et al. 2019; Srivastava et al. 2017; Zebda et al. 2018).

The photosynthetic microorganisms' metabolic pathways and genetic engineering enhancement can greatly boost biohydrogen production via organic feedstock bioconversion (Majidian et al. 2018). In addition, modifications to the synthesis and regulation of certain enzymes and the deletion of potentially harmful pathways can improve cyanobacterial hydrogen production. Hence, hydrogen metabolic pathways, genetic engineering techniques of certain hydrogen-producing enzymes, more excellent photobioreactors and synthetic biology may effectively increase hydrogen production (Nyberg et al. 2015; Unni et al. 2017). However, various limiting considerations, including high production costs, storage and transportation constraints and a lack of hydrogen infrastructure, make the fuel source economically unviable (Chu and Majumdar 2012; Oey et al. 2016; Rashid et al. 2013).

This chapter aims to evaluate the current knowledge on cyanobacterial hydrogen production. This chapter also summarises current concepts regarding the role of cyanobacteria in hydrogen production, explains all important points of the process, summarises the problems and improvement and suggests future research goals.

7.2 Bioprocesses of Hydrogen Production by Cyanobacteria

Microorganisms produce hydrogen by-products of their metabolic processes, called biological hydrogen generation. Biohydrogen is a clean, sustainable and high-yielding energy source that can become a significant source of energy in the future (Kumar et al. 2014; Nagarajan et al. 2017). Hydrogen gas has the maximum gravimetric energy density, making it suitable for electricity generation and transportation fuel. Recently, biohydrogen has primarily been produced through water electrolysis, microbial electrolysis cells (EME) and electrochemical photobioreactor (EPBR) and also via other mechanisms (Fig. 7.1) (Saratale et al. 2015). Biohydrogen generation from biomass or waste is gaining popularity due to its renewable and sustainable nature, low energy requirements and high efficiency (Hallenbeck 2005).

Cyanobacteria are Gram-negative prokaryotic oxygenic photoautotrophs developed during the Precambrian epoch and transformed the earth from anoxic to an oxic environment (Kaushik et al. 2017). Many of them can produce hydrogen on their own. Nitrogenases and hydrogenases are two enzymes involved in cyanobacteria hydrogen metabolism (Fig. 7.2a). The enzyme hydrogenase is found in cyanobacteria such as *Anabaena*, *Nostoc*, *Oscillatoria*, *Phormidium*, *Synechocystis* and *Synechococcus* (Eroglu and Melis 2016). According to the earlier reports, nitrogen-fixing cyanobacterial strains of the *Anabaena* have been extensively examined in hydrogen-producing potentiality. Among the strains belonging to this genus, *Anabaena variabilis* ATCC 29413 had the highest hydrogen-yielding activity of

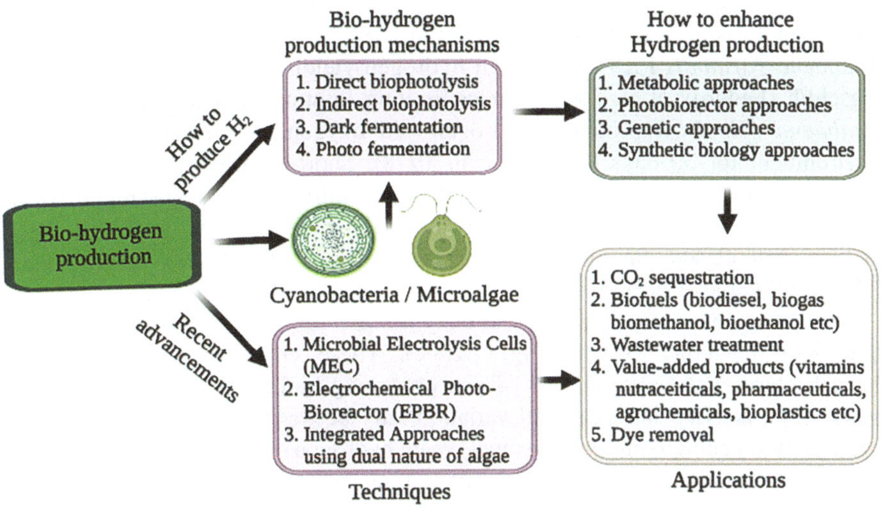

Fig. 7.1 General scheme of biohydrogen production and their applications

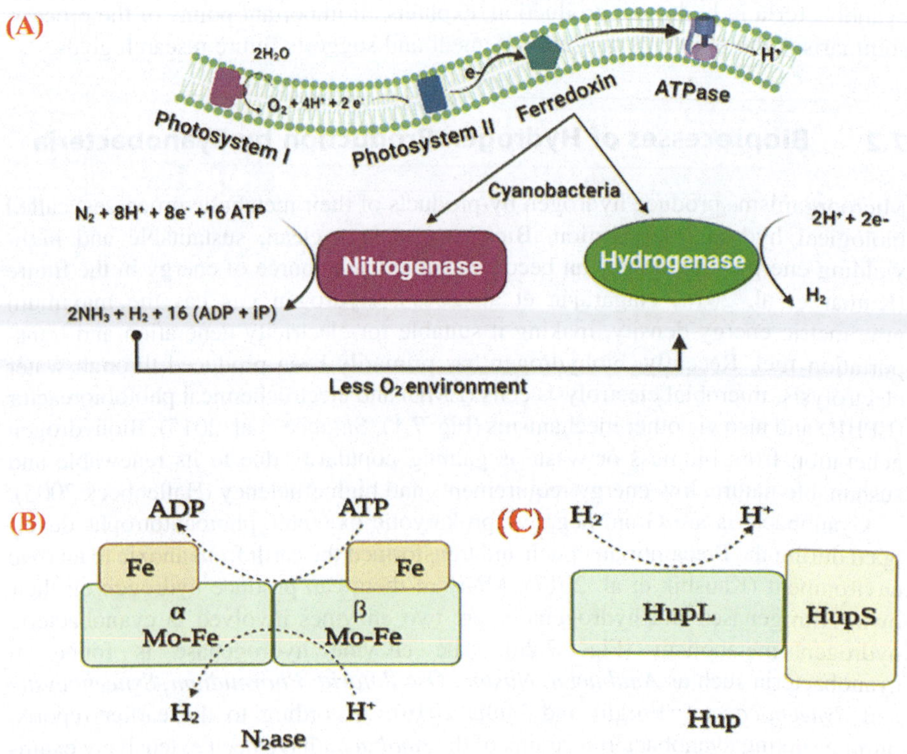

Fig. 7.2 Enzymes involved in the hydrogen production metabolism. (**a**) Nitrogenase and hydrogenase involved in biological hydrogen generation. (**b**) Nitrogenase. (**c**) Hup-hydrogenase

45.16 μmol H_2/mg chl/h (Sveshnikov et al. 1997). It is reported that the strain *Anabaena variabilis* PK84 has a high hydrogen-yielding capacity of 167.6 mmol/mg chl/h (Fedorov et al. 2001). Under limiting lighting circumstances, the strain *Anabaena cylindrica* produces hydrogen and oxygen concurrently in an argon environment for 30 days (Jeffries et al. 1978). Short-term hydrogen release was seen in members of the genus *Nostoc* with a gas output of 0.17–0.60 mmol H_2/mg chl/h. The *Oscillatoria* sp. strain Miami BG7 is the most widely investigated species lacking heterocysts having a maximal H_2 release rate of 260 mol/mg chl/h (Kumazawa et al. 1994).

The nitrogenase and hydrogenase enzymes are involved in the cyanobacterial hydrogen metabolic system (Fig. 7.2a–c). These enzymes are found in all cyanobacteria and play a critical role in producing biological hydrogen, but their maturation, activity and structural variation vary between species. Nitrogen-fixing cyanobacteria contain nitrogenases that create hydrogen due to nitrogen fixation and ammonia reduction. Bidirectional hydrogenase and uptake hydrogenase are two forms of [NiFe]-hydrogenases found in cyanobacteria (Lindblad 2018; Puggioni et al. 2016). Bidirectional hydrogenase recycles hydrogen produced by nitrogenase and produces hydrogen. All hydrogen-metabolising enzymes are oxygen-sensitive; hence, cyanobacteria have evolved spatial and temporal approaches to protect themselves. When cyanobacteria with this enzyme are exposed to light, they usually generate hydrogen for a short time before absorbing it. Oxygen sensitivity of bidirectional hydrogenase is a significant biotechnological hurdle to efficient hydrogen production.

Many cyanobacteria are genetically capable of producing biological hydrogen utilising sunlight and water in certain physiological circumstances. More than a hundred differently expressed proteins were involved in metabolic reprogramming in *Anabaena* sp.'s hydrogen-producing mutant (HupL) PCC 7120 (Ekman et al. 2011). Additionally, proteomics was used to spot targets for enhancing hydrogen generation yields in the diazotrophic unicellular cyanobacterium *Cyanothece* sp. (Bandyopadhyay et al. 2010; Aryal et al. 2013). These findings pave the way for metabolic engineering and growth in the future for efficient and long-term biological hydrogen production. Despite the widespread acceptance of cyanobacteria for multi-omics and in silico methods, attempts to enhance hydrogen production are still rare. One of the barriers to using systems biology as a platform for bioengineering is the lack of genome-scale metabolic models. *Anabaena* sp. PCC 7120 and *Anabaena* sp. ATCC 29413 has recently mutated to anticipate hydrogen producing proteins (Malatinszky et al. 2017; Malek Shahkouhi and Motamedian 2020). The in silico analyses mostly confirmed the important targets for enhancing heterocyst-based hydrogen production highlighted in the quantitative proteomic research. As a result, the preliminary simulations and modelling attempts show that these methodologies have the potential for further development in the field of biological hydrogen production.

7.3 Characteristics and Role of Hydrogen-Producing Enzymes

Hydrogen is the future fuel because of its high conversion efficiency, recyclability and non-polluting nature. In addition, biological hydrogen production processes are more economically friendly and energy-efficient than thermochemical and electrochemical procedures. The biological hydrogen production in photosynthetic cyanobacteria is greatly influenced by two enzymes, i.e. hydrogenase and nitrogenase enzymes (Poudyal et al. 2015; Bothe et al. 2010; Ghirardi et al. 2007). Furthermore, the condensation of the proton (H^+) to hydrogen (H_2) is aided by the enzymes hydrogenase and nitrogenase, and these enzymes are sensitive to oxygen (Anwar et al. 2019).

7.3.1 Hydrogenases

Hydrogenase is a metal enzyme that is involved in energy metabolism in a variety of microbial communities. There are three forms of hydrogenases identified in microorganisms based on metal catalytic activities: [FeFe]-hydrogenase,

[Fe]-hydrogenase and [NiFe]-hydrogenase (Mona et al. 2020; Bolatkhan et al. 2019).

[FeFe]-hydrogenases are found in both anaerobic prokaryotes and eukaryotes and are involved in the removal of excess reductant created during anaerobic fermentation as well as the use of hydrogen in sulphate reduction (Pohorelic et al. 2002). Only metal iron can be found in [Fe-Fe]-hydrogenases (Ghirardi et al. 2009). Although they have an iron-sulphur core with carbon monoxide and cyanide on the enzyme active site, their turnover rate is faster (Dasgupta et al. 2010). The activity of [Fe]-hydrogenase is being studied further. The [Fe]-hydrogenase active site contains only one Fe atom, one cysteine sulphur, two cis-oriented carbon monoxides and a bidentate guanylyl pyridinol ligand.

The [NiFe]-hydrogenase enzyme produces a considerable amount of hydrogen. It is a bifunctional enzyme identified in cyanobacteria that operates as an absorption hydrogenase. It comprises small and huge subunits of 34 kDa and 64 kDa, respectively. The catalytic core of these hydrogenases is connected to a large subunit HoxH and the small subunit HoxY, which is necessary for electron transport to the large subunit and has a [4Fe-4S] cluster. In addition, the diaphorase moiety in the catalytic site is made up of three additional subunits: HoxE, HoxF and HoxU. [NiFe]-hydrogenases are oxygen-tolerant and typically involved in hydrogen absorption processes, but they can also produce hydrogen (Eroglu and Melis 2011).

7.3.2 Nitrogenases

Nitrogenase is a nitrogen-fixing enzyme that can only be found in cyanobacterial strains and a few other prokaryotes. Nitrogenase catalyses biohydrogen production as a by-product of nitrogen fixation, which requires a high amount of energy in the form of ATP (Seefeldt et al. 2009; Wang et al. 2020). As a result, nitrogenase is a potential enzyme for producing biological hydrogen by cyanobacterial species. Nitrogenase requires a substantial quantity of electrons, reductant and at least 16 ATP molecules from photosynthesis in the cell to generate hydrogen (Zhang et al. 2016). Furthermore, nitrogenase catalyses the reaction under the most favourable environment for nitrogen fixation, as seen in the following equation.

$$N_2 + 8e^- + 8H^+ + 16\,ATP \rightarrow H_2 + 2NH_3 + 16\,ATP + iP$$

Nitrogenase is a two-protein enzyme complex consisting of dinitrogenase (MoFe protein) and dinitrogenase reductase (Fe protein) (Fig. 7.2b). Nitrogenase is heterotetramer subunits and encoded by the *nifD* and *nifK* genes, respectively. The nitrogenase reaction is catalysed by the interface of Fe and MoFe protein. The homodimeric Fe protein gets electrons from reduced ferredoxin and transfers it to the MoFe protein requiring ATP. The nitrogenase enzyme catalyse hydrogen production process. Thus, cells lacking hydrogenase activity can accumulate hydrogen for an extended period even when oxygen is present.

7.4 Biohydrogen Production Mechanisms

At this time, large-scale hydrogen production is carried out by water electrolysis and coal gasification. Both are destructive and harmful to the environment; therefore, certain photosynthetic microorganisms are being considered to produce more hydrogen energy. Biohydrogen is produced by cyanobacterial biomass via a variety of methods. A system-level mechanism analysis is required to understand better how cyanobacterial cell metabolism and bioenergetics navigate during biohydrogen production and identify possible targets for additional metabolic engineering. Most of the processes have been devoted to deciphering the nitrogenase-based hydrogen production metabolism (Kourpa et al. 2019; Sadler et al. 2016). Photosynthesis and fermentation are two strategies for cyanobacterial hydrogen production. The biohydrogen production from photosynthesis can be categorised into two categories: (i) direct biophotolysis and (ii) indirect biophotolysis. The use of light energy received by the photosynthetic apparatus for water photolysis is known as direct biophotolysis. The proton reduction and hydrogen production are followed by oxygen evolution and the development of low-potential reducing agents (Fig. 7.3a). The results of photosynthetic water photolysis and subsequent ferredoxin reduction are utilised to fix CO_2 during indirect biophotolysis, and the reduced carbon molecule can be used to trigger hydrogen release in a separate process (Fig. 7.3b). The fermentation process can be carried out in two ways: (i) photofermentation and (ii) dark fermentation. Chlorophyceae microalgae can only produce biohydrogen by direct biophotolysis due to oxygen evolution; however, cyanobacteria can perform direct and indirect biophotolysis because of their cell differentiation. Cyanobacteria have no oxygen evolution because they have specialised cells called heterocysts; therefore, they undergo indirect biophotolysis, whereas vegetative cells undergo direct biophotolysis (Kumar and Chowdhary 2016; Deviram et al. 2020). Biophotolysis is the most suitable among the different biological processes for producing clean hydrogen. There are four mechanisms for biological hydrogen production through microorganisms (Fig. 7.3).

7.4.1 Direct Biophotolysis

Direct biophotolysis converts solar energy into chemical energy in molecular hydrogen by phototrophic microorganisms to split water into oxygen and protons via photosynthesis (Azwar et al. 2014; da Silva Veras et al. 2017). Cyanobacteria can use solar energy to liberate protons and electrons during water cleavage. Hydrogen is produced by direct light absorption and electron transfer via two different enzymes, hydrogenase and nitrogenase (Manish and Banerjee 2008). With the use of [Fe]-hydrogenase, the photochemical oxidation of water in direct biophotolysis results in the production of hydrogen (Sharma and Arya 2017; Show et al. 2019). Photosystem I (PSI) and photosystem II (PSII) are the two photosynthetic reaction centres found in cyanobacteria chloroplasts. PSI generates the carbon dioxide reduction reductant, while PSII is concerned with water splitting and O_2 evolution.

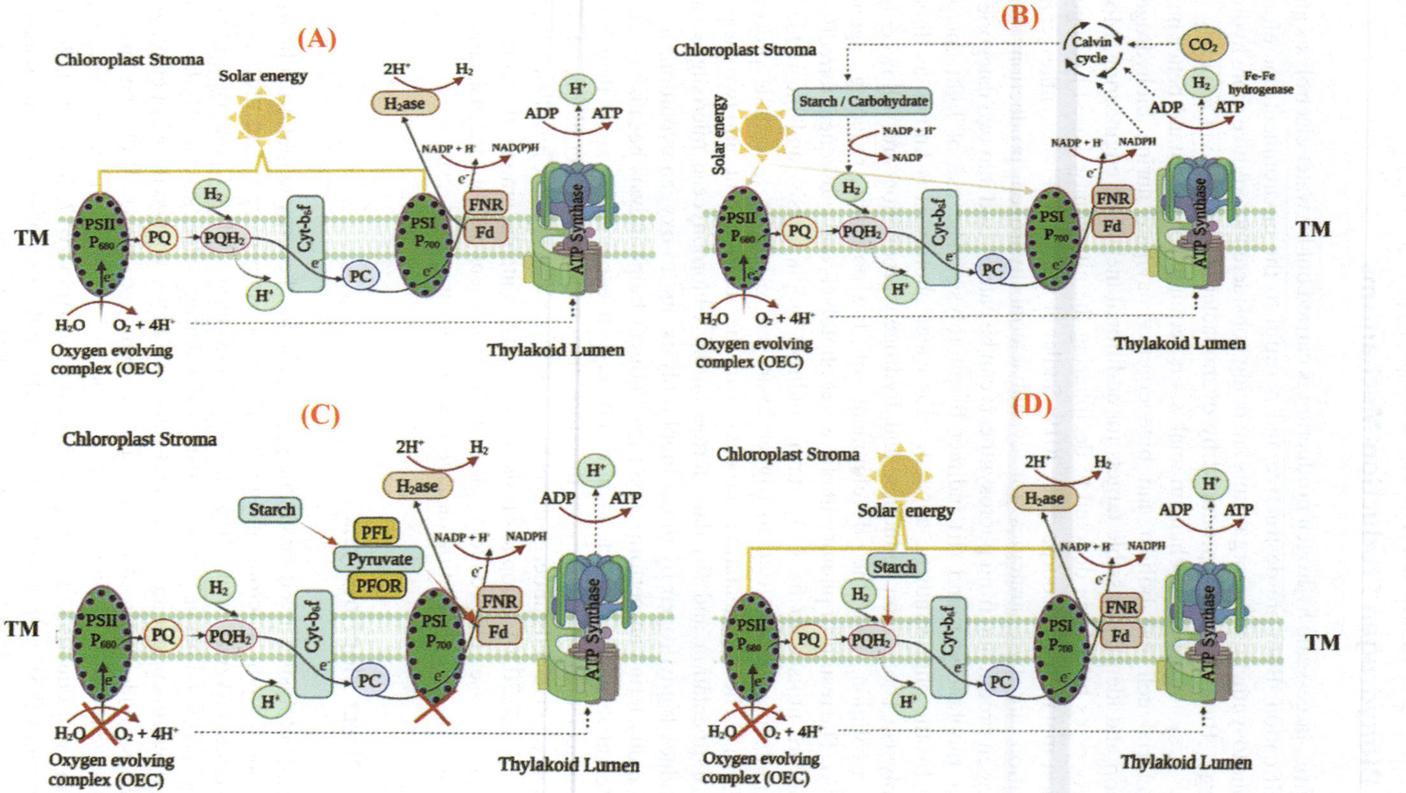

Fig. 7.3 Schematic representations of hydrogen-producing mechanisms under different conditions. (**a**) Direct biophotolysis. (**b**) Indirect biophotolysis. (**c**) Dark fermentation. (**d**) Photofermentation. Thylakoid membrane (TM), photosystem (PS), plastoquinones (PQ), cytochrome b6f complex (Cyt-b6f), plastocyanin (PC), ferredoxin (Fd), ferredoxin NAD(P) reductase (FNR), hydrogenase (H_2ase), pyruvate formate-lyase (PFL) and pyruvate ferredoxin oxidoreductase (PFOR). This figure was modified from Wang et al. (2020)

PSII, which is located in the thylakoid membrane, obtains electrons from water dissipation, which are then transferred to [Fe]-hydrogenase via a ferredoxin intermediate, resulting in hydrogen production (Fig. 7.3a) (Nagarajan et al. 2017). These electrons reduce plastoquinone (PQ) to plastoquinol (PQH2), which reduces $NADP^+$ to NADPH during cyclic photophosphorylation. In addition, the protons are used by the ATP synthase enzyme to establish a gradient for ATP synthesis. As a result, both oxygen and hydrogen are produced simultaneously. The researchers discovered that the electrons and protons emitted during water splitting processes were re-mixed with hydrogenase to produce highly pure hydrogen (up to 98%) (Hankamer et al. 2007; Horiuchi et al. 2001; Show et al. 2019). The direct biophotolysis reaction is depicted below:

$$2H_2O + \text{solar energy} \rightarrow 2H_2 + O_2 \uparrow$$

In theory, direct biophotolysis is an appealing process since it converts a readily available substrate, water, to oxygen and hydrogen using solar energy. However, it has been constrained in practice by several difficulties. First, because it is involved in the direct biophotolysis approach, hydrogenase activity (H_2ase) is particularly sensitive to oxygen and hinders hydrogen generation (Table 7.1). However, this constraint can be overcome by boosting respiration by utilising exogenous or endogenous substrates and developing oxygen-tolerant hydrogenases (Razu et al. 2019).

7.4.2 Indirect Biophotolysis

Cyanobacteria produce hydrogen from starch and glycogen during indirect biophotolysis. It consists of two steps: the first is the use of light energy to produce carbohydrates, and the second is the production of hydrogen from the carbohydrates through fermentation (Voloshin et al. 2016). Heterocystous cyanobacteria are used in indirect biophotolysis, and they split water and organic molecules to produce hydrogen (Vargas et al. 2018). Another approach for indirect biophotolysis is to perform two independent reactions: first producing oxygen (with CO_2 fixation) and then producing hydrogen (with CO_2 release). *Anabaena* strains have been studied extensively due to their greater hydrogen production rates (Levin et al. 2004). *Anabaena variabilis* mutant strains have been shown for hydrogen generation in indirect biophotolysis (Sveshnikov et al. 1997). Hydrogen may be generated from microalgae via indirect biophotolysis if photon conversion could be improved for industrial uses. The above section reports that nitrogenase and hydrogenase are two important enzymes found in cyanobacteria that perform metabolic tasks to produce hydrogen (Lindberg et al. 2012). Indirect biophotolysis has the disadvantage that uptake hydrogenase enzymes must be eliminated to prevent hydrogen breakdown, and around 30% O_2 is present in the gas mixture. The indirect biophotolysis reaction is depicted below:

Table 7.1 Advantages and disadvantages of different hydrogen production methods (Modified from Sharma and Arya 2017; Yin and Wang 2018; Shobana et al. 2017; Batista et al. 2018)

H_2 production mechanisms	Advantages	Disadvantages
Direct photolysis	1. Raw ingredients like water and sunlight are readily available	1. Hydrogenase enzyme sensitivity to oxygen
	2. More solar light conversion efficiency	2. High-intensity light is required
	3. It is not necessary to use carbon sources	3. Hydrogen output is being reduced
Indirect photolysis	1. Separate hydrogen and oxygen evolution	1. Adenosine triphosphate is required on a large scale
	2. Produces a large amount of hydrogen	2. A constant supply of light is required
		3. The efficiency of photosynthetic conversion is low
Dark fermentation	1. Carbon sources can come from a variety of places	1. Conditions must be nitrogen-free
	2. No light required	2. Low yield
	3. Acetic acid, butyric acid and lactic acid are examples of by-products	3. Substrate waste as a result of insufficient use
Photofermentation	1. Organic compounds are made from industrial waste	1. Nitrogenase oxidative inhibition
	2. Organic acid from dark fermentation is used, and wastewater can also be used	2. The enzyme is also inhibited by high levels of ammonia
	3. Reduces pollution	3. Working with high-intensity light is difficult
		4. Reactors are expensive
	4. There is no oxygen produced	
	5. Works in a non-sterile environment	

$$C_6H_{12}O_{16} + 6H_2O \rightarrow 6CO_2 + 12H_2$$

7.4.3 Dark Fermentation

It is a multistep process that does not necessitate sunlight, water or oxygen (Fig. 7.3c). Anaerobes and certain microalgae, such as green algae, perform this function. Several carbon-rich wastes can be handled via dark fermentation, which produces hydrogen and other important by-products such as acetic acid, butyric acid and volatile fatty acids (Bolatkhan et al. 2019; Elbeshbishy et al. 2017). Dark fermentation is an environmentally beneficial and cost-effective way of producing biohydrogen. The use of dark biological fermentation to produce hydrogen gas is particularly appealing because it is both renewable and carbon-free (Lukajtis et al.

2018). Under dark conditions, starch and glycogen found in cyanobacteria are transformed to pyruvate from carbohydrates and hydrogen. Pyruvate formate-lyase (PFL) and pyruvate ferredoxin oxidoreductase (PFOR) are two enzymes that act on pyruvate to produce hydrogen. The dark fermentation process has the advantage of not requiring the presence of light energy or the continual consumption of carbohydrates. However, the dark fermentation mechanism has not been widely adopted (Bundhoo and Mohee 2016). During the hydrogen-producing process, CO_2 must be separated. Therefore, the rate of light-independent hydrogen synthesis via dark fermentation usually is high. The dark fermentation reaction is depicted below:

$$C_{16}H_{12}O6 + 6H_2 \rightarrow 2H_2 + 6CO_2$$

The benefits of dark fermentation include removing environmental contaminants and synthesising organic acids such as acetic acid, butyric acid and lactic acid (Table 7.1). To improve hydrogen production in the dark fermentation process, temperature, nutrient supply and pH must be adjusted. The pH of the enzyme hydrogenase during hydrogen generation is one of the most critical factors that affect its activity.

7.4.4 Photofermentation

It is a process that uses light as an energy source to generate hydrogen and carbon dioxide from complicated organic molecules utilising ferredoxin and hydrogenase (Fig. 7.3d) (Rashid et al. 2013). Endogenous substrates provide electrons to the plastoquinone (PQ), which results in the production of organic molecules. Photosystem I (PSI) assimilation activity equalises the ferredoxin redox potential and [Fe]-hydrogenase. The presence of PSII inhibitor 3-(3,4-chlorophenyl)-1,1-dimethylurea (DCMU) has also been discovered to hinder electron passage from PSII to PQ; as a result, molecular hydrogen and carbon dioxide are formed (Show et al. 2019). Nitrogenase activity is inhibited by oxygen, high ammonia concentrations and increasing light intensity (Kumar et al. 2013, 2014). Organic substances such as glucose, fructose, acetic acid and malic acid are used by purple non-sulphur (PNS) bacteria to produce hydrogen. Some photosynthetic PNS bacteria, including *Rhodobacter* and *Rhodopseudomonas*, are less sensitive to oxygen and use the nitrogenase enzyme to produce hydrogen (Rashid et al. 2013). The disadvantages of photofermentation include the need for partial nitrogen conditions and the processing of harmful industrial waste.

7.5 Strategies to Improve Cyanobacterial Hydrogen Production

To be commercially viable, the rate of hydrogen production must be increased. Cyanobacteria are now being explored in various ways to improve their hydrogen generation. All of these approaches fall into three major categories.

7.5.1 Physiochemical Strategies

Many factors influenced the effectiveness of the hydrogen production process utilising cyanobacterial cells, which was especially crucial for large-scale hydrogen production. The molecular oxygen, salinity and nutritional media composition were the most relevant parameters and the temperature, light intensity and pH of the medium. Various parameters have been shown to have distinct effects throughout the hydrogen generation process by different strains of cyanobacteria. Therefore, many distinct factors influence hydrogen yield in cyanobacterial cells, and by regulating or modifying these factors, considerable outcomes could be improved. Thus, using cyanobacterial strains to generate enormous volumes of hydrogen played a crucial role in resolving and realising this problem. This indicator controls the effectiveness of the cyanobacterial enzymatic process and significantly impacts the redox potential of the cells. It is worth noting that enzyme activity is influenced by the content of nutritional media such as macro- and microelements essential for growth. The addition of specific trace elements to the culture medium greatly enhanced hydrogen generation due to their role in enzyme composition. The impact of the shortage of microelements on hydrogen generation by cyanobacteria was investigated using *Gloeocapsa alpicola* CALU 734 and *Synechocystis* sp. PCC 6803 (Troshina et al. 2002). According to these findings, a considerable increase in hydrogenase activity was found in the microelement-deficient culture, increasing hydrogen release. More than 14 species of cyanobacteria have been identified as producing hydrogen under diverse growth conditions.

Using various electron transport inhibitors is another successful method for boosting cyanobacterial hydrogen production. Currently, there are about 15 different types of inhibitors in use; some are 3-(3,4-dichlorophenyl)-1,1-dimethylurea (DCMU), carbonyl cyanide m-chlorophenyl hydrazone (CCCP), potassium cyanide (KCN), methyl viologen (MV), chloramphenicol, pentachlorophenol (PCP) and malonate being the most widely used (Table 7.2). DCMU is the most extensively utilised electron transport inhibitor because the molecular structure is highly comparable to reduced plastoquinone. This explains why these molecules attach so tightly to the PSII reaction centre quinone-binding site. The goal of using DCMU was to stop this photosystem II from working and create anaerobic conditions conducive to the synthesis of molecular hydrogen. DCMU binds to the plastoquinone (Q_B) of PSII and prevents electron transfer from plastoquinone (Q_A). In the presence of DCMU, unicellular freshwater cyanobacteria *Synechocystis* sp. PCC 6803 cells cultivated in dark anaerobic conditions produced more hydrogen (Cournac et al. 2004).

Table 7.2 The most common electron transport inhibitors employed in cyanobacteria to boost hydrogen production (Modified from Sadvakasova et al. 2020)

Electron transport inhibitor	Cyanobacteria	H_2 production	Mechanism of action
3-(3,4-Dichlorophenyl)-1,1-dimethylurea (DCMU)	*Anabaena siamensis* TISTR 8012 *Anabaena cylindrica*	22.0 μmol H_2/mg chl/h 30.8 μl H_2/mg dry wt/h	An inhibitor of PSII that prevents electron transfer from PSII to the reduced plastoquinone (PQH2) pool
Carbonyl cyanide *m*-chlorophenyl hydrazone (CCCP)	*Aphanothece halophytica* *Nostoc muscorum*	39.5 μmol H_2/g dry wt/h 5 μmoL H_2/ml pcv/h	Inhibits the photochemical activity of PSII
Potassium cyanide (KCN)	*Synechocystis* sp. PCC 6803 *Synechocystis* sp. PCC 6803	147 μmol H_2/mg chl/h 3.1 μmol H_2/1010 cells	A cytochrome oxidase inhibitor that inhibits all terminal oxidases
2,5-dibromo-3-methyl-6-isopropyl-p-benzoquinone (DBMIB)	*Synechocystis* sp. PCC 6803 *Nostoc muscorum*	22.0 μmol H_2/mg chl/h 15 moL H_2/ml pcv/h	A cytochrome b6f inhibitor promotes the chemical reduction of the plastoquinone (PQ) pool
Metribuzin	*Nostoc muscorum*	145 μmoL H_2/ml pcv/h	Inhibiting photosystem II
Atrazine	*Nostoc muscorum*	145 μmoL H_2/ml pcv/h	Inhibits photosynthesis

Carbonyl cyanide m-chlorophenyl hydrazone (CCCP) is another common PSII inhibitor in cyanobacteria. The CCCP, like DCMU, hindered the photochemical activity of PSII, resulting in a decrease in O_2 release. In the cyanobacteria *Synechocystis* sp. PCC 6803, *Synechococcus* sp. PCC 7942, *Nostoc* sp. and *Lyngbya* sp., CCCP decreased the photochemical activity of PSII in the presence of light (Torimura et al. 2001). Furthermore, CCCP has been demonstrated to reduce ATP synthesis in the filamentous nitrogen-fixing cyanobacteria *Anabaena variabilis* and *Anacystis nidulans*, increasing the rate of dark respiration, which facilitates hydrogen production (Pansook et al. 2019).

7.5.2 Photobioreactor Strategies

Aside from the physiological and biochemical strategies mentioned above, developing novel culture methods to improve biohydrogen generation has made major contributions, specifically by enhancing microalgae cultivation conditions and developing efficient photobioreactor systems. Temperature, light intensity, pH, oxygen, carbon dioxide concentration, types of medium and carbon-to-nitrogen ratio are the primary physicochemical factors influencing hydrogen yield. Therefore, it is possible to produce outstanding results in biohydrogen production by changing these factors.

The ideal pH values for various cyanobacterial cultures are 7–8, while the temperature range is 25–35 °C. The following requirements must be met in photobioreactors utilised for the large-scale production of biohydrogen from cyanobacterial cultures: (1) the photobioreactors must be a closed system to collect the hydrogen produced without loss; (2) the design of the reactors should ensure ease of operation, cleaning and sterilisation; (3) the materials used in the structure should be transparent (for light penetration) as well as durable; and (4) photobioreactors must have a high surface-to-volume ratio and the ability to mix culture nutrients quickly. Large-scale microalgal cultivation is frequently carried out in an open/raceway pond; however, this method of algal cultivation is unfavourable for hydrogen gas production and collection. To increase the process of biohydrogen production, many types of closed photobioreactors have been designed for commercial-scale microalgal cultivation (Skjånes et al. 2016; Geada et al. 2017). All photobioreactors require adequate light, which is frequently provided by sunshine. However, different artificial light sources are used in some photobioreactors to offer controlled illumination. Light penetration into the deeper parts of the photobioreactor is a major limiting factor in large-scale hydrogen production. Therefore, photobioreactor performance is hampered by area/volume ratio, temperature, agitation and gas exchange factors.

Various photobioreactors (PBRs) have been used to produce biohydrogen in which vertical column, tubular and flat-panel are primary types of PBRs (Fig. 7.4). Vertical PBRs are made up of transparent, high-quality glass surrounded by a water jacket (Fig. 7.4a) that allows appropriate light to enter while keeping the temperature stable with circulating water (Mirón et al. 1999). The cyanobacterial culture is disseminated using airlift pumps in a tubular photobioreactor (Fig. 7.4b) (Oncel and Kose 2014). The flat-panel photobioreactor comprises two compartments separated by a stainless steel sheet and three polycarbonate panels (Fig. 7.4c). The pH, temperature, optical density and dissolved oxygen content are all monitored and controlled by a specifically designed control system (Skjånes et al. 2016).

Furthermore, flat-panel photobioreactors can achieve high photosynthetic activity and high gas pressure management than other PBRs. However, maintaining the proper culture temperature and agitation during hydrogen production is challenging to inflate-panel PBR (Dasgupta et al. 2010). Overall, it was determined that the flat-panel photobioreactor is better than others for hydrogen production because the backpressure of collected hydrogen can be avoided in flat-panel photobioreactors

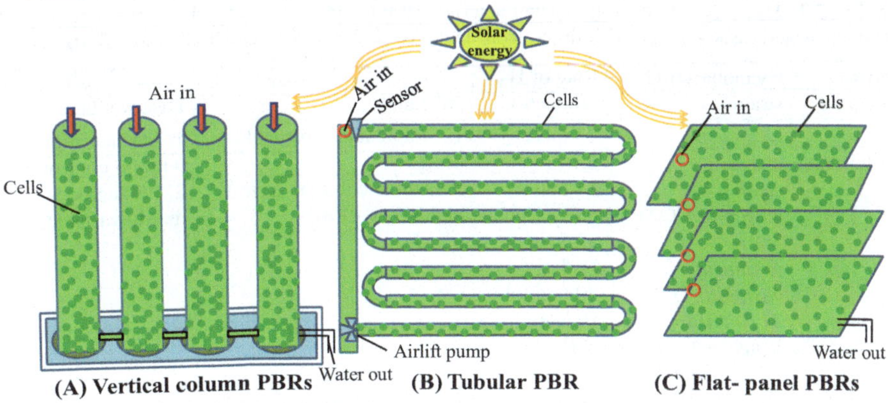

Fig. 7.4 The most common types of photobioreactors used to produce biohydrogen

(Oncel and Kose 2014; Nyberg et al. 2015). Table 7.3 shows the hydrogen yield produced by several cyanobacteria strains in various types of photobioreactors.

Developing a practical photobioreactor for cyanobacterial hydrogen production remains a challenge (Sevda et al. 2017; Kroumov et al. 2017). Study on microalgal-based hydrogen production has been successfully implemented in recent years. However, more depth research is still needed to improve the performance of photobioreactors for commercial cyanobacterial culture with lower biohydrogen production costs.

7.5.3 Genetic Engineering Strategies

Studying genetic engineering to develop effective cyanobacterial strains is the best way to optimise biohydrogen production. The genetic technique aims to create genetic mutants that improve photohydrogen production by blocking, adding new or overexpressing existing cyanobacterial genes. According to the earlier research, cyanobacterial genetic alterations are necessary to resolve the following significant issues: the uptake hydrogenase activity, less frequent formation of heterocyst, competition for reductants and low efficiency of light-capturing antennas. This section focuses on current and innovative advances in cyanobacterial engineering for long-term hydrogen production.

In the literature, there are numerous cases of interrupting or removing a structural gene or a gene that codes for a protein that is part of the HupSL hydrogenase maturation mechanism in cyanobacteria (Lindblad 2018). For example, the genes for large and small subunits of HupSL hydrogenase and the regulator (HupR protein) are knocked out to reduce hydrogenase expression and hence increase hydrogen yield (Kars et al. 2008). On the other hand, inactivation of uptake hydrogenase reprograms metabolism of the cyanobacterial cell and hydrogen production process; it may reduce the duration (Kourpa et al. 2019; Kosourov et al. 2014). Another

Table 7.3 Hydrogen-producing cyanobacterial strains in different photobioreactors with advantages and disadvantages (Modified from Dutta et al. 2005; Sadvakasova et al. 2020)

PBR type	Cyanobacterial strains	Rate of H_2 production	Advantage	Disadvantage
Vertical column	*Anabaena azollae* *Anabaena variabilis* CCAP 1403 *Anabaena variabilis* PCC 7120 ΔHup *Synechocystis* PCC sp. 6803	13 mL H_2/L/h 14 mL H_2/g dry wt/h 13.8 mL H_2/L/h 40.6 μmol H_2/mg chl/h	1. Simple and affordable design 2. Bubble columns have a higher rate of mass transfer	1. Irradiant light with no control 2. Productivity is highly variable
Tubular	*Anabaena variabilis* PK84 *Anabaena variabilis* PK84 *Anabaena* sp. PCC 7120 *Anabaena* AMC 414	8 mL H_2/L/h 0.82 mol H_2/m^3/h 14.9 mL H_2/L/h 4.1 mL H_2/g dry wt/h	1. Flexibility in the ratio of volume to the surface area 2. The ability to change the location where light is received 3. Internal static mixture provides more biomass	1. Poor mass transfer 2. Internal static mixtures require more mixing time
Flat-panel	*Nostoc* PCC 7120 ΔhupW *Anabaena variabilis* (ATCC 29413)	6.5 mL H_2/L/h 4.12 mL H_2/g dry wt/h	1. Better control of irradiant light 2. Gas pressure is effectively controlled	1. High production cost 2. More maintenance and complicated design

option is to change the electron flow in the HupSL hydrogenase from hydrogen absorption to hydrogen evolution. The large subunit has a NiFe containing active site, while the small subunit (HupS) has three Fe-S clusters. The HupS Fe-S cluster was altered by replacing cysteine 12 with proline (C12P) and expressed in *N. punctiforme*, producing more hydrogen. All of these scenarios resulted in higher hydrogen yield.

Although cyanobacteria are the most interesting and emerging organisms for biotechnological purposes, generating cyanobacterial strains that lack a functioning nitrate assimilation pathway boosts photobiological hydrogen generation in *Synechocystis* sp. C 6803 (Baebprasert et al. 2011). A current study found that the heterocyst-specific Flv3B enzyme removes O_2, allowing for a microaerobic environment inside the heterocyst that enables nitrogenase enzyme activity when the heterocyst is eliminated (Ermakova et al. 2014). As a result of overexpression of the Flv3B, protein induces hydrogen photoproduction even though this phenotype was not associated with increased nitrogenase activity (Roumezi et al. 2020). Solving the problem of hydrogenase oxygen sensitivity is a significant genetic technique for enhancing cyanobacterial hydrogen productivity. Many investigations have been conducted on developing an O_2-tolerant hydrogenase active enzyme for the long-term release of hydrogen in cyanobacteria.

Furthermore, immobilising enzymes and microorganisms within polymer matrices is a promising method for prolonging catalyst lifetime and enabling a continuous and straightforward hydrogen production platform. Moreover, cyanobacteria and green microalgae immobilised in thin films have increased cell fitness and biohydrogen production activity (Kosourov et al. 2017; Wutthithien et al. 2019). In conclusion, each of the presented genetic techniques needs special attention and has new potential for developing biohydrogen production technology due to solving several fundamental challenges. However, the current development of genetic engineering technologies has yet to solve all challenges, manipulating cyanobacterial metabolism and opening up new possibilities for enhancing efficient biohydrogen production.

7.5.4 Synthetic Biology Strategies

Hydrogenases have a significant potential for inter-conversion of H^+ to H_2; nevertheless, the enzymes perform this activity using first-row transition metals and hence can be inexpensively synthesised in large quantities (Kleinhaus et al. 2021; Land et al. 2020). As a result, they have become more common in hydrogen production applications, particularly for [FeFe]-hydrogenases. They have the maximum catalytic activity of all hydrogenases and a very high maturation process that involves three enzymes known as HydE, HydG and HydF (Posewitz et al. 2004). To produce the active enzyme, the [FeFe]-hydrogenase receives the pre-catalyst, which contains [4Fe-4S]-hydrogenase pre-assembled cluster (Britt et al. 2020). On the other hand, the requirement for the co-expression of the [2Fe]-hydrogenase maturation enzyme to generate active [FeFe]-hydrogenases impeded large-scale screening and extensive analysis of these biotechnologically essential enzymes. In biotechological application the following techniques are included in relation to biohydrogen production:

(1) Separation of the active enzyme on a large scale.
(2) Synthetic chemistry is used to manipulate the enzyme.
(3) New [FeFe]-hydrogenases are being screened.
(4) Hydrogen-producing cyanobacteria are screened.

Increased hydrogen evolution activity of [FeFe]-hydrogenase in the *Synechocystis* sp. PCC 6803 shows how a manufactured metalloenzyme can modify the efficiency of cyanobacterial cell systems and exceed native catalytic pathways (Wegelius et al. 2018). In this regard, the idea of synthetic hydrogenases offers a viable avenue for developing non-natural metabolic ways for hydrogen generation.

7.6 Challenges and Improvements of Hydrogen Production from Cyanobacteria

Cyanobacteria have a lot of potentials and appear to be an excellent choice for producing biohydrogen on a long-term basis. Nonetheless, due to the numerous research obstacles involved with cyanobacterial-based hydrogen production systems, doubts have been raised about their viability. The great O_2 sensitivity of cyanobacterial hydrogenases is one of the major problems to biohydrogen production. Because O_2 inhibits the activity of [FeFe]-hydrogenases and proteins, transcription and maturation are also suppressed (Oey et al. 2016). Very low amounts of O_2 have been shown to render catalytically inactive hydrogenases by damaging the hydrogen-[4Fe-4S] cluster domain (Stripp et al. 2009). Random and targeted mutagenesis and recombinant systems have also been used to change the O_2 sensitivity for long-term hydrogen generation (Ghirardi et al. 2000; Xu et al. 2005). Protein engineering research could be a hopeful path to develop an oxygen-insensitive hydrogenase for long-term biohydrogen production. The key obstacles in biological hydrogen production under the influence of certain conditions are summarised in Fig. 7.5.

Hydrogenases with a limited supply of electrons are another difficulty linked with biophotolysis during biohydrogen production. Other metabolic processes like ferredoxin and NAD(P)H compete for the principal electron donors. Ferredoxin-NADP$^+$ reductase (FNR), glutamate synthase, nitrate reductase and sulphite reductase are among the enzymes that catalyse these competitive metabolic processes (Das et al.

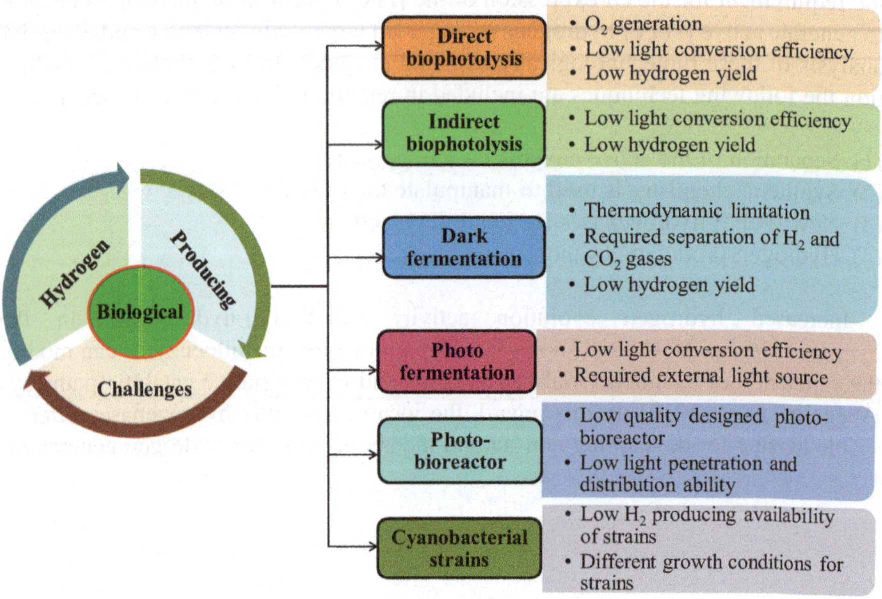

Fig. 7.5 Different types of major challenges in biological hydrogen production

2014). The downregulation of FNR, ferredoxin and hydrogenase fusion and ferredoxin with PSI have all been used to boost electron flow towards the hydrogenase and increase hydrogen production (Rumpel et al. 2014; Wittenberg et al. 2013).

The low photon conversion efficiency of the microalgal hydrogen production system is another significant restriction. Cyanobacteria build massive light-harvesting complex (LHC) systems to capture the most photons under natural environmental conditions. As a survival strategy, 90% of these photons are dissipated in the form of heat through a technique known as non-photochemical quenching (NPQ). NPQ is found on the upper surface of dense cyanobacterial cultures. A self-shading effect occurs when light does not pass effectively to the lower surface. Hence, the photon conversion efficiency is low due to NPQ and the self-shading effect. Modifying the light-harvesting complex and proper designing of PBRs have been offered as a possible solution. A PBR with improved light penetration and distribution is required to achieve high photon conversion efficiency. The PBRs surface area/volume (A/V) ratio is the most crucial factor affecting photosynthetic photon flux density (Das et al. 2014). Tredici and Zittelli (1998) used helical tubular PBR with a high A/V ratio to obtain a photosynthetic efficiency using cyanobacteria *Arthrospira platensis*. Light-scattering nanoparticles doped in planar waveguides were recently inserted into the flat-panel PBR for the homogeneous distribution of light energy throughout the microalgal culture. However, major improvements in process efficiency and economic feasibility are required before this technology can be extensively adopted.

Another interesting method for biological hydrogen production is dark fermentation with cyanobacterial feedstock. In this method, the hydrogenase enzyme is never exposed to O_2 because this is a method of anaerobic fermentation. Furthermore, unlike biophotolysis, this system efficiency is unaffected by sun radiation. Despite these obvious advantages, dark fermentation faces severe problems that must be overcome to acquire commercial significance. The limited energy recovery from the substrate is the most significant disadvantage. A considerable portion of the organic matter in dark fermentation is not converted to biohydrogen and remains in the liquid fraction as soluble metabolites such as acetate, alcohols and butyrate. As a result, this reduces the overall energy recovery process, and the organic acid-rich effluent may harm the environment. The separation/purification of biohydrogen is the other dark fermentation obstacle that must be addressed. In contrast to biophotolysis, which produces hydrogen, dark fermentation creates a mixture of hydrogen, CO_2 and moisture. As a result, hydrogen must be purified and separated.

Furthermore, before the microalgal hydrogen production technique is released for commercialisation, large-scale research is required. Various operational and technological issues may occur at a large scale that is not apparent at the laboratory stage. As a result, realistic pilot-scale research is required to convert all the methods into useful technology.

7.7 Future Perspective

Cyanobacterial systems are quickly establishing themselves as one of the most promising platforms for producing renewable biofuels. Various researches are required for enhancing cyanobacterial hydrogen production because cyanobacteria still produce insufficient hydrogen to make hydrogen a commercially viable energy carrier. Therefore, there is a requirement to identify various ways to increase the ability of cyanobacteria to produce hydrogen. The cyanobacterial network of anoxic metabolic pathways must be well studied and understood to meet this requirement. The current largest difficulty is to upgrade the existing fermentation system to produce hydrogen more efficiently. With the advancement of synthetic biology, genetically engineered cyanobacterial cells will manufacture specific products, such as biohydrogen and biofuels. The development of synthetic biology and microbial genetic engineering will be helpful in producing hydrogen in the future.

7.8 Conclusion

According to an analysis of relevant literature, many aspects of using cyanobacteria as a source of hydrogen have been the topic of several scientific pieces of research in recent years, which have a growing emphasis on renewable energy development. Researchers are concentrating their efforts on understanding the basic principles of cyanobacterial photosynthesis for hydrogen production, promising to obtain fundamental knowledge and their practical application in biotechnology. The majority of research focuses on applying various metabolic methods to boost hydrogen yield in various cyanobacterial strains. There is a lot of information about genetic engineering and synthetic biology to boost the hydrogen yield of cyanobacterial strains. Biohydrogen generation on a commercial scale can only be achieved by overcoming the most significant barriers using current biotechnology and a revolutionised industry. To facilitate the synthesis of biohydrogen, entire new pathways can be introduced into cyanobacteria.

Acknowledgements The authors are thankful to the Head, Department of Botany Centre of Advanced Study, Institute of Science, Banaras Hindu University, Varanasi, and DST-FIST for providing the necessary research facilities. The University Grants Commission, Government of India, New Delhi, is also acknowledged for providing Senior Research Fellowship (SRF) and ISLS, ISC, BHU, for financial assistance.

Conflict of Interest The authors declare that they have no conflict of interest.

References

Anwar M, Lou S, Chen L, Li H, Hu Z (2019) Recent advancement and strategy on bio-hydrogen production from photosynthetic microalgae. Bioresour Technol 292:121972

Aryal UK, Callister SJ, Mishra S, Zhang X, Shutthanandan JI, Angel TE, Sherman L (2013) Proteome analyses of strains ATCC 51142 and PCC 7822 of the diazotrophic cyanobacterium *Cyanothece* sp. under culture conditions resulting in enhanced H_2 production. Appl Environ Microbiol 79:1070–1077

Azwar MY, Hussain MA, Abdul-Wahab AK (2014) Development of biohydrogen production by photobiological, fermentation and electrochemical processes: a review. Renew Sust Energ Rev 31:158–173

Baebprasert W, Jantaro S, Khetkorn W, Lindblad P, Incharoensakdi A (2011) Increased H_2 production in the cyanobacterium *Synechocystis* sp. strain PCC 6803 by redirecting the electron supply via genetic engineering of the nitrate assimilation pathway. Metab Eng 13:610–616

Bandyopadhyay A, Stöckel J, Min H, Sherman LA, Pakrasi HB (2010) High rates of photobiological H_2 production by a cyanobacterium under aerobic conditions. Nat Commun 1:1–7

Batista AP, Gouveia L, Marques PA (2018) Fermentative hydrogen production from microalgal biomass by a single strain of bacterium *Enterobacter aerogenes*–effect of operational conditions and fermentation kinetics. Renew Energy 119:203–209

Bolatkhan K, Kossalbayev BD, Zayadan BK, Tomo T, Veziroglu TN, Allakhverdiev SI (2019) Hydrogen production from phototrophic microorganisms: reality and perspectives. Int J Hydrog Energy 44:5799–5811

Bothe H, Schmitz O, Yates MG, Newton WE (2010) Nitrogen fixation and hydrogen metabolism in cyanobacteria. Microbiol Mol Biol Rev 74:529–551

Britt RD, Rao G, Tao L (2020) Biosynthesis of the catalytic H-cluster of [FeFe] hydrogenase: the roles of the Fe–S maturase proteins HydE, HydF, and HydG. Chem Sci 11:10313–10323

Bundhoo MZ, Mohee R (2016) Inhibition of dark fermentative bio-hydrogen production: a review. Int J Hydrog Energy 41:6713–6733

Chatzitakis A, Nikolakaki E, Sotiropoulos S, Poulios I (2013) Hydrogen production using an algae photoelectrochemical cell. Appl Catal B Environ 142:161–168

Chu S, Majumdar A (2012) Opportunities and challenges for a sustainable energy future. Nature 488:294–303

Cournac L, Guedeney G, Peltier G, Vignais PM (2004) Sustained photoevolution of molecular hydrogen in a mutant of *Synechocystis* sp. strain PCC 6803 deficient in the type I NADPH-dehydrogenase complex. J Bacteriol Res 186:1737–1746

Das D, Khanna N, Dasgupta CN (2014) Biohydrogen production: fundamentals and technology advances. CRC Press/Taylor and Francis Group, Boca Raton

Dasgupta CN, Gilbert JJ, Lindblad P, Heidorn T, Borgvang SA, Skjanes K, Das D (2010) Recent trends on the development of photobiological processes and photobioreactors for the improvement of hydrogen production. Int J Hydrog Energy 35:10218–10238

Deviram G, Mathimani T, Anto S, Ahamed TS, Ananth DA, Pugazhendhi A (2020) Applications of microalgal and cyanobacterial biomass on a way to safe, cleaner and a sustainable environment. J Clean Prod 253:119770

Dutta D, De D, Chaudhuri S, Bhattacharya SK (2005) Hydrogen production by cyanobacteria. Microb Cell Factories 4:1–11

Ekman M, Ow SY, Holmqvist M, Zhang X, van Wagenen J, Wright PC, Stensjo K (2011) Metabolic adaptations in a H_2 producing heterocyst-forming cyanobacterium: potentials and implications for biological engineering. J Proteome Res 10:1772–1784

Elbeshbishy E, Dhar BR, Nakhla G, Lee HS (2017) Critical review on inhibition of dark biohydrogen fermentation. Renew Sust Energ Rev 79:656–668

Ermakova M, Battchikova N, Richaud P, Leino H, Kosourov S, Isojärvi J, Aro EM (2014) Heterocyst-specific flavodiiron protein Flv3B enables oxic diazotrophic growth of the filamentous cyanobacterium *Anabaena* sp. PCC 7120. Proc Natl Acad Sci 111:11205–11210

Eroglu E, Melis A (2011) Photobiological hydrogen production: recent advances and state of the art. Bioresour Technol 102:8403–8413

Eroglu E, Melis A (2016) Microalgal hydrogen production research. Int J Hydrog Energy 41: 12772–12798

Fedorov AS, Tsygankov AA, Rao KK, Hall DO (2001) *Anabaena variabilis* production of hydrogen by an mutant in a photobioreactor under aerobic outdoor conditions. In: Biohydrogen II. Pergamon, Amsterdam, pp 223–228

Geada P, Vasconcelos V, Vicente A, Fernandes B (2017) Microalgal biomass cultivation. In: Rastogi RP, Madamwar D, Pandey A (eds) Algal green chemistry recent progress in biotechnology. Elsevier, Amsterdam, pp 257–284

Ghirardi ML, Dubini A, Yu J, Maness PC (2009) Photobiological hydrogen-producing systems. Chem Soc Rev 38:52–61

Ghirardi ML, Posewitz MC, Maness PC, Dubini A, Yu J, Seibert M (2007) Hydrogenases and hydrogen photoproduction in oxygenic photosynthetic organisms. Annu Rev Plant Biol 58:71–91

Ghirardi ML, Zhang L, Lee JW, Flynn T, Seibert M, Greenbaum E, Melis A (2000) Microalgae: a green source of renewable H_2. Trends Biotechnol 18:506–511

Hallenbeck PC (2005) Fundamentals of the fermentative production of hydrogen. Water Sci Technol 52:21–29

Hankamer B, Lehr F, Rupprecht J, Mussgnug JH, Posten C, Kruse O (2007) Photosynthetic biomass and H_2 production by green algae: from bioengineering to bioreactor scale-up. Physiol Plant 131:10–21

Horiuchi JI, Kikuchi S, Kobayashi M, Kanno T, Shimizu T (2001) Modeling of pH response in continuous anaerobic acidogenesis by an artificial neural network. Biochem Eng J 9:199–204

Hosseini SE, Wahid MA (2016) Hydrogen production from renewable and sustainable energy resources: promising green energy carrier for clean development. Renew Sust Energ Rev 57: 850–866

Jeffries TW, Timourian H, Ward RL (1978) Hydrogen production by *Anabaena cylindrica*: effects of varying ammonium and ferric ions, pH, and light. Appl Environ Microbiol 35:704–710

Kars G, Gündüz U, Rakhely G, Yücel M, Eroğlu İ, Kovacs KL (2008) Improved hydrogen production by uptake hydrogenase deficient mutant strain of *Rhodobacter sphaeroides* OU 001. Int J Hydrog Energy 33:3056–3060

Kaushik MS, Srivastava M, Singh A, Mishra AK (2017) NtcA transcriptional factor: a global nitrogen regulator and connecting link between nitrogen metabolism and other crucial metabolisms. In: Plant and microbes in ever changing environment. Nova Science, Hauppauge, pp 101–127

Khetkorn W, Rastogi RP, Incharoensakdi A, Lindblad P, Madamwar D, Pandey A, Larroche C (2017) Microalgal hydrogen production–a review. Bioresour Technol 243:1194–1206

Kleinhaus JT, Wittkamp F, Yadav S, Siegmund D, Apfel UP (2021) [FeFe]-Hydrogenases: maturation and reactivity of enzymatic systems and overview of biomimetic models. Chem Soc Rev 50:1668–1784

Kosourov S, Leino H, Murukesan G, Lynch F, Sivonen K, Tsygankov AA, Allahverdiyeva Y (2014) Hydrogen photoproduction by immobilized N_2-fixing cyanobacteria: understanding the role of the uptake hydrogenase in the long-term process. Appl Environ Microbiol 80:5807–5817

Kosourov S, Murukesan G, Seibert M, Allahverdiyeva Y (2017) Evaluation of light energy to H_2 energy conversion efficiency in thin films of cyanobacteria and green alga under photoautotrophic conditions. Algal Res 28:253–263

Kourpa K, Manarolaki E, Lyratzakis A, Strataki V, Rupprecht F, Langer JD, Tsiotis G (2019) Proteome analysis of enriched heterocysts from two hydrogenase mutants from *Anabaena* sp. PCC 7120. Proteomics 19:1800332

Kroumov AD, Scheufele FB, Trigueros DEG, Modenes AN, Zaharieva M, Najdenski H (2017) Modeling and technoeconomic analysis of algae for bioenergy and coproducts. In: Algal green chemistry. Elsevier, Amsterdam, pp 201–241

Kumar GR, Chowdhary N (2016) Biotechnological and bioinformatics approaches for augmentation of biohydrogen production: a review. Renew Sust Energ Rev 56:1194–1206

Kumar GS, Kumari S, Reddy K, Bux F (2013) Trends in biohydrogen production: major challenges and state-of-the-art developments. Environ Technol 34:1653–1670

Kumar P, Pant DC, Mehariya S, Sharma R, Kansal A, Kalia VC (2014) Ecobiotechnological strategy to enhance efficiency of bioconversion of wastes into hydrogen and methane. Indian J Microbiol 54:262–267

Kumazawa T, Sato S, Kanenari D, Kunimatsu A, Hirose R, Matsuba S, Onodera JI (1994) Precursor of carthamin, a constituent of safflower. Chem Lett 23:2343–2344

Land H, Senger M, Berggren G, Stripp ST (2020) Current state of [FeFe]-hydrogenase research: biodiversity and spectroscopic investigations. ACS Catal 10:7069–7086

Levin DB, Pitt L, Love M (2004) Biohydrogen production: prospects and limitations to practical application. Int J Hydrog Energy 29:173–185

Lindberg P, Devine E, Stensjö K, Lindblad P (2012) HupW protease specifically required for processing of the catalytic subunit of the uptake hydrogenase in the cyanobacterium *Nostoc* sp. strain PCC 7120. Appl Environ Microbiol 78:273–276

Lindblad P (2018) Hydrogen production using novel photosynthetic cell factories. Cyanobacterial hydrogen production: design of efficient organisms. In: Microalgal hydrogen production: achievements and perspectives. The Royal Society of Chemistry, London, pp 323–334

Lukajtis R, Hołowacz I, Kucharska K, Glinka M, Rybarczyk P, Przyjazny A, Kamiński M (2018) Hydrogen production from biomass using dark fermentation. Renew Sust Energ Rev 91:665–694

Majidian P, Tabatabaei M, Zeinolabedini M, Naghshbandi MP, Chisti Y (2018) Metabolic engineering of microorganisms for biofuel production. Renew Sust Energ Rev 82:3863–3885

Malatinszky D, Steuer R, Jones PR (2017) A comprehensively curated genome-scale two-cell model for the heterocystous cyanobacterium *Anabaena* sp. PCC 7120. Plant Physiol 173:509–523

Malek Shahkouhi A, Motamedian E (2020) Reconstruction of a regulated two-cell metabolic model to study biohydrogen production in a diazotrophic cyanobacterium *Anabaena variabilis* ATCC 29413. PLoS One 15:0227977

Manish S, Banerjee R (2008) Comparison of biohydrogen production processes. Int J Hydrog Energy 33:279–286

Mirón AS, Gomez AC, Camacho FG, Grima EM, Chisti Y (1999) Comparative evaluation of compact photobioreactors for large-scale monoculture of microalgae. In: Progress in industrial microbiology. Elsevier, Amsterdam, pp 249–270

Mona S, Kumar SS, Kumar V, Parveen K, Saini N, Deepak B, Pugazhendhi A (2020) Green technology for sustainable biohydrogen production (waste to energy): a review. Sci Total Environ 728:138481

Mostafaeipour A, Khayyami M, Sedaghat A, Mohammadi K, Shamshirband S, Sehati MA, Gorakifard E (2016) Evaluating the wind energy potential for hydrogen production: a case study. Int J Hydrog Energy 41:6200–6210

Nagarajan D, Lee DJ, Kondo A, Chang JS (2017) Recent insights into biohydrogen production by microalgae–from biophotolysis to dark fermentation. Bioresour Technol 227:373–387

Nyberg M, Heidorn T, Lindblad P (2015) Hydrogen production by the engineered cyanobacterial strain *Nostoc* PCC 7120 ΔhupW examined in a flat panel photobioreactor system. J Biotechnol 215:35–43

Oey M, Sawyer AL, Ross IL, Hankamer B (2016) Challenges and opportunities for hydrogen production from microalgae. Plant Biotechnol J 14:1487–1499

Oncel S, Kose A (2014) Comparison of tubular and panel type photobioreactors for biohydrogen production utilizing *Chlamydomonas reinhardtii* considering mixing time and light intensity. Bioresour Technol 151:265–270

Pansook S, Incharoensakdi A, Phunpruch S (2019) Effects of the photosystem II inhibitors CCCP and DCMU on hydrogen production by the unicellular halotolerant cyanobacterium *Aphanothece halophytica*. Sci World J 2019:1030236

Pohorelic BK, Voordouw JK, Lojou E, Dolla A, Harder J, Voordouw G (2002) Effects of deletion of genes encoding Fe-only hydrogenase of *Desulfovibrio vulgaris* Hildenborough on hydrogen and lactate metabolism. J Bacteriol 184:679–686

Posewitz MC, King PW, Smolinski SL, Zhang L, Seibert M, Ghirardi ML (2004) Discovery of two novel radical S-adenosylmethionine proteins required for the assembly of an active [Fe] hydrogenase. J Biol Chem 279:25711–25720

Poudyal RS, Tiwari I, Koirala AR, Masukawa H, Inoue K, Tomo T, Veziroğlu TN (2015) Hydrogen production using photobiological methods. In: Compendium of hydrogen energy. Woodhead Publishing, Cambridge, pp 289–317

Puggioni V, Tempel S, Latifi A (2016) Distribution of hydrogenases in cyanobacteria: a phylum-wide genomic survey. Front Genet 7:223

Rashid N, Rehman MSU, Memon S, Rahman ZU, Lee K, Han JI (2013) Current status, barriers and developments in biohydrogen production by microalgae. Renew Sust Energ Rev 22:571–579

Razu MH, Hossain F, Khan M (2019) Advancement of bio-hydrogen production from microalgae. In: Microalgae biotechnology for development of biofuel and wastewater treatment. Springer, Singapore, pp 423–462

Roumezi B, Avilan L, Risoul V, Brugna M, Rabouille S, Latifi A (2020) Overproduction of the Flv3B flavodiiron, enhances the photobiological hydrogen production by the nitrogen-fixing cyanobacterium *Nostoc* PCC 7120. Microb Cell Factories 19:1–10

Rumpel S, Siebel JF, Farès C, Duan J, Reijerse E, Happe T, Winkler M (2014) Enhancing hydrogen production of microalgae by redirecting electrons from photosystem I to hydrogenase. Energy Environ Sci 7:3296–3301

Sadler NC, Bernstein HC, Melnicki MR, Charania MA, Hill EA, Anderson LN, Wright AT (2016) Dinitrogenase-driven photobiological hydrogen production combats oxidative stress in *Cyanothece* sp. strain ATCC 51142. Appl Environ Microbiol 82:7227–7235

Sadvakasova AK, Kossalbayev BD, Zayadan BK, Bolatkhan K, Alwasel S, Najafpour MM, Allakhverdiev SI (2020) Bioprocesses of hydrogen production by cyanobacteria cells and possible ways to increase their productivity. Renew Sust Energ Rev 133:110054

Saratale GD, Kshirsagar SD, Saratale RG, Govindwar SP, Oh MK (2015) Fermentative hydrogen production using sorghum husk as a biomass feedstock and process optimization. Biotechnol Bioprocess Eng 20:733–743

Schipper K, Al Muraikhi M, Alghasal GSH, Saadaoui I, Bounnit T, Rasheed R, Barbosa MJ (2019) Potential of novel desert microalgae and cyanobacteria for commercial applications and CO_2 sequestration. J Appl Phycol 31:2231–2243

Seefeldt LC, Hoffman BM, Dean DR (2009) Mechanism of Mo-dependent nitrogenase. Annu Rev Biochem 78:701–722

Sekoai PT, Ouma CNM, Du Preez SP, Modisha P, Engelbrecht N, Bessarabov DG, Ghimire A (2019) Application of nanoparticles in biofuels: an overview. Fuel 237:380–397

Sevda S, Bhattacharya S, Reesh IMA, Bhuvanesh S, Sreekrishnan TR (2017) Challenges in the design and operation of an efficient photobioreactor for microalgae cultivation and hydrogen production. In: Biohydrogen production: sustainability of current technology and future perspective. Springer, New Delhi, pp 147–162

Sharma A, Arya SK (2017) Hydrogen from algal biomass: a review of production process. Biotechnol Rep 15:63–69

Shobana S, Kumar G, Bakonyi P, Saratale GD, Nemestóthy N, Bélafi-Bakó K, Chang JS (2017) A review on the biomass pretreatment and inhibitor removal methods as key-steps towards efficient macroalgae-based biohydrogen production. Bioresour Technol 244:1341–1348

Show KY, Yan Y, Zong C, Guo N, Chang JS, Lee DJ (2019) State of the art and challenges of biohydrogen from microalgae. Bioresour Technol 289:121747

da Silva Veras T, Mozer TS, da Silva CA (2017) Hydrogen: trends, production and characterization of the main process worldwide. Int J Hydrog Energy 42:2018–2033

Skjånes K, Andersen U, Heidorn T, Borgvang SA (2016) Design and construction of a photobioreactor for hydrogen production, including status in the field. J Appl Phycol 28: 2205–2223

Srivastava N, Srivastava M, Kushwaha D, Gupta VK, Manikanta A, Ramteke PW, Mishra PK (2017) Efficient dark fermentative hydrogen production from enzyme hydrolyzed rice straw by *Clostridium pasteurianum* (MTCC116). Bioresour Technol 238:552–558

Stripp ST, Goldet G, Brandmayr C, Sanganas O, Vincent KA, Haumann M, Happe T (2009) How oxygen attacks [FeFe] hydrogenases from photosynthetic organisms. Proc Natl Acad Sci 106: 17331–17336

Sveshnikov DA, Sveshnikova NV, Rao KK, Hall DO (1997) Hydrogen metabolism of mutant forms of *Anabaena variabilis* in continuous cultures and under nutritional stress. FEMS Microbiol Lett 147:297–301

Torimura M, Miki A, Wadano A, Kano K, Ikeda T (2001) Electrochemical investigation of cyanobacteria *Synechococcus* sp. PCC7942-catalyzed photoreduction of exogenous quinones and photoelectrochemical oxidation of water. J Electroanal Chem 496:21–28

Tredici MR, Zittelli GC (1998) Efficiency of sunlight utilization: tubular versus flat photobioreactors. Biotechnol Bioeng 57:187–197

Troshina O, Serebryakova L, Sheremetieva M, Lindblad P (2002) Production of H_2 by the unicellular cyanobacterium *Gloeocapsa alpicola* CALU 743 during fermentation. Int J Hydrog Energy 27:1283–1289

Unni JK, Govindappa P, Das LM (2017) Development of hydrogen fuelled transport engine and field tests on vehicles. Int J Hydrog Energy 42:643–651

Vargas SR, dos Santos PV, Zaiat M, do Carmo Calijuri M (2018) Optimization of biomass and hydrogen production by *Anabaena* sp. (UTEX 1448) in nitrogen-deprived cultures. Biomass Bioenergy 111:70–76

Voloshin RA, Rodionova MV, Zharmukhamedov SK, Veziroglu TN, Allakhverdiev SI (2016) Biofuel production from plant and algal biomass. Int J Hydrog Energy 41:17257–17273

Wang Y, Yang H, Zhang X, Han F, Tu W, Yang W (2020) Microalgal hydrogen production. Small Methods 4:1900514

Wegelius A, Khanna N, Esmieu C, Barone GD, Pinto F, Tamagnini P, Lindblad P (2018) Generation of a functional, semisynthetic [FeFe]-hydrogenase in a photosynthetic microorganism. Energy Environ Sci 11:3163–3167

Wittenberg G, Sheffler W, Darchi D, Baker D, Noy D (2013) Accelerated electron transport from photosystem I to redox partners by covalently linked ferredoxin. Phys Chem Chem Phys 15: 19608–19614

Wutthithien P, Lindblad P, Incharoensakdi A (2019) Improvement of photobiological hydrogen production by suspended and immobilized cells of the N_2-fixing cyanobacterium *Fischerella muscicola* TISTR 8215. J Appl Phycol 31:3527–3536

Xu Q, Yooseph S, Smith HO, Venter JC (2005) Development of a novel recombinant cyanobacterial system for hydrogen production from water. In: Contractor-grantee workshop III, p 63

Yin Y, Wang J (2018) Pretreatment of macroalgal *Laminaria japonica* by combined microwave-acid method for biohydrogen production. Bioresour Technol 268:52–59

Zebda A, Alcaraz JP, Vadgama P, Shleev S, Minteer SD, Boucher F, Martin DK (2018) Challenges for successful implantation of biofuel cells. Bioelectrochemistry 124:57–72

Zhang Y, Yang H, Guo L (2016) Enhancing photo-fermentative hydrogen production performance of *Rhodobacter capsulatus* by disrupting methylmalonate-semialdehyde dehydrogenase gene. Int J Hydrog Energy 41:190–197

Molecular Biology of Non-ribosomal Peptide (NRP) and Polyketide (PK) Biosynthesis in Cyanobacteria

8

Laxmi, Sweksha Singh, Avinash Singh, and Ravi K. Asthana

Abstract

Secondary metabolites produced by cyanobacteria are numerous and structurally diverse, with a focus on non-ribosomal peptide and polyketide structures. These compounds could have a variety of bioactivities, and some of them could be valuable in the creation of commercial medications or as biochemical research tools. Natural product gene clusters can be identified and characterized via genome mining. This has allowed for the clustering of biosynthetic genes encoding diverse substances, including non-ribosomal peptides (NRPs), polyketides (PKs), and hybrids thereof, which span a wide range of bioactive metabolites of human significance. Various bioinformatics techniques have only just been created to anticipate the outcomes of orphan gene clusters, although the cyanobacterial pathways are complex in some cases, making prediction difficult. However, this backdrop can be avoided through the heterologous expression and/or mass spectrometry. Several cyanobacterial genomes have been found to contain peptide synthetase and polyketide synthase genes, according to recent research. This information could be extremely useful in future screening initiatives aiming at detecting new bioactive chemicals.

Keywords

Cyanobacteria · Bioactive molecules · Peptide synthetase · Polyketide synthase · Antimicrobial · Antiviral compounds

Laxmi · S. Singh · A. Singh · R. K. Asthana (✉)
R. N. Singh Memorial Laboratory, Centre of Advanced Study in Botany, Institute of Science, Banaras Hindu University, Varanasi, Uttar Pradesh, India
e-mail: rasthana@bhu.ac.in

B. Neilan et al. (eds.), *Cyanobacterial Biotechnology in the 21st Century*,
https://doi.org/10.1007/978-981-99-0181-4_8

8.1 Introduction

Cyanobacteria are a diverse group of the oldest oxygenic photosynthetic gram-negative prokaryotes, and their fossilized remnants have been found in rocks that were formed approximately 3.5 billion years ago (Singh et al. 2021). They can be found in any ecosystem, from salty to fresh water, such as seas, estuaries, lakes, rivers, and streams, as well as on land (e.g., rocks, soils, estuarine, and even deserts). One way to classify them is according to the environmental conditions that are favorable for their growth, such as high pH, high salinity, low nutrients, low or high temperatures, and high or moderate levels of radiation (Huang et al. 2021). Cyanobacteria have, over the course of billions of years of evolution, developed a wide variety of intricate molecular, physiological, and metabolic characteristics that allow them to thrive in their specific environments.

Natural products, also known as specialized metabolites, are organic molecules with low molecular weight that perform a variety of biological roles, many of which are quite potent (Newman and Cragg 2020). Typically, they have complicated chemical structures with peculiar chemical moieties, enabling them to react with particular molecular targets, thereby killing or impeding the development of other species. Secondary metabolites are not required for an organism's growth and development, but they do assist the organism in dealing with the survival strategies of the system. From 39 genera of cyanobacteria, approximately 1100 secondary metabolites with complex and unique chemical structures have been found (Dittmann et al. 2015). Antibacterial and antifungal medications and anticancerous drugs are just a few of the secondary metabolites that have proven useful. Polyketides and non-ribosomal peptides are important secondary metabolites, as well as alkaloids, terpenoids, shikimate-derived chemicals, and aminoglycosides. Even though there has been a great deal of bioinformatics research indicating a highly diverse biochemical route for the formation of secondary metabolite, relatively few natural compounds can presently be linked to those strains whose genome has been sequenced. Each microorganism has its own intrinsic characteristics to produce any of the secondary metabolites, which are products of the NRPS and PKS genes or their hybrids. As these genes are modular in nature, they have an exceptionally high capacity to mutate, resulting in the production of various types of secondary metabolites. Many current pharmaceuticals are built on NRPs and PKs, which operate as a molecular/chemical platform for the extraction of novel biologically active molecules. Thus, identification of novel compounds with therapeutic uses is the focus of natural product research.

8.2 Cyanobacterial Secondary Metabolites

Despite the fact that the majority of metabolites are classed as primary and secondary, there is some overlap between the two categories. By offering protection and assisting in survival in response to various stresses, viz., biotic or abiotic, in their niche, cyanobacteria frequently create secondary metabolites, giving them an edge over the other species directly or indirectly (Burja et al. 2003). Due to their adaptability to a variety of environments, cyanobacteria constitute an abundant source of secondary metabolites. According to the stress environment, several suites of secondary metabolites can be created with a great deal of structural diversity among the other chemical classes. The peptides, polyketides, alkaloids, terpenoids, and UV-absorbing metabolites are part of this group. These qualities can be used to create medicines, cosmeceuticals, and nutraceuticals in industrial biotechnology. Characterization of novel natural compounds provides new active structures as a starting point for drug design. Secondary metabolism of cyanobacteria is so complicated that myriads of molecules can be generated, yet only a small proportion of them have been described so far. These substances might exhibit a range of bioactivities, and some of them might be useful in the development of pharmaceutical products or as instruments for biochemical studies. The non-ribosomal production of the cyanobacterial hepatotoxin microcystin was established by knock-out mutagenesis as the very first metabolite.

Besides producing toxins, cyanobacteria also produce compounds that may be useful in medicine. Cryptophycins, which are produced by the *Nostoc* GSV 224 strain, are potent microtubule assembly inhibitors with potent anticancer effects on a variety of tumors (Panda et al. 1998). Also, dolastatins, curacin A, and lynbyabellins are a few of the antimitotic and cytotoxic metabolites that the *Lyngbya majuscula* produces (Mitchell et al. 2000). *L. majuscula* also produces tanikolide and majusculamide, whereas *Anabaena* was the first species to synthesize laxaphycins (Singh et al. 1999). Dihydro-2H-pyran-2-one derivative from the cyanobacteria *Geitlerinema* CCC728 and *Arthrospira* CCC729 was also shown to exhibit antifungal and antitumor activity (Srivastava et al. 2017). Thus, in addition to its well-studied marine counterpart, freshwater cyanobacteria are prolific makers of secondary metabolite(s).

8.3 Non-ribosomal Peptide Synthetases, Polyketides, and Hybrid Pathway

The secondary metabolites produced by cyanobacteria are quite diverse. Approximately 5% of cyanobacteria's genome has been attributed to non-ribosomal peptide synthatases (NRPS) and polyketide synthase (PKS) pathways. With the exception of peptides and polyketides, the great majority of secondary metabolites produced by cyanobacteria are synthesized by one of the four metabolic pathways: the shikimate pathway, the malonate pathway, the mevalonate pathway, or the methylerythritol phosphate (MEP) system. The biosynthesis of NRPs and PKs is aided by a collection of multimodular enzymes, each of which comprises a core set of three domains required for activating and condensing a single amino or carboxylic component. Condensation of amino acid (proteinogenic and/ or non-proteinogenic amino acid) monomers in a predetermined sequence yields non-ribosomal peptides. The production of polyketides and non-ribosomal peptides is carried out by polyketide synthases (PKSs) and non-ribosomal peptide synthetases (NRPSs), two multi-enzymatic, multi-domain megasynthases. Polyketides, in contrast, are made by continually combining two carbon ketide units made from thioesters of acetate or other short carboxylic acids. Many of these secondary metabolites (discussed later) are effective antibacterial, antifungal, antiparasitic, antitumor, and immunosuppressive medications in clinical practice (Verma et al. 2022).

8.3.1 Non-ribosomal Peptide Synthetases (NRPSs)

As previously mentioned, NRPS possesses many modules (Fig. 8.1) where each module possesses a collection of domains, separately which controls a separate enzymatic process. Each module, with the exception of the starter one, typically consists of three core biosynthetic domains which are, namely, the adenylation (A) domain, whose assigned role is of selecting the substrate, and then the substrate is covalently linked to the phosphopantetheinyl cofactor (PPT), which is attached to the neighboring peptidyl carrier protein (PCP) domain after adenylation. At the end of the process, condensation (C) domains, which are thought to operate as gatekeepers, help in regulating substrate activation and peptide bond formation by catalyzing the repeated coupling of amino acid substrates to the nascent peptide chain (Meyer et al. 2016). Various domains, viz., thioesterase, epimerization, N-methylation, heterocyclization, formylation, and oxidation, work as optional auxiliary domains. The A domain binding pocket's ten essential amino acids can be utilized to forecast which amino acid will be selected and incorporated into the A domain (Challis et al. 2000). Proteins with one or more modules are coded for by non-ribosomal peptide biosynthetic gene clusters, which form complexes that function like assembly lines, and following the co-linearity rule quite oftenly (Marahiel

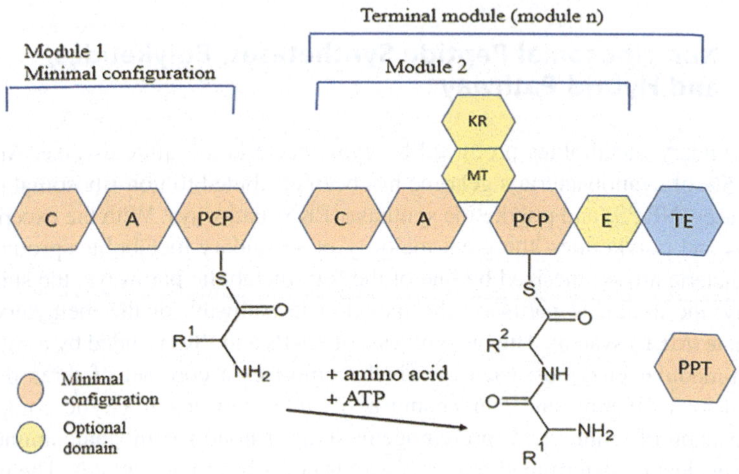

Fig. 8.1 The NRPS along with their domains. (For abbreviations, see text) (Modified from Kehr et al. 2011)

et al. 1997). For example, *L. majuscula* produces lyngbyatoxin-a through the NRPS pathway. The binding and activation of protein kinase C (PKC) are responsible for the toxicity of the dermatoxic alkaloid lyngbyatoxin (LTX; 437 Da). Furthermore, a linalyl group is attached to the C-7 position of the indolactam ring (Cullen et al. 2019).

8.3.2 Polyketide Synthases (PKSs)

Polyketide synthases (PKSs) are versatile enzymes that biosynthesize a wide range of natural chemicals, many of which are now exploited as medication sources. These are both structurally and enzymatically connected to fatty acid synthases; they evolve in tandem and create polyketides, a class of secondary metabolites ranging in size from 3 to 60 kb, from carboxylic acid (Nivina et al. 2019).

The PKSs have been classified into three groups according to their structure and catalytic mechanism. Each functional domain in modular type I PKSs catalyzes one step in polyketide biosynthesis. Several single-module proteins, each with its own unique set of enzymatic activity, work together to synthesise polyketides in iterative type II PKSs, whereas type III PKSs are homodimeric, composed of a single active site enzyme that operates in a loop to produce the final polyketide product. Type I PKSs are usually found in bacteria, followed by type II in fungi and type III in plants (Kaluzhnaya and Itskovich (2022).

These biosynthesize them by sequentially assembling polyketides from acyl-CoA. PKS modules, like NRPS, are made up of cis-AT domain, an ACP domain, a KS domain, and other domains for extra modification (Micallef et al. 2015). The co-linearity rule (Fig. 8.2), which determines the sequence of peptide or polyketide chains by the order of modules (and hence by the demand of certain substrates), is a common method for doing this.

Similar to how fatty acid synthesis is carried out, PKSs catalyze the decarboxylative condensation of malonyl-coenzyme A (CoA) and its related counterparts onto a substrate molecule to produce polyketides (Lai and Cronan 2003). Recent research has indicated that the nosperin pathway is an exception to this rule. The majority of the nosperin backbone is composed of a modular trans-AT PKS devoid of integrated AT domains and working in collaboration with a free-standing AT, with one PKS module containing an integrated AT domain (Kampa et al. 2013). For example, nosperin was obtained from *Nostoc* sp. (Kampa et al. 2013). A neurotoxin, anatoxin-a, synthesized by three PKS modules, was obtained from *Anabaena* sp. (Kehr et al. 2011).

8.3.3 Hybrid Pathway (NRPS/PKS)

The cyanotoxins microcystin and cylindrospermopsin, which are made by both NRPSs and type I PKSs, are among the many such "hybrid" clusters that have been discovered in cyanobacteria (Tillett et al. 2000). The common coupling of

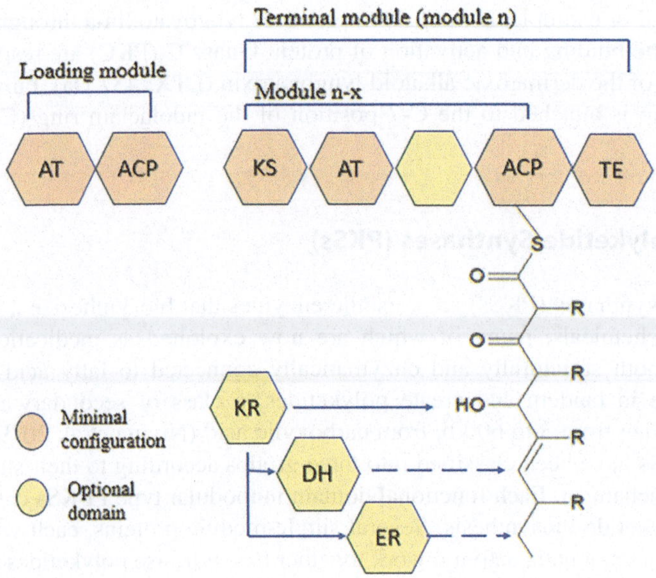

Fig. 8.2 PKS pathway along with their domains. (Abbreviations: see text) (Modified from Jenke-Kodama et al. 2005)

NRPS and PKS modules inside a single reading frame in the secondary metabolite biosynthesis pathways of cyanobacteria is one of their most defining characteristics (Welker and Von Döhren 2006). Because these two pathways have so many structural and catalytic similarities, it's only inevitable that hybrids of them exist. Hybrid routes have been identified in both linear and cyclic peptides, and they have been found in freshwater, saltwater, and terrestrial cyanobacteria. The first NRPS/PKS hybrid route in cyanobacteria was discovered in *Microcystis aeruginosa*, a freshwater cyanobacterium that produces the well-known hepatotoxin (Verma et al. 2022). Cryptophycins, a compound produced by NRPS/PKS hybrid route, have been found in free-living or lichen-symbiotic terrestrial *Nostoc* strains (Dittmann et al. 2015). Further, examples of peptides from hybrid routes include the recently found jamaicamide and barbamide from the *Lyngbya majuscula*. In order to create a variety of chemical structures with distinct functions and bioactivity, polyketides or fatty acids are typically attached to non-ribosomal peptides utilizing PKS to form hybrid metabolites.

8.4 Genome Mining in Search of New Natural Products

Most secondary metabolite biosynthetic gene clusters in cyanobacteria have been found by comparing their chemical structure to that of recognized natural compounds. Biosynthetic gene clusters were previously discovered by time-consuming manual screening of gene libraries. Genome initiatives have been launched in recent years to identify the biochemical routes for natural products (Kellmann et al. 2008). Here, the chemical structure and projected biosynthetic route were used to pinpoint biosynthetic gene clusters in microbial genomes. A new phenomenon, the accidental identification of biosynthetic routes for known natural products via genome projects, is expected to continue at a rapid rate as the number of accessible genomes grows.

Genome mining has also led to the identification of natural products from cyanobacteria, including trichamides, viridisamides, anacyclamides, and prochlorosins, when no previously known natural product was available to serve as a starting point (Dittmann et al. 2015). Using a targeted genome mining strategy, researchers isolated nosperin, a novel macrolide, from lichens harboring symbiotic strains of *Nostoc commune* (previously mentioned). Genome mining also uncovered novel mechanisms in cyanobacteria that insert different d-amino acid sequences into all-L peptides, exhibiting strong regioselectivity, substrate heterogeneity, and action which are irreversible (Morinaka et al. 2014). These finding are very encouraging because these suggest the potential of genomics as a means of locating novel natural compounds and because it opens up promising avenues for peptide engineering.

8.5 Bioactive Compounds from Cyanobacteria

Cyanobacteria provide a wealth of natural goods, and a vast variety of bioactive substances, including pharmaceuticals, enzymes, colored pigments, bioplastics, toxins, and a variety of other compounds, has been identified. As previously mentioned, the bulk of bioactive compounds recognized with established antibacterial, antifungal, antiviral, and anti-HIV effect have been obtained from the cyanobacteria. To further add to this, they produce several significant bioactive compounds, including amides (9%), lipopeptides (40%), amino acids (5.6%), fatty acids (4.2%), and macrolides (4.2%) (Burja et al. 2001). Such bioactive substances can kill living cells, tissue, or organisms at very low concentrations in a solution.

8.5.1 Antimicrobial Compounds

Antimicrobial compounds work by interfering with the physiological processes of microorganisms to either kill or hinder their growth. Since multidrug-resistant (MDR) bacterial strains are becoming more prevalent, significant efforts are being made to find and develop novel antibiotics from other organisms. Today, it is known that a number of cyanobacteria produce intracellular and extracellular bioactive substances with strong antibacterial and antifungal properties (Singh et al. 2021). According to the reports, members of the order *Nostocales* generate the highest rates (nearly 81%) of antibacterial metabolites, followed by the order *Oscillatoriales* (13%). 70% or more of the cyanobacterially generated antimicrobial chemicals were alkaloids and polyketides (Carpine and Sieber 2021). Inhibition of human pathogenic bacteria by the polyketide antibiotic malyngolide, which was initially isolated from *L. majuscula*, has been demonstrated. Many cyanobacterial metabolites have been studied for their potential to inhibit fungal growth. Scytophycin C produced from *Scytonema pseudohofmanni* was found to be efficient against *Pythium ultimum*, *Candida albicans*, *Sclerotia homoeocarpa*, and *Rhizoctonia solani*. Hassallidins, a well-known group of antifungal metabolites, were initially discovered in the epilithic cyanobacterium *Hassallia* sp. and were effective against *Ustilago maydis*, *Cryptococcus neoformans*, *Penicillium* sp., *Fusarium* sp., *Aspergillus* sp., and *Acremonium strictum*. *Aphanizomenon*, *Anabaena* sp., *Cylindrospermopsis raciborski*, *Tolypothrix*, and *Nostoc* are also reported to produce hassallidins. Hassallidins were also reported to be produced by the cyanobacteria *Anabaena* sp., *Aphanizomenon*, *Cylindrospermopsis raciborskii*, *Nostoc*, and *Tolypothrix* (Vestola et al. 2014).

8.5.2 Antiviral Compounds

Several cyanobacterial species have been demonstrated to produce antiviral compounds. *Nostoc ellipsosporum* has yielded a 101 amino acid protein named cyanovirin-N (CV-N) which irreversibly inactivates various main HIV-1 strains and concomitantly prevents cell-to-cell transmission. CV-N is surprisingly efficient against all forms of HIV, including the M- and T-tropic strains of HIV-1, HIV-2, SIV (simian), and FIV (feline). Scytovirin, a compound discovered in *Scytonema varium*, possesses potent anti-HIV activity because it interacts with the viral coat proteins gp120, gp160, and gp41. (Bokesch et al. 2003). Furthermore, nostoflan, a bioactive compound, was isolated from *Nostoc flagelliforme* and showed a significant antiviral activity against herpes simplex virus type 1 (HSV-1), herpes simplex virus type 2 (HSV-2), human cytomegalovirus, and influenza A (Verma et al. 2022).

8.5.3 Anticancer Compounds

Because of the lack of effective therapies and medicines, cancer kills millions of people every year. Synthesis of a bioactive compound with anticancer characteristics is facilitated by cyanobacteria. There is evidence that compounds produced by marine cyanobacteria can interact with certain molecular targets, leading to the regulated death of tumor cells. Cell cycle arrest, mitochondrial dysfunction, oxidative damage, and alterations in caspase cascade, noncaspase proteases, the Bcl-2 protein family, and membrane sodium channel dynamics are most strongly influenced by these chemicals (Costa et al. 2019). A different study found that homogenous leukoplakia might be avoided with the help of *Spirulina fusiformis* extract taken orally (Patil et al. 2015). A subset of spirulina extract components inhibited chemically induced carcinogenesis in the buccal pouches of model hamsters.

8.5.4 Antiprotozoal Compounds

Cyanobacteria produce many bioactive compounds that have inhibitory effects on protozoans including *Plasmodium* (causal agent of malaria), *Trypanosoma* (causal agent of sleeping sickness), *Leishmania* (causal agent of leishmaniasis), etc. Calothrixins A and B, which were isolated from two isolates of the *Calothrix* species, and cyclic hexapeptides, which were recovered from the marine cyanobacterium *Oscillatoria* sp., have all been reported to display antiplasmodial action (Verma et al. 2022). When tested against *Plasmodium falciparum*, the quaternary carbolinium alkaloid nostocarboline, isolated from the cyanobacterium *Nostoc*, was highly inhibitive even at micromolar concentrations (Barbaras et al. 2008). The ribosomal cyclic peptide aerucyclamide B was extracted from *P. aeruginosa*, and it was found to have potent antiplasmodial activity. Aerucyclamide was also very effective against *Trypanosoma*. According to the recent reports, cyanobacteria produce antiplasmodial substances (Saad et al. 2022), but many more strains need to be screened before a powerful molecule can be produced on a wide scale and used to treat malarial disorders.

8.6 Natural Product Biosynthetic Gene Cluster Prediction via Bioinformatics

The intricate chemical structures of the metabolites and the domain organization of these gene clusters must be correlated using substantial bioinformatics analysis in order to discover the metabolic products of NRPS and PKS clusters. However, several multifunctional NRPS and PKS proteins have sequences that are searchable in multiple sequence databases, but their domain and module organization as well as their substrate features have not yet been extensively annotated. Domain organization in multifunctional proteins is typically difficult to capture with conventional domain identification techniques as the Conserved Domain Database (CDD) (Marchler-Bauer et al. 2002) and InterPro (Mulder et al. 2003). These proteins' domain arrangements and substrate preferences cannot currently be predicted, and it takes a lot of work to match the sequences of these proteins to their metabolites. In addition, in contrast to the limited starting pitcher precursors that are involved in polyketide biosynthesis, NRPS adenylation domains are known to select starter/extender precursors from a reservoir of at least 50 natural or artificial amino acids. This is in contrast to the limited starting pitcher precursors that are involved in polyketide biosynthesis. Therefore, it is difficult to make accurate predictions regarding the substrate sensitivity of NRPS proteins (Ansari et al. 2004). Automated computational methods need to be developed in order to correctly identify NRPS and/or PKS domains and estimate the substrate specificity of these domains based on the identification of potential specificity-determining residues. Because the multiple PKS domains have a high degree of homology within a functional family, pairwise comparisons of each domain can be recognized by a single template sequence. However, because of the significant sequence diversity across members of a certain domain family, it has been difficult to distinguish different NRPS domains with accurate boundaries. SEARCHNRPS is a web-based software application that was developed following an in-depth analysis of the sequence and structural characteristics of a large number of experimentally reported NRPS clusters. Its purpose is to make predictions regarding the domain organization and substrate specificity of NRPS proteins. They have merged SEARCHNRPS with their previously developed tool SEARCHPKS despite the fact that NRPS domains and modular PKS domains can coexist in a single open reading frame (ORF) and that such hybrid NRPS/PKS clusters have also been found in a large number of microbial genomes. This was done in order to improve the precision of their search results, and it was made available through a single interface known as NRPS-PKS. This interface is able to evaluate hybrid NRPS/PKS sequences in addition to NRPS and PKS sequences individually. In addition to modular PKSs, iterative equipment is utilized in the production of polyketides and type III polyketide synthases (Yadav et al. 2003).

The majority of gene clusters uncovered by bioinformatic genome mining techniques are NRPS, PKS, or hybrid NRPS/PKS. This is because the genes involved in the NRPS/PKS pathways are extremely well conserved and defined and because they tend to cluster together in predictable patterns across the genome. Kalaitzis et al. (2009) screened the genome for replicated genomic elements essential to NRPS/PKS and hybrid biosynthetic pathways, including PF00501 (AMP-binding enzyme), PF00608 (condensation domain), and PF00550 (phosphopantetheine attachment site), as well as PF00109/PF02801 (phosphopantetheine attachment site) (-ketoacyl synthase N- and C-terminal) Other applications capable of searching whole genomes include NP (Tae et al. 2007). Instead of searching for individual gene clusters, NaPDoS (Natural Product Domain Seeker) analyzes entire genomes to identify all C and KS domains inside a query genome (Ziemert et al. 2012). Then, NaPDoS uses BLAST and phylogeny to assess if the domain is a member of a distinct cluster and, consequently, is responsible for a distinct natural product. In order to compute the number of gene clusters present, the user must manually link each discovered domain to its appropriate gene cluster.

8.7 Exploiting the Biosynthesis of Secondary Metabolites in Cyanobacteria

Despite the discovery of a vast array of natural compounds in cyanobacteria, its commercialization is not on track. The slow development of these bacteria, as well as the need for a photobioreactor, is undoubtedly limiting factors. *Synechococcus* sp. and *Synechocystis* sp., two unicellular organisms known for their rapid growth and extensive genetic-level research, do not make any of the NRPS or PKS products, which is an interesting fact. Still, there are a number of approaches that may be used to build different cyanobacterial PKSs. Module swapping to generate novel chimeric PKSs, domain swapping to introduce novel AT domains into the acyl extender, and site-directed mutagenesis to alter substrate specificity at individual amino acid sites are all examples. The development of efficient modular PKSs relies on enhancing the catalytic activity of specific domains while retaining the vast majority of their protein-protein interactions (Neves et al. 2022). However, full biosynthetic gene clusters may eventually need to be expressed heterologously, either in cyanobacteria that grow quickly or in other heterotrophic hosts (Dittmann et al. 2001). Using OMICS (omics) data (genomics, metabolomics, proteomics, and transcriptomics) and computational analysis, system biology delves deep into the intricate regulatory networks and metabolic pathways of microbes in an effort to develop more metabolically efficient strains (Xu et al. 2020). These aspects of engineering could be quite useful in further research.

8.8 Future Prospects and Challenges of Cyanobacterial Drugs

Recent advances have significantly sped up the process of creating new medicines. Many different kinds of cyanobacteria are responsible for synthesizing these novel metabolites, but they only produce trace amounts, and their extraction and purification are laborious and time-consuming. Large-scale reacquisition or fermentation is required to produce sufficient amounts for preclinical research, which would significantly lengthen the development cycle. The gathering of cyanobacterial strains from uncharted regions is one of the main methods for the field-collected strains' research, and the development of novel medications' suitability for laboratory culture is a crucial element in the process of finding new drugs using them. Once they have been introduced to pure culture, simple methods are available for growing them in greater numbers, chemically analyzing the natural products to identify the chemicals. In addition to selecting the best strain and developing the ideal physicochemical conditions for manufacturing, the next piece of the technical puzzle will be genomics of a single cell employing amplification by multiple displacement which will enable the investigation of the full biochemical potential of isolated, uncultivated bacteria through intricate microbial communities. Furthermore, developing new NRPS and associated megasynthase enzymes to function as biosynthetic assembly lines is a potent approach to developing novel antibiotics as well as other biologically active natural compounds that are too complicated to synthesize chemically (Thong et al. 2021). As a result of their large size, their complexity, and the need to protect several functional groups, medicinal and combinatorial chemists avoid working with these natural products. Preparing several natural product analogues as synthetic compounds simultaneously is challenging. The development of drugs from cyanobacteria strains is moving very slowly as a result of all these difficulties.

Gathering cyanobacterial strains from unexplored locations is one of the key strategies for field-collected strains' research, and developing newer pharmaceuticals suitable for lab growth is a crucial aspect in finding new drugs employing them. Once introduced to pure culture, simple ways exist for growing them in bigger numbers and chemically analyzing natural products to identify compounds. The next piece of the technical puzzle will be genomics of a single cell using amplification by multiple displacements, which will allow the examination of the complete biochemical potential of isolated, uncultivated bacteria through complicated microbial communities (Thong et al. 2021). This new approach makes it easier to reach rare microbiota, which could yield novel enzymes and natural compounds from cyanobacterial research. Metagenomics can indicate which genes exist in a given environment, but not which are active. Metatranscriptomics-based techniques for analyzing microbial communities must only access transcriptionally active genes to avoid this issue (Gilbert and Dupont 2011). Utilizing metaproteomics in-depth analysis of the environment's proteins is an alternate method of accessing the metagenome. However, these innovative and promising techniques haven't yet been used to study how cyanobacteria produce secondary metabolites, despite the fact that they hold great promise for the creation of novel medications using them. Future drug development efforts are aimed at creating innovative medications with anticancer or antibacterial properties that are conjugated with compound derived from cyanobacteria (Swain et al. 2017). It may be possible to transfer appropriate genes from a cyanobacterium to a cultivable strain making in order to produce desired bioactive compounds in large quantities. Therefore, the creation of synthetic analogues and/or conjugate medications is necessary in order to effectively utilize cyanobacterial compounds for the control of severe drug-resistant diseases.

8.9 Conclusions

There have been a lot more cyanobacterial genome sequences published recently. The following produced valuable information about the number, clusters of known natural product genes that are diverse and widely distributed, as well as orphan gene cluster identification. While there are a variety of bioinformatic methods that may be used to analyze particularly with regard to clusters of genes from natural products, many of them encounter difficulties when used with cyanobacterial pathways. During the analysis of gene clusters encoding cyanobacterial natural products, this should be taken into account, as additional manual analysis would be necessary. To avail use of the potential for new chemical sources in cyanobacterial natural product routes, heterologous expression (along with the OMICS approaches) will be required. Despite the modest progress, recent advances in the expression of crucial pathways and the cloning of huge gene clusters lay the groundwork for additional study and the discovery of current natural outcome.

Acknowledgments We greatly acknowledge Head and Programme Coordinator, CAS in Botany, DST-FIST, ISLS, B.H.U., for providing research facilities. The financial supports are also acknowledged to the University Grants Commission and CSIR (Council of Scientific and Industrial Research), Government of India, New Delhi, for providing Senior Research Fellowship (SRF) and Junior Research Fellowship (JRF).

References

Ansari MZ, Yadav G, Gokhale RS, Mohanty D (2004) NRPS-PKS: a knowledge-based resource for analysis of NRPS/PKS megasynthases. Nucleic Acids Res 32(suppl_2):W405–W413

Barbaras D, Kaiser M, Brun R, Gademann K (2008) Potent and selective antiplasmodial activity of the cyanobacterial alkaloid nostocarboline and its dimers. Bioorg Med Chem Lett 18(15):4413–4415

Bokesch HR, O'Keefe BR, McKee TC, Pannell LK, Patterson GM, Gardella RS et al (2003) A potent novel anti-HIV protein from the cultured cyanobacterium Scytonema varium. Biochemistry 42(9):2578–2584

Burja AM, Banaigs B, Abou-Mansour E, Burgessd JG, Wrighta PC (2001) Marine cyanobacteria: a prolific source of natural products. Tetrahedron 57:9347–9377

Burja AM, Dhamwichukorn S, Wright PC (2003) Cyanobacterial postgenomic research and systems biology. Trends Biotechnol 21(11):504–511

Carpine R, Sieber S (2021) Antibacterial and antiviral metabolites from cyanobacteria: their application and their impact on human health. Curr Res Biotechnol 3:65–81

Challis GL, Ravel J, Townsend CA (2000) Predictive, structure-based model of amino acid recognition by nonribosomal peptide synthetase adenylation domains. Chem Biol 7(3):211–224

Costa M, Sampaio-Dias IE, Castelo-Branco R, Scharfenstein H, Rezende de Castro R, Silva A, Leão PN (2019) Structure of hierridin c, synthesis of hierridins b and c, and evidence for prevalent alkylresorcinol biosynthesis in picocyanobacteria. J Nat Prod 82(2):393–402

Cullen A, Pearson LA, Mazmouz R, Liu T, Soeriyadi AH, Ongley SE, Neilan BA (2019) Heterologous expression and biochemical characterisation of cyanotoxin biosynthesis pathways. Nat Prod Rep 36(8):1117–1136

Dittmann E, Neilan B, Börner T (2001) Molecular biology of peptide and polyketide biosynthesis in cyanobacteria. Appl Microbiol Biotechnol 57(4):467–473

Dittmann E, Gugger M, Sivonen K, Fewer DP (2015) Natural product biosynthetic diversity and comparative genomics of the cyanobacteria. Trends Microbiol 23(10):642–652

Gilbert JA, Dupont CL (2011) Microbial metagenomics: beyond the genome. Annu Rev Mar Sci 3: 347–371

Huang IS, Pietrasiak N, Gobler CJ, Johansen JR, Burkholder JM, D'Antonio S, Zimba PV (2021) Diversity of bioactive compound content across 71 genera of marine, freshwater, and terrestrial cyanobacteria. Harmful Algae 109:102116

Jenke-Kodama H, Sandmann A, Müller R, Dittmann E (2005) Evolutionary implications of bacterial polyketide synthases. Mol Biol Evol 22(10):2027–2039

Kalaitzis JA, Lauro FM, Neilan BA (2009) Mining cyanobacterial genomes for genes encoding complex biosynthetic pathways. Nat Prod Rep 26(11):1447–1465

Kaluzhnaya OV, Itskovich VB (2022) Features of diversity of polyketide synthase genes in the community of freshwater sponge Baikalospongia fungiformis. Russ J Genet 58(3):336–346

Kampa A, Gagunashvili AN, Gulder TA, Morinaka BI, Daolio C, Godejohann M et al (2013) Metagenomic natural product discovery in lichen provides evidence for a family of biosynthetic pathways in diverse symbioses. Proc Natl Acad Sci 110(33):E3129–E3137

Kehr JC, Picchi DG, Dittmann E (2011) Natural product biosyntheses in cyanobacteria: a treasure trove of unique enzymes. Beilstein J Org Chem 7(1):1622–1635

Kellmann R, Mihali TK, Jeon YJ, Pickford R, Pomati F, Neilan BA (2008) Biosynthetic intermediate analysis and functional homology reveal a saxitoxin gene cluster in cyanobacteria. Appl Environ Microbiol 74(13):4044–4053

Lai CY, Cronan JE (2003) β-Ketoacyl-acyl carrier protein synthase III (FabH) is essential for bacterial fatty acid synthesis. J Biol Chem 278(51):51494–51503

Marahiel MA, Stachelhaus T, Mootz HD (1997) Modular peptide synthetases involved in nonribosomal peptide synthesis. Chem Rev 97(7):2651–2674

Marchler-Bauer A, Panchenko AR, Shoemaker BA, Thiessen PA, Geer LY, Bryant SH (2002) CDD: a database of conserved domain alignments with links to domain three-dimensional structure. Nucleic Acids Res 30(1):281–283

Meyer S, Kehr JC, Mainz A, Dehm D, Petras D, Süssmuth RD, Dittmann E (2016) Biochemical dissection of the natural diversification of microcystin provides lessons for synthetic biology of NRPS. Cell Chem Biol 23(4):462–471

Micallef ML, D'Agostino PM, Al-Sinawi B, Neilan BA, Moffitt MC (2015) Exploring cyanobacterial genomes for natural product biosynthesis pathways. Mar Genomics 21:1–12

Mitchell SS, Faulkner DJ, Rubins K, Bushman FD (2000) Dolastatin 3 and two novel cyclic peptides from a palauan collection of Lyngbya majuscula. J Nat Prod 63(2):279–282

Morinaka BI, Vagstad AL, Helf MJ, Gugger M, Kegler C, Freeman MF et al (2014) Radical S-adenosyl methionine epimerases: regioselective introduction of diverse d-amino acid patterns into peptide natural products. Angew Chem Int Ed 53(32):8503–8507

Mulder NJ, Apweiler R, Attwood TK, Bairoch A, Barrell D, Bateman A, Zdobnov EM (2003) The InterPro Database, 2003 brings increased coverage and new features. Nucleic Acids Res 31(1): 315–318

Neves RP, Ferreira P, Medina FE, Paiva P, Sousa JP, Viegas MF et al (2022) Engineering of PKS megaenzymes—a promising way to biosynthesize high-value active molecules. Top Catal 65(1):544–562

Newman DJ, Cragg GM (2020) Natural products as sources of new drugs over the nearly four decades from 01/1981 to 09/2019. J Nat Prod 83(3):770–803

Nivina A, Yuet KP, Hsu J, Khosla C (2019) Evolution and diversity of assembly-line polyketide synthases: focus review. Chem Rev 119(24):12524–12547

Panda D, DeLuca K, Williams D, Jordan MA, Wilson L (1998) Antiproliferative mechanism of action of cryptophycin-52: kinetic stabilization of microtubule dynamics by high-affinity binding to microtubule ends. Proc Natl Acad Sci 95(16):9313–9318

Patil S, Al-Zarea BK, Maheshwari S, Sahu R (2015) Comparative evaluation of natural antioxidants spirulina and aloe vera for the treatment of oral submucous fibrosis. J Oral Biol Craniofac Res 5(1):11–15

Saad MH, El-Fakharany EM, Salem MS, Sidkey NM (2022) The use of cyanobacterial metabolites as natural medical and biotechnological tools. J Biomol Struct Dyn 40(6):2828–2850

Singh IP, Milligan KE, Gerwick WH (1999) Tanikolide, a toxic and antifungal lactone from the marine cyanobacterium Lyngbya majuscula. J Nat Prod 62(9):1333–1335

Singh U, Singh P, Singh AK, Kumar D, Tilak R, Shrivastava SK, Asthana RK (2021) Identification of antifungal and antibacterial biomolecules from a cyanobacterium, Arthrospira platensis. Algal Res 54:102215

Srivastava A, Singh VK, Patnaik S, Tripathi J, Singh P, Nath G, Asthana RK (2017) Antimicrobial assay and genetic screening of selected freshwater Cyanobacteria and identification of a biomolecule dihydro-2H-pyran-2-one derivative. J Appl Microbiol 122(4):881–892

Swain SS, Paidesetty SK, Padhy RN (2017) Antibacterial activity, computational analysis and host toxicity study of thymol-sulfonamide conjugates. Biomed Pharmacother 88:181–193

Tae H, Kong EB, Park K (2007) ASMPKS: an analysis system for modular polyketide synthases. BMC Bioinform 8(1):1–9

Thong WL, Zhang Y, Zhuo Y, Robins KJ, Fyans JK, Herbert AJ et al (2021) Gene editing enables rapid engineering of complex antibiotic assembly lines. Nat Commun 12(1):1–10

Tillett D, Dittmann E, Erhard M, Von Döhren H, Börner T, Neilan BA (2000) Structural organization of microcystin biosynthesis in Microcystis aeruginosa PCC7806: an integrated peptide–polyketide synthetase system. Chem Biol 7(10):753–764

Verma S, Thapa S, Siddiqui N, Chakdar H (2022) Cyanobacterial secondary metabolites towards improved commercial significance through multiomics approaches. World J Microbiol Biotechnol 38(6):1–22

Vestola J, Shishido TK, Jokela J, Fewer DP, Aitio O, Permi P et al (2014) Hassallidins, antifungal glycolipopeptides, are widespread among cyanobacteria and are the end-product of a nonribosomal pathway. Proc Natl Acad Sci 111(18):E1909–E1917

Welker M, Von Döhren H (2006) Cyanobacterial peptides—nature's own combinatorial biosynthesis. FEMS Microbiol Rev 30(4):530–563

Xu X, Liu Y, Du G, Liu L (2020) Systems biology, synthetic biology, and metabolic engineering. In: Systems and synthetic metabolic engineering. Academic Press, London, pp 1–31

Yadav G, Gokhale RS, Mohanty D (2003) Computational approach for prediction of domain organization and substrate specificity of modular polyketide synthases. J Mol Biol 328(2): 335–363

Ziemert N, Podell S, Penn K, Badger JH, Allen E, Jensen PR (2012) The natural product domain seeker NaPDoS: a phylogeny based bioinformatic tool to classify secondary metabolite gene diversity. PLoS One 7(3):e34064

Cyanobacteria as Bioindicator of Water Pollution

9

Shivam Yadav, Amit Kumar Singh, and Ekta Verma

Abstract

Water pollution is a global problem in recent times, and its remedy has become imperative to safeguard the health of ecosystems and humans. To solve the water pollution problem, detecting the presence of various pollutants in water is the first critical step. Although many physical and chemical techniques have been developed to date to measure the concentrations of different types of contaminants in the water, the biological methods are the most feasible due to their easy operational process and cost-effectiveness. The cyanobacteria, a unique group of photoautotrophs known to occupy various aquatic environments, are the potent candidates for detecting the presence of different types of pollutants in water. Due to their common occurrence in water bodies, toxicological responses of cyanobacteria have widely been investigated, which suggests that toxicological research on cyanobacteria may help us evaluate the levels of various contaminants in water bodies because they respond to different pollutants in distinct manners, either by modulating their morphological, ultrastructural, and physiological indices or by changing their community composition and altering molecular biological phenomena. However, water pollution is becoming very complex daily because new pollutants are being produced and discharged daily. Therefore, the toxicological responses of cyanobacteria must be assessed against both existing and emerging contaminants to be used as bioindicators for both kinds of pollutants. In the present chapter, we initially provided glimpses of the status of global water pollution and the types of major water pollutants, followed by discussing bioindicators' advantages over other traditional methods in

S. Yadav
Department of Botany, T.P.S. College, Patna, Bihar, India

A. K. Singh · E. Verma (✉)
Post Graduate Department of Botany, Magadh University, Bodh-Gaya, India

B. Neilan et al. (eds.), *Cyanobacterial Biotechnology in the 21st Century*,
https://doi.org/10.1007/978-981-99-0181-4_9

determining water pollution. Besides, the chapter also included information on the potential of cyanobacteria as bioindicators of water pollution and detailed descriptions of the cyanobacterial bioindicator characteristics for specific pollutants. Eventually, the chapter suggested that investigating the toxicological impacts of emerging contaminants like pharmaceuticals and aromatic pollutants on cyanobacteria and the selection of unique and distinct bioindicator characteristics will be helpful to better predict the presence of these pollutants in aquatic systems.

Keywords

Water pollution · Cyanobacteria · Bioindicator · Pharmaceuticals · Aromatic pollutants · Aquatic systems

9.1 Introduction

The continuously increasing magnitude of environmental distortion and the advent of new pollutants not only raised concern among the environmental scientists but also necessitated the designing of monitoring programs to identify and manage the impacts of these factors on the environment (Cooper et al. 2009). The physical and chemical methods directly measure various environmental quality parameters at a particular point and provide information related to environmental health. However, hampered or low sampling frequency may result in the loss of important information. In addition, data obtained from physical and chemical analysis sometimes give a very narrow overview of the environmental conditions, hence are not reliable for addressing the issues related to environmental health. As a result, due to external factors, bioindicators and biomonitoring have gained huge attention for analyzing and recording chemical and physical alterations in any environment over time and space (Holt and Miller 2010). While the living organisms like microbes, cyanobacteria, algae, plants, and animals that respond to environmental changes are the representatives of bioindicators, biological monitoring comprises biomarkers (measurable parameters of biological entities) and biosensors (signal produced in response to the concentration of a specific substance) (Markert 2007; Britstein et al. 2016). Since bioindication assesses the collective impacts of alterations in pollutants and habitat, it has advantages over chemical-physical methods, biomarkers, and biosensors.

Furthermore, it requires low cost and less time and gives detailed and reliable descriptions of a specific environment (Girotti et al. 2008). Bioindicators are those organisms which exhibit a sharp response to environmental stress. The bioindicator organisms are generally ubiquitous and can be sampled and identified easily. Moreover, the long life cycle displayed by them and harmless nature are the additional advantages.

The abovementioned standards suggest that cyanobacteria, an ancient group of photosynthetic microbes, would be the efficient indicators of water pollution.

Cyanobacteria have a huge range of physiological adaptations for colonizing various ecosystems. Cyanobacteria are known to form mats on the water's surface (Teta et al. 2015) and are supposed to help evaluate water grade due to their nutrient demands, faster rate of reproduction, and life cycle. However, the suitability of cyanobacteria for analyzing the quality of water in different aquatic systems has long been a matter of discussion (Soltani et al. 2012). Although the unicellular forms like *Synechococcus* spp. and *Prochlorococcus* spp. show profuse growth in open ocean, tropical, and subtropical oligotrophic aquatic environments, the occurrence of other cyanobacterial species with a high relative abundance has been found in response to eutrophication. This eutrophication of marine and terrestrial aquatic ecosystems results from excess input of nutrients from municipal wastewater discharges, agricultural runoff (WHO 1999), and release of organic matters and fecal wastes into the water systems (Loza et al. 2013a, b, c). The ecological distribution of cyanobacteria in eutrophicated aquatic systems may vary depending upon their physiological tolerances (Mateo et al. 2015). Cyanobacterial eutrophication in different aquatic systems leads to a large-scale fish mortality due to oxygen depletion in surface waters.

Moreover, cyanobacterial blooms in eutrophicated water also produce harmful secondary metabolites that cause fish mortality and increase public health risks. Hence, the occurrence and composition of cyanobacterial bloom serve as outstanding indicators of nutrient pollution or water eutrophication (Richardson et al. 2014). The cyanobacterial blooms can thoroughly be studied either through the manual sampling of environmental sampling and laboratory-based experimentation or by applying the satellite remote sensing technique. Satellite remote sensing has emerged as an outstanding technique to assess the cyanobacterial bloom formation followed by its progression and related environmental parameters because of its impeccable efficiency in monitoring spatial and temporal distribution of cyanobacterial blooms (Cracknell et al. 2001). Since different satellites contain different sensors to capture a range of wavelengths, the amount of light reflected from the ocean's surface can be measured with the help of this technique (Teta et al. 2017). Furthermore, it has been reported that the measurement of cyanobacteria's photosynthetic pigments (chlorophyll, carotenoid, phycobiliproteins) through satellite imaging is very useful in monitoring cyanobacterial blooms (Mishra and Mishra 2014; Teta et al. 2017). Moreover, phycocyanin (PC), due to its distinctive reflectance profile, provides added advantages over chlorophyll and carotenoids to distinguish cyanobacteria from other algal species and accurately decipher their roles as water pollution bioindicators (Teta et al. 2017).

To provide the maximum information about water pollution and the usefulness of cyanobacteria as a bioindicator of water pollution, three different segments have been incorporated in this book chapter. The first section gives glimpses of water pollution, bioindicators, and advantages of bioindicators over the traditional methods in determining water pollution, whereas the second section elaborately describes the bioindicator characteristics of cyanobacteria. Finally, in the third section of the chapter, the techniques used to study the bioindicator characteristics of cyanobacteria are discussed.

9.2 Ecosystem Pollution

Ecosystem pollution is defined as the introduction of some specific substances into an ecosystem which generally exerts various harmful effects on that ecosystem by degrading its health, structure, and biota. The major reason behind ecosystem pollution is uncontrolled human or anthropogenic activities. Nearly all the important components of ecosystems, such as air, water, and soil, are deteriorating because of this continuously increasing pollution problem. This has not only imposed a serious constraint on the existence of various biological communities but also raised the question of the human race's survival (Butterworth et al. 2001; Holt and Miller 2010). The classification of pollutants is generally done based on their physical state and chemical composition and sometimes depending upon their solubility, reactivity, and biodegradability. Additionally, pollutants are often classified based on their sources (fuel combustion, industry, or domestic) and the sectors (air and water) they pollute. In the present chapter, our concern is water pollution; hence, further discussions are dedicated to this topic.

9.3 Water Pollution

Water, being the source of life and an important natural resource, is essentially required by almost all groups of organisms. However, most of the water resources (fresh, brackish, and sea) are becoming unusable because of anthropogenic activities, called water pollution (https://www.hsph.harvard.edu/ehep/82-2). The deterioration of freshwater resources is a common environmental problem in most countries worldwide. Human activities or activities in the manufacturing sectors of different industries are regarded as the major reason for freshwater pollution (Fig. 9.1). Since removing pollutants from polluted water bodies is tough and expensive, a load of contaminants is continuously increasing in the freshwater resources. Despite this growing problem of freshwater pollution, 80% of untreated

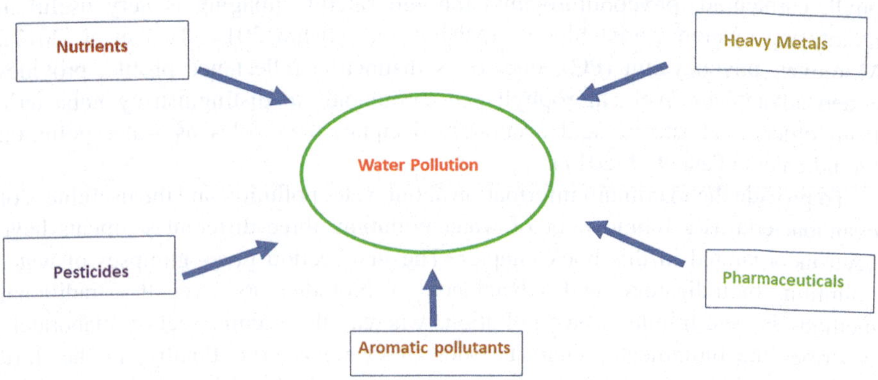

Fig. 9.1 The major water pollutants

global wastewater comprising human waste and toxic industrial effluents is discharged into water bodies and makes water unusable for humans (https://www.unep.org/explore-topics/water/what-we-do/tackling-global-water-pollution). Substances that cause pollution in freshwater ecosystems generally include pathogens (from septic tanks, untreated sewage), organic matter (mainly nitrogen or phosphorus from agricultural runoff), pesticides, heavy metals, detergents, and salts (Kumar Reddy and Lee 2012). Apart from the above pollutants, aromatic pollutants and pharmaceuticals have also emerged as new contaminants in waterways. However, the exact impacts of these substances on freshwater remain unknown. Therefore, identifying pollutants' nature and their impacts and developing suitable monitoring systems are essential for managing water bodies.

The traditional methods for assessing water pollution rely on the measurement of physicochemical properties, but this is not much convenient for managing water bodies at a large scale. Biological methods have gained a significant attention from scientific communities because the final point of pollution in any ecosystem is represented by the extent of well-being of biotic communities in that system (Norris and Thoms 1999). Hence, functional and compositional changes in biodiversity in response to the presence of pollutants are the best indicators of water pollution.

9.4 Biological Indicators and Their Advantages over Traditional Methods in Determining Water Pollution

The biological communities in any ecosystem depict the characteristics of that ecosystem (Gaston 2000; Joanna 2006), and organisms that quickly respond to minute alterations in the ecosystems are generally referred to as biological indicators (Gaston 2000; Marques 2001). The term "biological indicators" is often used broadly to describe nearly all biotic and abiotic reactions associated with the alterations in a specific ecosystem. Among the 1.7 million species documented on earth, no single species has yet been found that satisfactorily indicates all types of changes in every ecosystem (Holt and Miller 2010). Therefore, utilization of already established organisms or biological communities and localized changes in a specific ecosystem appear to be the best approach to select efficient biological indicators. It is considered that all components of a biological system may serve as biological indicators for its native ecosystem. Biological indicators are often used to determine the synergetic and antagonistic effects of different pollutants on biota. The sensitivity range of a biological indicator provides information about pollution rates in each ecosystem.

The traditional methods of detecting water pollution generally rely on assessing the environment's physical and chemical parameters (such as nutrients, pollutants, salinity, temperature, light availability, and levels of gases) (Mateo et al. 2015). In contrast, bioindicator-based analyses measure the combined effects of pollutants and changes in habitats on biota. Bioindicators offer various advantages over chemical and physical methods because their utilization differs entirely from the earlier ones. In addition, to the monitoring of biota, bioindicators provide information about the present, past, or future conditions of a specific ecosystem (Teta et al. 2017). On the contrary, measurements of chemical and physical parameters only provide insight into conditions prevailing during sampling resulting in the loss of sporadic pulses of pollutants.

Furthermore, since pollutants may occur at low concentrations, their detection requires employing costly compassionate technologies. Only after successfully measuring such low concentrations of contaminants that researchers would be able to connect the biological responses with these pollutants, and these connections may be vague. However, the extents of tolerance of bioindicators, irrespective of the pollutant concentrations, provide information about the pollutant levels that are meaningful from the biological viewpoint. Apart from this, bioindicators have another important advantage over physical and chemical methods because of their capability to indicate the indirect impacts of pollutants on biota (Zaghloul et al. 2020). As a result, various organizations worldwide, e.g., the International Union for Conservation of Nature and the World Conservation Union, are recommending the use of biological indicators for handling the process of biomonitoring in polluted ecosystems.

9.5 Bioindicator Selection Criteria

The prompt and accurate response to pollution and efficiency in diagnosing ecosystem changes caused by inappropriate management, incorrect land-use patterns, pollution, and climatic modulations in a specific ecosystem are critical for selecting a potent bioindicator (Jain et al. 2010). In addition, the bioindicator must have a causal relationship to a significant ecological endpoint and specific dose responsiveness to a particular pollutant. Apart from this, bioindicators should have a wide temporal and spatial distribution and be available throughout the year. Furthermore, the low variability to noise in the system and the ability to give transparent and reproducible results are other two essential criteria that bioindicators must fulfil. Besides, accessible collection and inexpensive nature are also crucial for selecting bioindicators.

9.6 Bioindicator Types

Bioindicators may be categorized under two sections, i.e., accumulation bioindicators, which accumulate pollutants without changing their metabolic activities, and response bioindicators, which exhibit changes in cellular properties or symptoms of cellular damages upon exposure to even minute concentration of toxic pollutants (Jain et al. 2010). While bioindicators are used experimentally, responses that may be observed include ecological changes, behavioral changes, and physiological changes. Ecological changes may involve modulations in species diversity, key species, and population density, whereas behavioral changes may alter bacterial mobility and feeding activities. Moreover, the physiological changes may include heavy metal accumulation, BOD, production of CO_2, and microbial activity. To sense or measure the extent of water pollution, organisms, predominantly plants, animals, and microorganisms, are frequently used as bioindicators.

9.6.1 Plant Indicators

Plants are considered very sensitive tools for detecting environmental pollution among various bioindicators. Marine plants may be used as bioindicators to detect pollution in the oceanic systems. This is because marine plants are capable of obtaining equilibrium rapidly with their natural surroundings and providing ample information about the changes in this system (Plafkin et al. 1989; Klemm 1990; Kennish 1992; Phillips and Rainbow 1993; De Lange 1994; Jain et al. 2010). Alterations in the diversity of phytoplankton such as *Euglena clastica*, *Phacus tortus*, and *Trachelomonas* are generally used as an indicator of pollution in the marine ecosystems (Plafkin et al. 1989; Phillips and Rainbow 1993; Jain et al. 2010). Another example of plant indicator is lichens, which are composed of algae and fungi and grow on tree barks and rocks. Lichens are highly sensitive to ecological changes in the forests. They specifically respond to changes in the forest structure, air quality of the forest ecosystems, and climate change. As lichens are sensitive to elevated levels of sulfur dioxide (SO_2) and nitrogen (N_2), the increased level of sulfur dioxide (SO_2) and nitrogen (N_2) can be indicated by the disappearance of lichen in forests (Walsh 1978; Gerhardt 2002; Holt and Miller 2010; Khatri and Tyagi 2015). Besides, *Wolffia globosa*, which exhibits high cadmium sensitivity, is essential for indicating cadmium contamination.

9.6.2 Animal Indicators

Like plant indicators, animals are also used to detect environmental pollution. Their altered population may indicate the presence of harmful pollutants. To measure the number of toxins in the animal tissues, animal indicators may be used (Joanna 2006; Khatri and Tyagi 2015). Among the animal indicators, frogs are known to be the impeccable indicators of environmental quality both in the aquatic and terrestrial ecosystems. Moreover, other animals like zooplankton, for example, *Alona guttata*, *Mesocyclops edax*, *Cyclops*, and *Aheyella*, are considered pollution indicators of specific zones in the environment (Underwood and Shapiro 1999; Hans et al. 2003; Ramchandra et al. 2006; Pradhan et al. 2008; Zannatul and Muktadir 2009; Jain et al. 2010; Hosmani 2014). Aquatic invertebrates that generally thrive in the bottom regions of water bodies may be powerful indicators of watershed health.

9.6.3 Microbial Indicators

Apart from plants and animals, microbes are another important group of bioindicators. Microbes are frequently used to indicate the health of aquatic and terrestrial ecosystems. The easy handling and the high availability of microbes are the major reasons behind their usage as bioindicators. Few microbes synthesize specific novel proteins when exposed to pollutants like cadmium and benzene (Khatri and Tyagi 2015). This property may be used to detect cadmium and benzene pollution. Since microorganisms constitute a significant part of oceanic biomass and respond to even significantly fewer amounts of contaminants and small alterations in physicochemical and biological entities, they can be used to detect pollution in the marine systems (Underwood and Shapiro 1999; Gerhardt 2002; Hans et al. 2003; Markert et al. 2003; Ramchandra et al. 2006; Pradhan et al. 2008; Zannatul and Muktadir 2009; Nkwoji et al. 2010; Hosmani 2014). Microbes may also serve as a powerful tool for indicating heavy metal pollution in aquatic systems. The bacterium *Vogesella indigofera*, which quantitatively responds to heavy metal pollution, is the most exceptional example of microbial indicator. It has been shown that this bacterium produces blue pigmentation without metal pollutants. However, with the availability of hexavalent chromium, the production of blue pigment is blocked in the bacterium (Oberholster et al. 2009; Jain et al. 2010; Malik and Bharti 2012). Therefore, this bioluminescence property of this bacterium may be exploited to find out chromium pollution in water.

Cyanobacteria, considered the intermediate of plant and microbes, are an interesting group of organisms that also rapidly respond to minute changes in their surrounding environments. Moreover, these organisms are the autotrophic and common entity of water systems. Therefore, they are likely to be the best suited ones for being used as bioindicators of water pollution.

9.7 Cyanobacteria as Bioindicators of Water Pollution

Despite being a problematic entity of eutrophicated water, the presence of cyanobacteria cannot always be connected to ecological deterioration. In contrast, few cyanobacterial species that strictly thrive in clean water bodies are emerging as tools for analyzing the grade of flowing or stagnant waters (Mateo et al. 2015). In addition, different cyanobacterial species exhibit specific responses to nitrogen and phosphorus, and these responses are often corroborated with the modulations in the dominance of species in eutrophicated water systems. Cyanobacteria display ubiquitous distribution, and the underlying reasons for such distribution are general and a few unique characteristics. Many species possess a natural capacity to tolerate multiple environmental constraints (Castenholz 2001). The ability of cyanobacteria to tolerate stressful conditions may vary with the genus and species. Few cyanobacterial species exhibit occurrences in a broad range of niches, whereas many are found in restricted niches (Koma'rek and Anagnostidis 1999). For example, some cyanobacterial species thrive only in unpolluted waters, while others inhabit highly polluted water systems (Fjerdingstad 1964; Palmer 1969; Fogg et al. 1973). The saprobic system was the first one which used cyanobacteria to indicate water quality (Kolkwitz and Marsson 1908). The attempt made utilizing this system suggested that the composition of the cyanobacterial community can be governed by water quality. Afterwards, various cyanobacteria-based methods were developed for monitoring water quality. Many nations around the world frequently use cyanobacteria for regular monitoring of water quality, including Poland, Austria, the Czech Republic, and Germany (Kelly 2013; Whitton 2013). Cyanobacteria are often combined with diatoms and benthic algae in Austria for water quality assessment purposes (Rott and Schneider 2014). In the last two decades, the research group of Barinova analyzed the cyanobacterial community present in Georgia, Israel, and Russia for selecting suitable species or members that can efficiently serve as bioindicators (Barinova et al. 2006, 2008, 2011; Barinova and Tavassi 2009). Moreover, the rigorous review of various literature disclosed that a few species of the cyanobacterial genus *Oscillatoria*, such as *O. limosa*, *O. princeps*, and *O. tenuis*, are potent indicators of eutrophicated water bodies (Palmer 1969; Walley et al. 2001; Gutowski et al. 2004; Sierra and Go'mez 2007). Apart from this, the associations of *Leptolyngbya foveolarum*, *Phormidium terebriforme*, and *P. ambiguum* with water systems containing high concentrations of growth-promoting nutrients have also been demonstrated by many workers (Gutowski et al. 2004; Gutowski and Foerster 2009; Stancheva et al. 2012; Kolkwitz and Marsson 1908; Fjerdingstad 1964; Sla'decek 1973; Loza et al. 2013b; Gutowski and Foerster 2009). Therefore, the above reports suggest that using cyanobacteria to detect water pollution is a safe and scientifically sound approach.

9.7.1 Cyanobacteria as Indicators of Organic Water Pollution

Cyanobacteria exhibit a wide range of variations in their morphology and physiology with respect to different genus and species. These morphophysiological characteristics of cyanobacteria, especially those which are noticed in multiple genera or species, may serve as potent rapporteurs of waters' nutrient status.

9.7.1.1 Morphological Characteristics

Heterocytes

A considerable number of filamentous cyanobacteria are frequently observed to form specialized nitrogen-fixing cells called heterocysts or heterocytes. These specialized cells differentiate from normal vegetative cells as a result of nitrogen deficiency and are larger in size than other vegetative cells. The heterocytes fix atmospheric dinitrogen gas into ammonia with the help of the multisubunit enzyme nitrogenase and allocate the fixed nitrogen to vegetative cells. In return, the vegetative cells provide fixed carbon and reductants to the heterocytes. Based on this theory, it can easily be said that the development of heterocyst and the expression of nitrogenase could be useful for measuring the level of nitrogenous nutrients in water. Horne and

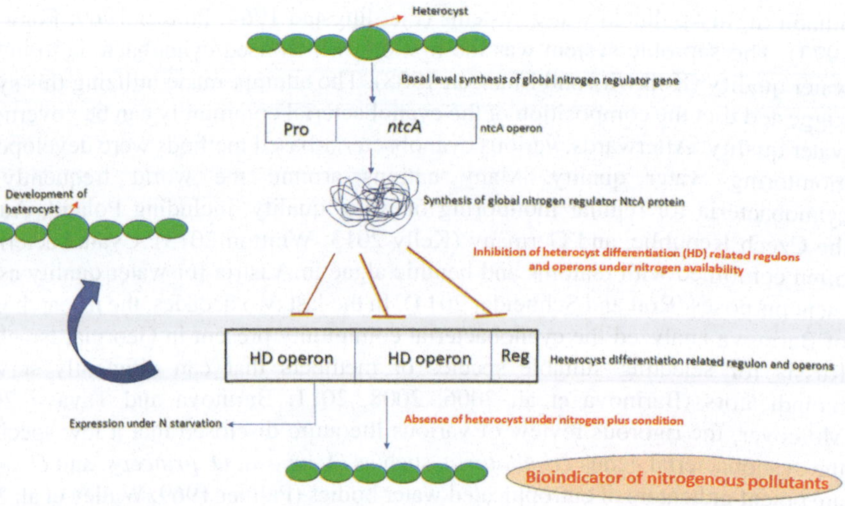

Fig. 9.2 Inhibition of heterocyst development as bioindicator of nitrogenous pollutants

Carnmiggelt (1975) and Grimm and Petrone (1997) experimentally showed a high nitrogen fixation rate in the streams located in the desert of northern California and Arizona due to the low levels of nitrate. Many years after these two reports, Stancheva et al. (2013) made a similar kind of observations in cyanobacteria. They analyzed the expression level of nitrogenase gene in two cyanobacterial genera, i.e., *Nostoc* and *Calothrix*, along a gradient of nitrate and were able to measure detectable amount of nitrogenase expression only at a low nitrate concentration. It is evident from various experimental studies that a high amount of nitrate and ammonium inhibits the differentiation of heterocysts from vegetative cells. Therefore, the absence of heterocytes in heterocytous cyanobacteria is a bioindicator of surplus nitrogen or nitrogen pollution (Fig. 9.2).

Polyphosphate Granules and Calyptra

Phosphorus is an essential macronutrient for cyanobacteria as this element is required for synthesizing nucleic acids and energy currencies like ATP and NADPH. Cyanobacteria take phosphate from their surrounding medium and store it inside the cells as polyphosphate granules; when phosphorus is available in the surplus amount in the surrounding environment, the abundance of polyphosphate bodies increases inside the cyanobacterial cells. Therefore, this accumulation of polyphosphate bodies inside the cells indicates P richness or pollution (Whitton and Mateo 2012; Muñoz-Martín et al. 2014) (Fig. 9.3). Apart from this, a few cyanobacterial strains (e.g., *Phormidium*) are also known to develop thickened caps, termed calyptra, on the outer wall of apical cells at high concentrations of phosphate (Whitton and Neal 2011; Whitton and Potts 2000; Muñoz-Martín et al. 2014). Thus, the accumulation of polyphosphate bodies and the development of

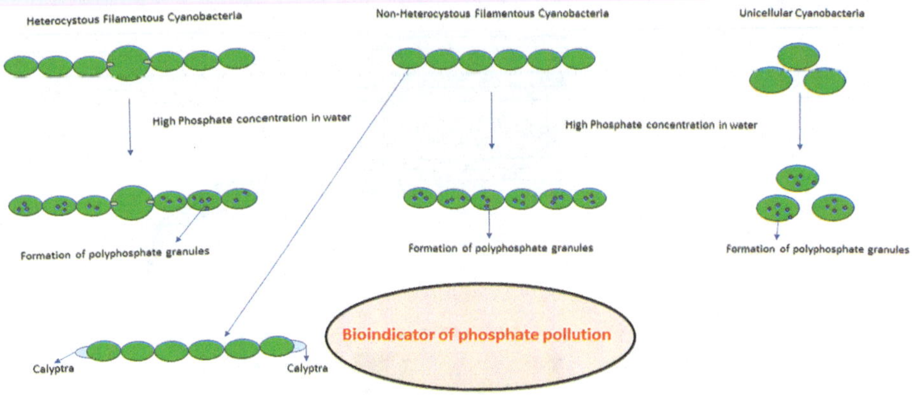

Fig. 9.3 Intracellular deposition of polyphosphate granules as bioindicator of phosphate pollution pollutants

calyptra in cyanobacteria can be used to detect the level of phosphorus in polluted water systems.

9.7.1.2 Physiological Characteristics

Chlorophyll-*a*

Chlorophyll-*a*, the major photosynthetic pigment of cyanobacteria, serves as a tool for evaluating the cyanobacterial responses at the community level to nutrient richness. With the availability of high amounts of growth-promoting nutrients, cyanobacteria increase their biomass, and along with that biomass chlorophyll-*a* content proportionately increases. Therefore, the value of chlorophyll-*a* content alone can indicate the growth responses of cyanobacterial communities in a water system polluted with excessive amounts of nutrients, especially N and P.

Phycobiliproteins

Besides chlorophyll-*a*, another pigment called phycobilin may also be used as a water pollution indicator property of cyanobacteria. Since this pigment is highly proteinaceous, its synthesis mainly depends on nitrogen availability. Therefore, a higher content of phycobilins may serve as an indicator of nitrogen status in water.

Both chlorophyll-*a* and phycobiliproteins have recently been used by Teta et al. (2017) to monitor the water quality status of Campania coasts, Italy. Hence, these two parameters appear to be potent bioindicators of water eutrophication (Fig. 9.4).

9.7.1.3 Altered Cyanobacterial Community Structure as Water Eutrophication Indicator

Cyanobacteria, based on their morphology and genomic constructions, are classified into various groups. The cyanobacterial members of an individual group share nearly similar morphology and basic genome structure. Many individuals of a single group form a community, and a single community displays distinct responses to their

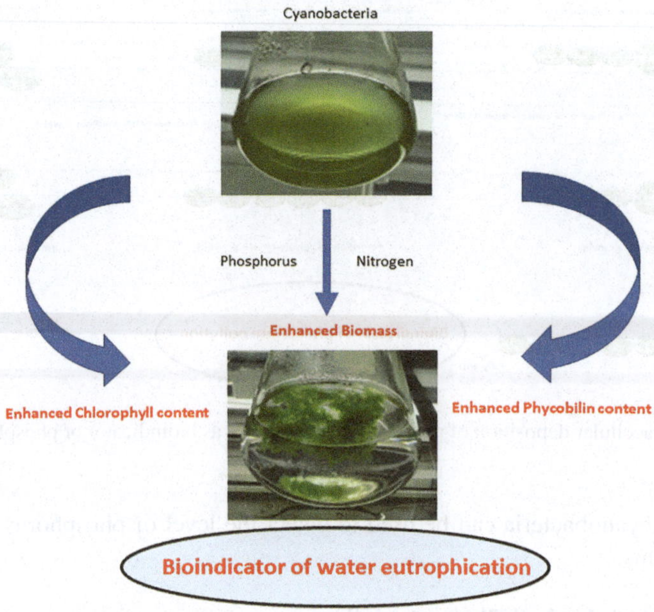

Fig. 9.4 Enhanced biomass, chlorophyll content, and phycobilin content of cyanobacteria as bioindicators of water eutrophication

surrounding environments. The occurrence of different cyanobacterial communities in different aquatic systems represents the nutrient status of those environments. For an instance, heterocystous cyanobacteria are generally found in the water bodies containing low levels of nitrogenous nutrients, whereas non-heterocystous communities strictly inhabit high-nitrogen-containing water. A considerable number of studies have been performed by several workers in many rivers of Spain (Perona et al. 1998; Douterelo et al. 2004; Rodriguez et al. 2007; Loza et al. 2013a, b, c, 2014). In these studies, they analyzed the diversity, abundance, and richness of cyanobacterial communities along the eutrophic gradients. The results of these studies reflected that the communities present in the less eutrophicated upstream regions of the rivers were completely different from the communities present in the highly eutrophicated downstream regions. Changes in the community structure were also observed by them with the alterations in water quality because many cyanobacterial species that were found to be absent in the river water of disturbed zones showed profuse growth in undisturbed zones (Douterelo et al. 2004; Perona and Mateo 2006; Loza et al. 2013a, b, c, 2014). In the upstream sites of the rivers where the water was less eutrophicated, the presence of heterocystous cyanobacterial communities especially *Nostocales* was noticed, whereas the highly eutrophicated downstream sites of the rivers were observed to be occupied by mainly non-heterocystous communities especially *Oscillatoriales* (Perona et al. 1998; Douterelo et al. 2004; Perona and Mateo 2006; Rodriguez et al. 2007). Branco and Pereira (2002) who performed the same study in Brazilian rivers obtained similar outcomes. They proposed that the abundance ratio of non-heterocytous and heterocytous cyanobacterial communities is an exceptional tool for determining the level of nitrogen pollution in tropical aquatic systems. However, the studies performed by Cantonati (Cantonati et al. 1996, 2006; Cantonati 2008) have demonstrated that the concentration of nitrate nitrogen decides the abundance of heterocytous and non-heterocytous communities of cyanobacteria and have shown the prevalence of non-heterocytous communities in the nitrate-enriched regions of the Southern Alps and the abundance of heterocytous cyanobacterial communities in the low-nitrate-containing flowing water of a national park in Northern Italy. Based on the negative correlation of nitrogen levels and nitrogen-fixing cyanobacterial richness, researchers from North America have also mentioned the heterocytous cyanobacterial communities as indicators of nutrient status (Porter et al. 2008; Stancheva et al. 2012, 2013). Therefore, the above studies affirm that the community structure of cyanobacteria is an outstanding indicator of water eutrophication.

9.7.2 Cyanobacteria as Indicators of Heavy Metal Pollution in Water

9.7.2.1 The Morphophysiological and Ultrastructural Features of Cyanobacteria as Bioindicator of Heavy Metal Pollution

In the last few decades, a large number of reports related to the responses of cyanobacteria to heavy metal stress have been accumulated. These reports suggested that cyanobacteria can respond to even minute alterations in the concentrations of heavy metals. Since cyanobacteria are frequently found to thrive in aquatic environments and exhibit morphological, ultrastructural, and physiological modifications in response to heavy metal exposure, these responses may be considered indicators of heavy metal pollution in water. Various species from the genera *Anabaena*, *Nostoc*, *Fischerella*, *Spirulina*, *Calothrix*, *Synechococcus*, *Synechocystis*, *Oscillatoria*, etc. have been shown to modulate their morphophysiological properties in response to heavy metals like cadmium, copper, zinc, nickel, mercury, lead, and many more (Fernandez-Piñas et al. 1995; Heng et al. 2004; Surosz and Palinska 2004; Anjana et al. 2007; Ghorbani et al. 2022). The most prominent changes in the morphology and ultrastructure are reduced cell size, spherical or elongated cell shape, demolished exopolysaccharide sheath, distortion or degradation of thylakoid membranes, and dissolution of phycobilisomes, cyanophycin granules, polyphosphate bodies, etc. (Newby Jr. et al. 2017; Chakraborty and Mishra 2021).

Moreover, the physiological responses include reduced chlorophyll content and photosynthetic efficiency, production of free oxygen radicals, a decline in ATP and NADPH contents, etc. (Chakraborty and Mishra 2021). The schematic presentation of heavy metal-induced morphophysiological modulations in cyanobacteria has been shown in Figs. 9.5 and 9.6. Based on the above reports, it can be said that cyanobacteria would be outstanding bioindicators of heavy metal pollution in water.

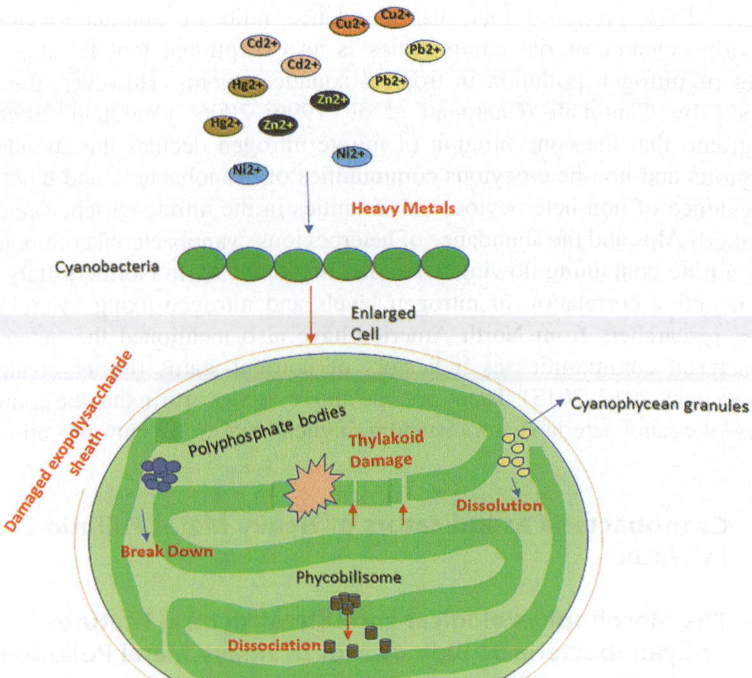

Fig. 9.5 Morphological and ultrastructural variations in cyanobacterial cell as bioindicators of heavy metal pollution

9.7.2.2 Altered Bioenergetics in Photosynthesis as Bioindicator of Heavy Metal Pollution

In photoautotrophic organisms including cyanobacteria, photosynthesis is one of the most important bioenergetics processes, in which the energy stored in the photons of sunlight is captured by photosynthetic pigments and used to convert atmospheric carbon dioxide gas into carbohydrates. In oxygenic photosynthesis, commonly observed in cyanobacteria, algae, and higher plants, two photosystems, i.e., PSI and PSII, located on the thylakoid membrane (grana of chloroplast in higher plants) coordinately perform the light reaction during which O_2 gas, NADPH, and ATP are produced. While the produced O_2 gas is liberated in the environment, ATP and NADPH serve as the energy and reducing power respectively to fix atmospheric CO_2 to form carbohydrates in the dark cycle (Garret and Grisham 2010). In this way, photosynthesis fulfils the energy demand of living organisms on the Earth by capturing energy stored in sun light. Of the total amount of trapped energy, a small portion is lost as heat and fluorescence emission, and the rest is channelized toward the carbohydrate synthesis pathway (Voet et al. 2012; Campbell et al. 1998). Although the electron transport chain of the photosynthetic apparatus requires the presence of a small amount of iron and copper, the high concentrations of these

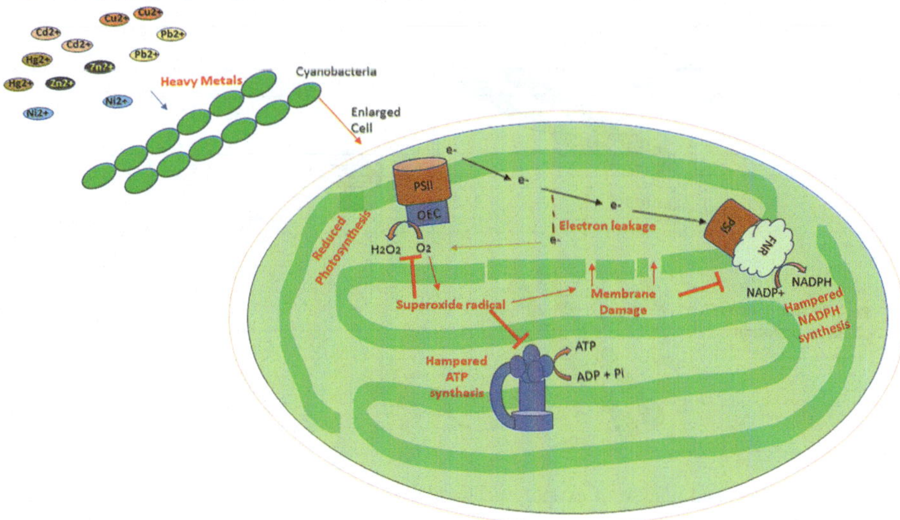

Fig. 9.6 Physiological modulations coupled with disturbed bioenergetics of photosynthesis as bioindicators of heavy metal pollution

metals and other toxic heavy metals dismantle the whole process of photosynthesis. The highly adverse impacts of heavy metals on the photosynthesis have been reported by many workers from different parts of the world (Clijsters and Assche 1985; Prasad and Strzałka 1999; Babu et al. 2010; Nagajyoti et al. 2010).

In cyanobacteria, the inhibited photosynthesis can easily be characterized by decreased oxygen evolution and triggered emission of fluorescence. Both of these altered physiological behaviors of cyanobacteria have been used by researchers to determine the environmental concentrations of heavy metals (Chay et al. 2005; Nguyen-Ngoc et al. 2009; Wong et al. 2008; Wong et al. 2012, 2013). These integrated biological indicators, generally referred to as biosensors, are the analytical tools combining biological component, transducer, and reader (Su et al. 2011). The biological components, which constitute both cells and biomolecules (deoxyribonucleic acids and amino acids), give feedback corresponding to specific target compounds, and the transducers translate that biological feedback to an electronic signal. This signal is eventually captured by the readers and displayed in the small screen monitor (Fig. 9.6).

9.7.2.3 Exopolysaccharide Production as Bioindicator of Heavy Metal Pollution

A few cyanobacterial species have been shown to produce exopolysaccharides in high concentrations of different metals. For example, while Chakraborty and Mishra (2020) have shown that the cyanobacterium *Anabaena oryzae* releases polysaccharides in the growth medium in response to 6 mg/L zinc (Zn), elevated production of capsular polysaccharides by the cyanobacterium *Nostoc*

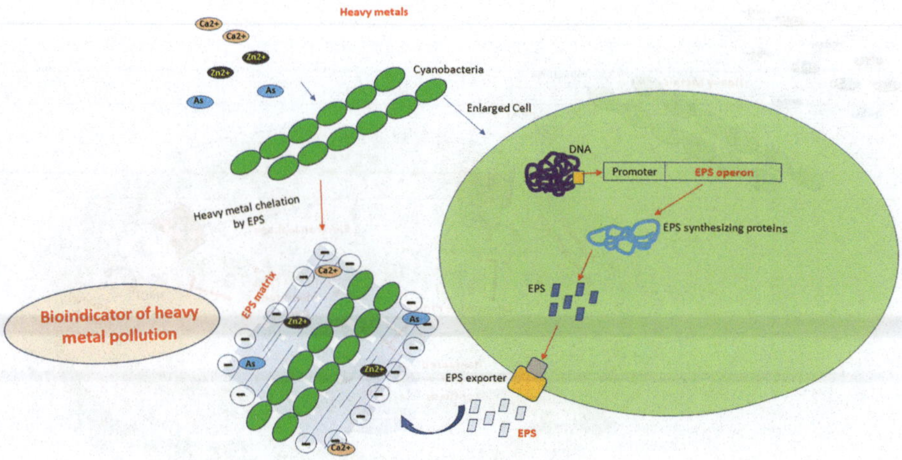

Fig. 9.7 Synthesis and release of exopolysaccharides by cyanobacteria as bioindicator of heavy metal pollution

ellipsosporum in response to 400 μM arsenic (As) has been experimentally demonstrated by Dutta and Bhadury (2020). Apart from this, a high concentration of calcium (Ca) has been found to induce exopolysaccharide production in *Anabaena* sp. PCC7120 (Singh et al. 2016). Hence, it can be stated that the abovementioned individual cyanobacterial strains and their exopolysaccharide-producing ability may be used to detect the presence of metals like Zn, As, and Ca (Fig. 9.7).

9.7.2.4 Metallothionein Synthesis as Bioindicator of Heavy Metal Pollution

In addition to the responses of cyanobacteria to metals mentioned earlier, the most common and prominent metal-responsive feature is the synthesis of cysteine-rich short peptides called metallothioneins (Turner and Robinson 1995). Many studies regarding the metal-induced synthesis of metallothioneins have so far been performed in various cyanobacteria. Most synthesize this typical low molecular weight protein as a defense strategy. The cysteine residues in this protein help cyanobacteria chelate excess metal ions at the intracellular level (Kojima 1991). Metallothionein synthesis has been observed in cadmium- and zinc-exposed *Synechococcus* PCC 7942 (Olafson et al. 1988; Takatera and Watanabe 1992).

Moreover, the resistance to Cd in this organism has been correlated with the increased production of metallothionein. Additionally, the induced expression of metallothionein has also been noticed in the cyanobacterium *Anabaena* PCC7120 under Cd, Zn, and Cu stress (Divya et al. 2018). Recently, the increased levels of metallothionein content have been experimentally shown in zinc-exposed *Anabaena doliolum* and *Anabaena oryzae* (Chakraborty and Mishra 2020). Thus, quantitative or expressional analysis of metallothionein in cyanobacteria growing in polluted

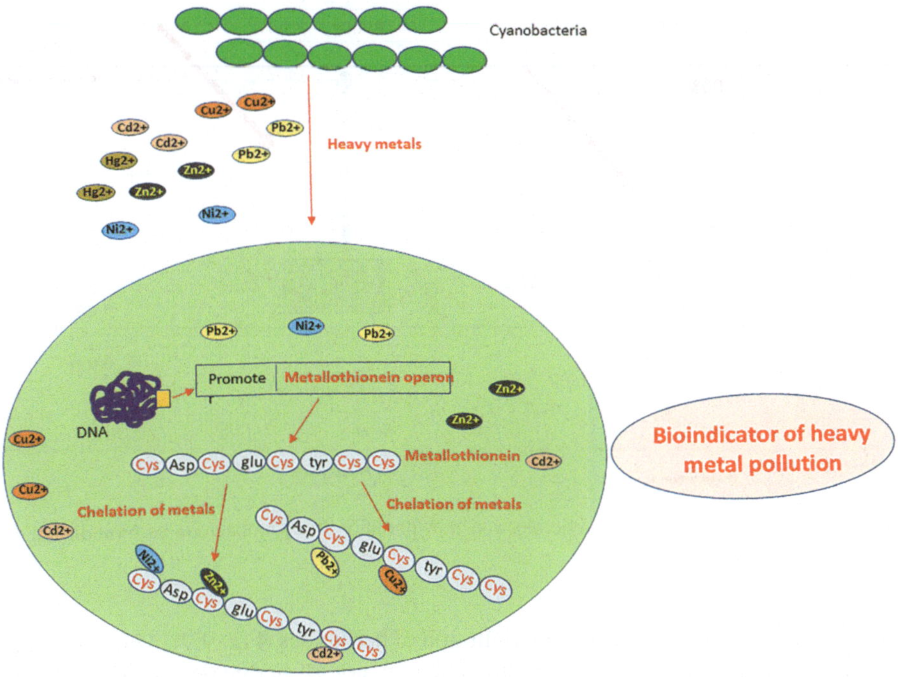

Fig. 9.8 Synthesis of metallothionein by cyanobacteria as bioindicator of heavy metal pollution

water could be useful for detecting contamination of heavy metals like Cd, Zn, and Cu (Fig. 9.8).

9.7.3 Cyanobacteria as Indicators of Pesticide Pollution

9.7.3.1 Inhibition of Photosystem II Activity in Cyanobacteria as Bioindicator of Herbicide Pollution

A large number of herbicides are generally used throughout the world to kill unwanted herbs which grow in agricultural fields and house gardens. However, these herbicides further enter the aquatic ecosystems through agrarian runoff water and exert toxic impacts on the nontarget aquatic organisms (Gupta 2018). Therefore, herbicides are one of the key pollutants of aquatic systems (Hollis et al. 2000), and their detection is of utmost importance. It is well known that one-half of the herbicides presently used in agriculture acts by inhibiting the light reactions in photosynthesis, mostly by targeting the photosystem II (PSII) complex (Draber et al. 1991). Besides the higher plants, the same pattern of photosynthetic inhibition is also observed in cyanobacteria. Since cyanobacteria are the environmentally important organisms in most aquatic systems, this phenomenon of photosynthetic inhibition of cyanobacteria in response to herbicides may be referred to as an

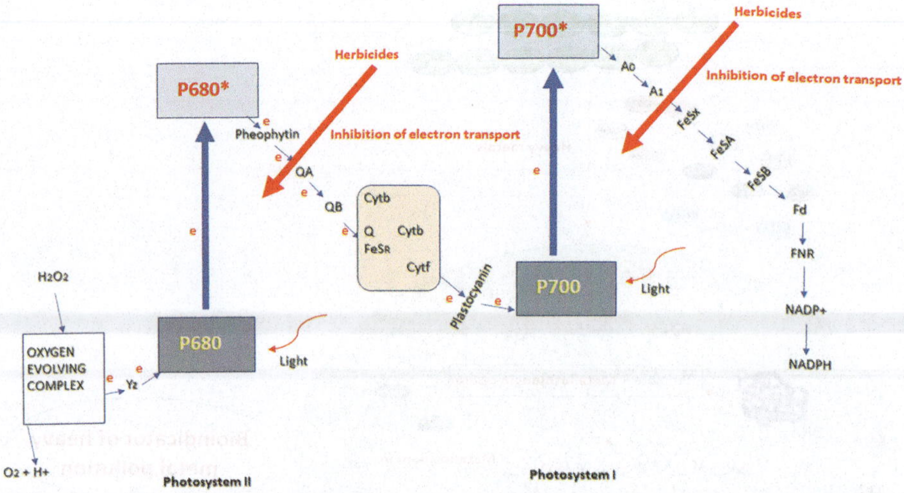

Fig. 9.9 Inhibition of photosynthetic electron transportation in cyanobacteria as bioindicator of herbicide pollution

indicator for monitoring the aquatic pollution of herbicides (Fig. 9.9). This referral is supported by the fact that by using this phenomenon of herbicide-mediated photosynthetic inhibition as a base, several whole-cell cyanobacterial biosensors have been developed for detecting herbicides which inhibit the photosynthetic electron transportation process (Arsalane et al. 1993; Brewster et al. 1995; Jay et al. 1997; Frense et al. 1998; Campanella et al. 2000).

9.7.3.2 Inhibition of Nitrogen Fixation as an Indicator of Herbicides Butachlor and Benthiocarb

Butachlor and benthiocarb are very commonly used herbicides in agricultural fields and are found to contaminate various ecosystems, including the aquatic ones. The impacts of these herbicides have been assessed in many cyanobacterial strains; while few strains have shown increased growth, others have displayed inhibited growth. Recently, in a study performed by Dash et al. (2017), the comparative effects of insecticides (quinalphos, carbofuran, and methyl parathion) and herbicides (benthiocarb, butachlor, and 2,4-D) have been evaluated in *Aphanothece*, *Anabaena*, *Aulosira*, *Nostoc*, and *Gloeotrichia* in terms of their growth, nitrogen fixation efficiency (or acetylene reduction activity (ARA)), and yield of nitrogen. This study has clearly shown that the pesticides carbofuran and methyl parathion and the herbicide 2,4-D promoted the growth and acetylene reduction ability of all the tested strains. Still, the pesticide quinalphos and other two herbicides, i.e., butachlor and benthiocarb, reduced the values of these measured parameters. In another study, Kaushik et al. (2018, 2019) demonstrated the unaltered growth but reduced acetylene reduction ability of the cyanobacterium *Nostoc linckia* in the presence of 0.5 ppm benthiocarb. Thus, apart from the inhibition of the photosynthetic electron

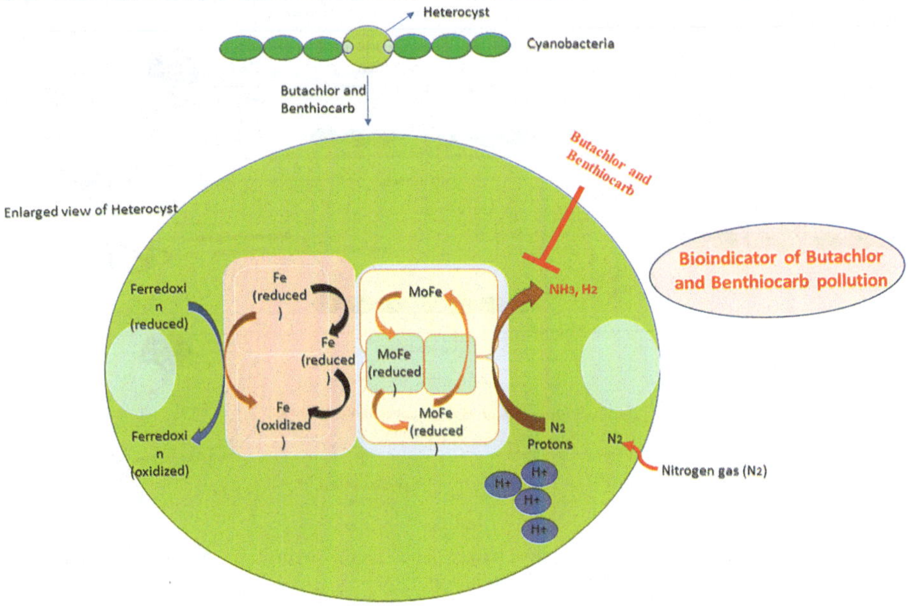

Fig. 9.10 Reduced nitrogen fixation efficiency in cyanobacteria as bioindicator of herbicides butachlor and benthiocarb

transport chain, reduction in nitrogen fixation efficiency or acetylene reduction ability could be used as an indicator for the detection of herbicides butachlor and benthiocarb (Fig. 9.10).

9.7.3.3 Cyanobacterial Alkaline Phosphatase Activity and Lectin Domain-Containing Hydrolase as Bioindicator of Organophosphorus Pesticide Methyl Parathion

Organophosphorus pesticides are another group of toxic environmental pollutants frequently observed in aquatic systems (Cho et al. 2002). These pesticides exert toxicity by inhibiting the enzyme acetylcholinesterase (AcHE) in animals (Elersek and Filipic 2011). However, the lower groups of organisms, especially bacteria, are comparatively less toxic because bacteria efficiently degrade them and utilize them as nutrients (Fulton et al. 1998). The degradation of organophosphorus pesticides in these organisms mainly occurs due to the activity of the enzyme organophosphorus hydrolase (Dumas et al. 1989, 1990). In cyanobacteria, organophosphorus hydrolase activity has not yet been reported, but the role of alkaline phosphatase has recently been shown by Tiwari et al. (2017a, b). In these studies, it has been found that the supplementation of methyl parathion in the inorganic phosphate-free BG11$_0$ growth medium resulted in the high alkaline phosphatase activity in two cyanobacterial genera, i.e., *Scytonema* sp. BHUS-5 and *Fischerella* LMG1, followed by elevated organophosphorus pesticide methyl parathion degradation. The same group has also shown that the cyanobacterium *Fischerella* LMG1, under phosphate-supplemented

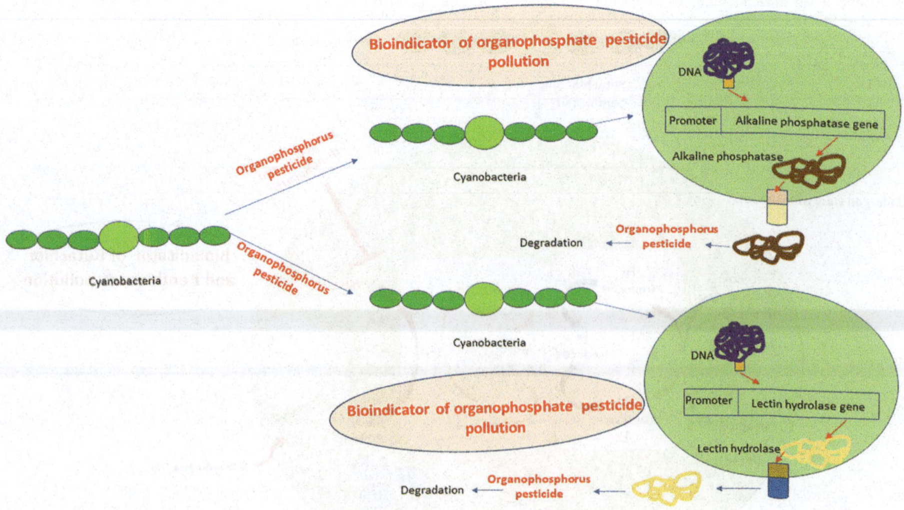

Fig. 9.11 Production of alkaline phosphatase and lectin hydrolase by cyanobacteria as bioindicators of organophosphate pesticide pollution

conditions, upregulates a carbohydrate-binding lectin domain-containing hydrolase upon methyl parathion exposure.

Further, based on a molecular docking study, they disclosed that this particular lectin possessed a high binding affinity for methyl parathion (Tiwari et al. 2018). Eventually, they concluded that the degradation of methyl parathion was possibly due to the expression and activity of this lectin hydrolase. These observations indicate that estimation of cyanobacterial alkaline phosphatase activity and expressional analysis of lectin hydrolases may help detect the levels of organophosphorus pesticides in polluted water (Fig. 9.11).

9.7.3.4 Cyanobacteria as Indicators of Different Carbamate-Type Pesticides

Besides organophosphorus pesticides, carbamate-type pesticides are a group of potentially hazardous chemicals frequently observed to contaminate various water systems. Several studies performed using cyanobacteria and carbamate pesticides reflected that these pesticides are highly toxic to these organisms. For example, it was found that carbaryl and carbofuran completely inhibited the growth of two nitrogen-fixing cyanobacteria, *Nostoc muscorum* and *Cylindrospermum* sp., respectively, at 120 and 100 μg/mL concentrations (Vaishampayan 1985). Similar growth inhibitory effects of carbaryl and carbofuran at 100 and 120 ppm have also been noticed in the N_2-fixing cyanobacterium, *Anabaena* PCC 7120 (Padhy 2001; Padhy and Mohapatra 2001). These researchers suggested that carbofuran is comparatively less toxic to cyanobacteria than carbaryl. Although many other toxic pollutants also exert growth inhibitory effects on cyanobacteria, reduction in growth may be a helpful parameter to distinguish between the concentrations of different carbamate pesticides.

9.7.4 Cyanobacteria as Indicators of Pharmaceuticals in Aquatic Systems

Apart from metals, nutrients, and pesticides, pharmaceuticals are another group of crucial water pollutants. The most common pharmaceutical pollutant is antibiotics (DoerrMacEwen and Haight 2006). The distribution of antibiotics in aquatic environments mainly depends on the source and hydrological conditions (Yang et al. 2017). In the environment, the sources of antibiotics are untreated hospital sewage, pharmaceutical plant effluent, and aquafarm discharge (Xu et al. 2014; Rodriguez-Mozaz et al. 2015; Dai et al. 2016; Lu et al. 2018; Zhang et al. 2018b). Antibiotics exert their adverse impacts on prokaryotes through the inhibition of cell envelope, protein, or nucleic acid (DNA/RNA) synthesis (Le Page et al. 2017) (Fig. 9.12). Dias et al. (2015) have performed a study using cyanobacterial strains belonging to different genera, i.e., *Microcystis aeruginosa*, *Aphanizomenon gracile*, *Chrysosporum bergii*, and *Planktothrix agardhii*, and nine different species of the genus *Microcystis* to evaluate their susceptibility toward antibiotics like amoxicillin, ceftazidime, ceftriaxone, kanamycin, gentamicin, tetracycline, trimethoprim, nalidixic acid, and norfloxacin. This study has reflected that the less amount of amoxicillin strongly inhibited the growth of all tested cyanobacteria.

In contrast, the high concentrations of tetracycline and norfloxacin severely decreased the viability of the cyanobacterial cells. Besides, the impacts of antibiotics

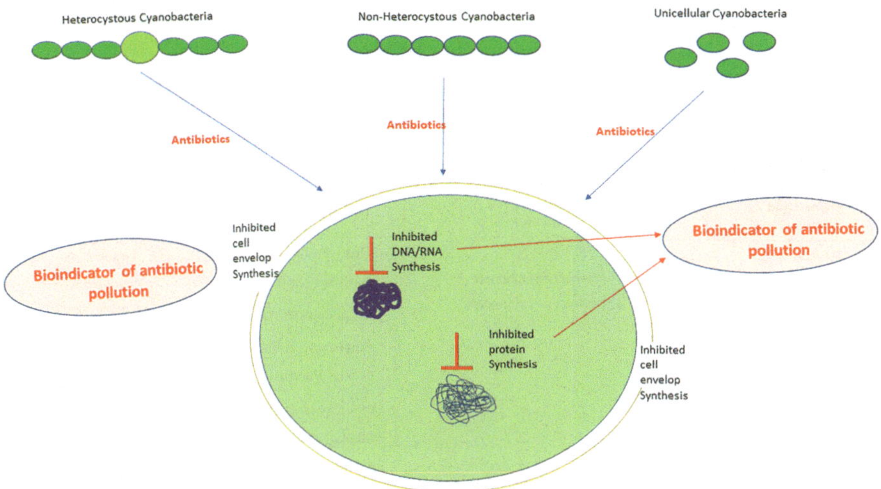

Fig. 9.12 Inhibited synthesis of cell envelope, nucleic acids, and proteins in cyanobacteria as bioindicators of antibiotic pollution

were also observed to vary with the cyanobacterial strains. For example, while *A. gracile* (LMECYA 40) and *C. berghii* (LMECYA 246) were identified as most susceptible to β-lactams, *M. aeruginosa* (LMECYA 7) was found to be vulnerable to norfloxacin. This species-specific antibiotic susceptibility of cyanobacteria may be considered as a bioindicator for assessing the extent of antibiotics pollution in water. Another recent study by Zhong et al. (2021) disclosed that the cyanobacterium *Anabaena cylindrica* exhibits elevated suppression of growth and increased production of ROS in response to the antibiotics spiramycin, tigecycline, and amoxicillin. This study suggested that the reduced growth coupled with ROS production is a potent bioindicator property of the cyanobacterium *A. cylindrica*, which may be used for detecting spiramycin, tigecycline, and amoxicillin in water.

9.7.5 Cyanobacteria as Indicators of Aromatic Pollutants in Water

Besides the abovementioned pollutants, another group of highly toxic water pollutants is aromatic pollutants. Aromatic pollutants represent a large number of chemicals which can be divided into various categories, i.e., (1) phenols and its derivatives, (2) aromatic industrial pollutants, (3) phthalate esters, and (4) polycyclic aromatic hydrocarbons (Fig. 9.13). The major sources of aromatic pollutants are oil and petroleum refinery effluents; coal processing; coking plants; resin, dye, paint, and petrochemical industries; paper mills; and textile industries. Moreover, using aromatic compounds as bactericidal and fungicidal compounds (Krastanov et al. 2013; Zhang et al. 2013) also significantly contributes to environmental pollution. Much literature on the toxic impacts of aromatic compounds on cyanobacteria has accumulated over the years (Ramakrishnan et al. 2010; Singh and Khattar 2016;

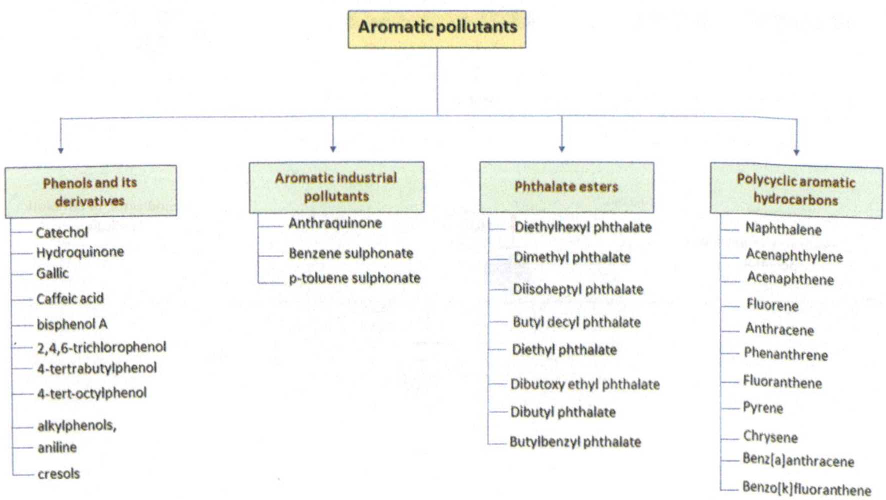

Fig. 9.13 Categorization of aromatic pollutants

Kaushik et al. 2018, 2019; Kumar et al. 2018; Singh et al. 2018). Cyanobacterial species like *Phormidium valderianum*, *Phormidium tenue*, *Microcystis aeruginosa*, *Nostoc muscorum*, *Synechococcus* PCC7942, *Arthrospira platensis*, *Anabaena cylindrica*, *Synechococcus lividus*, *Anabaena flos-aquae*, *Cyanothece* sp., and *Anabaena fertilissima* have so far been tested for their responses to various groups of aromatic pollutants (Acey et al. 1987; Shashirekha et al. 1997; Dziga et al. 2007; Kumar et al. 2009; Babu and Wu 2010; Sundaram et al. 2011; Sundaram and Soumya 2011; Ni et al. 2013; Patel et al. 2013; Kumar et al. 2014; Laue et al. 2014; Huang et al. 2016; Zhang et al. 2016; Cheng et al. 2017; Zhang et al. 2018). These studies have been performed mainly based on measuring growth behaviors, pigment contents, and antioxidant responses. For all these pollutants, elevated suppression of growth, decreased pigment content, and reduced photosynthetic activities have been the common responses except the cellular aggregation of *Synechococcus lividus* to di-n-butyl phthalate and enhanced level of glutathione in *Microcystis aeruginosa* in the presence of triclosan. Since these responses are common for other groups of pollutants, they cannot be explicitly used to detect aromatic pollutants. Hence, more intense studies on the cellular responses of cyanobacteria to aromatic pollutants are required to select specific bioindicator properties.

9.8 Conclusions

Anthropogenic activities adversely affect the ecological processes by exerting physical and chemical forces and interfering with biodiversity. In the cyanobacterial cells, the mechanisms of toxicity of different pollutants are determined by their characteristic features. Dynamic alterations in the community composition and morphophysiological and biochemical parameters of cyanobacteria can be the convenient signals of polluted water systems. While the shift in the cyanobacterial community, formation of blooms, abundance of polyphosphate bodies, and development of calyptra are the indicators of organic water pollution, elevated damages to the cell envelope, deterioration of ultrastructural organizations, hampered photosynthetic bioenergetics, and synthesis of exopolysaccharides and metallothionein are the indicators of heavy metal pollution. Moreover, the high activity of alkaline phosphatase and the production of lectin hydrolase are the responses specific to organophosphorus pesticides. In addition, blocking photosynthetic electron transport and nitrogen fixation are indicators of herbicide pollution in water.

In contrast, the available information on the ecotoxicological responses of cyanobacteria to emerging pollutants like antibiotics and aromatic compounds is still poor and fragmentary. This suggests that further studies are required on the toxic effects of these pollutants on cyanobacteria to use cyanobacteria as bioindicators of antibiotics and aromatic pollutants. Therefore, it can be concluded that cyanobacteria, a group of photoautotrophic organisms readily available throughout the year and easy to test, are excellent bioindicators of organics, heavy metals, organophosphorus, and carbamate-type pesticides and herbicides that act by inhibiting photosynthetic electron transportation.

Acknowledgments The Head of P.G. Department of Botany, Magadh University, Bodh Gaya, India, is gratefully acknowledged.

References

Acey R, Healy P, Unger TF, Ford CE, Hudson RA (1987) Growth and aggregation behavior of representative phytoplankton as affected by the environmental contaminant di-n-butyl phthalate. Bull Environ Contam Toxicol 39:1–6

Anjana K, Kaushik A, Kiran B, Nisha R (2007) Biosorption of Cr^{6+} by immobilized biomass of two indigenous strains of cyanobacteria isolated from metal contaminated soil. J Hazard Mater 148: 383–386

Arsalane W, Paresys G, Duval JC, Conrad R, Buechel C (1993) A new fluorometric device to measure the in vivo chlorophyll a fluorescence yield in microalgae and its use as a herbicide monitor. Eur J Phycol 28:247–252

Babu B, Wu JT (2010) Biodegradation of phthalate esters by cyanobacteria. J Phycol 46:1106–1113

Babu NG, Sarma PA, Attitalla IH, Murthy SDS (2010) Effect of selected heavy metal ions on the photosynthetic electron transport and energy transfer in the thylakoid membrane of the cyanobacterium *Spirulina platensis*. Acad J Plant Sci 3:46–49

Barinova S, Tavassi M (2009) Study of seasonal influences on algal biodiversity in the River Yarqon (Central Israel) by bio-indication and canonical correspondence analysis (CCA). Turk J Bot 33:353–372

Barinova S, Tavassi M, Nevo E (2006) Algal indicator system of environmental variables in the Hadera River basin, Central Israel. Plant Biosyst 140:65–79

Barinova S, Medvedeva L, Nevo E (2008) Regional influences on algal biodiversity in two polluted rivers of Eurasia (Rudnaya River, Russia, and Qishon River, Israel) by bioindication and canonical correspondence analysis. Appl Ecol Environ Res 6:29–59

Barinova S, Kukhaleishvili L, Nevo E, Janelidze Z (2011) Diversity and ecology of algae in the Algeti National Park as a part of the Georgian system of protected areas. Turk J Bot 35:729–774

Branco LHZ, Pereira JL (2002) Evaluation of seasonal dynamics and bioindication potential of macroalgal communities in a polluted tropical stream. Arch Hydrobiol 155:147–161

Brewster JD, Lightfield AR, Bermel PL (1995) Storage and immobilization of photosystem II reaction centers used in an assay for herbicides. Anal Chem 67:1296–1299

Britstein M, Devescovi G, Handley KM, Malik A, Haber M (2016) A new N-Acyl homoserine lactone synthase in an uncultured symbiont of the red sea sponge *Theonella swinhoei*. Appl Environ Microbiol 82(4):1274–1285

Butterworth FM, Gunatilaka A, Gonsebatt ME (2001) Biomonitors and biomarkers as indicators of environmental change, vol 2. Springer Science & Business Media, Boston (MA)

Campanella L, Cubadda F, Sammartino MP, Saoncella A (2000) An algal biosensor for the monitoring of water toxicity in estuarine environments. Water Res 35:69–76

Campbell D, Hurry V, Clarke AK, Gustafssson P, Oquist G (1998) Chlorophyll fluorescence analysis of cyanobacterial photosynthesis and acclimation. Microbiol Mol Biol Rev 62:667–683

Cantonati M (2008) Cyanoprokaryotes and algae other than diatoms in springs and streams of the Dolomiti Bellunesi National Park (Northern Italy). Arch Hydrobiol Algol Stud 126:113–136

Cantonati M, Rott E, Pipp E (1996) Ecology of cyanophytes in mountain springs of the river Sarca catchment (Adamello-Brenta Regional Park, Trentino, Northern Italy). Arch Hydrobiol Algol Stud 83:145–162

Cantonati M, Gerecke R, Bertuzzi E (2006) Springs of the Alps–sensitive ecosystems to environmental change: from biodiversity assessments to long-term studies. Hydrobiologia 562:59–96

Castenholz RW (2001) Phylum BX cyanobacteria. Oxygenic photosynthetic bacteria. In: Bergey's manual of systematic bacteriology, vol 1, 2nd edn. Springer, New York, pp 473–487

Chakraborty S, Mishra AK (2020) Mitigation of zinc toxicity through differential strategies in two species of the cyanobacterium *Anabaena* isolated from zinc polluted paddy field. Environ Pollut 263:114375

Chakraborty S, Mishra AK (2021) Effects of zinc toxicity on the nitrogen-fixing cyanobacterium *Anabaena sphaerica*- ultrastructural, physiological and biochemical analyses. Environ Sci Pollut Res 28:33292–33306

Chay TC, Surif S, Heng LY (2005) A copper toxicity biosensor using immobilized cyanobacteria, *Anabaena Torulosa*. Sens Lett 3:49–54

Cheng L, He Y, Tian Y et al (2017) Comparative biotoxicity of N-Phenyl-1-naphthylamine and N-Phenyl-2-naphthylamine on cyanobacteria Microcystis aeruginosa. Chemosphere 176:183–191

Cho YA, Lee H, Park EU, Lee YT, Hammock BD, Ahn KC, Lee JK (2002) Development of an ELISA for the organophosphorus insecticide Chlorpyrifos. Bull Korean Chem Soc 23:481–487

Clijsters H, Assche F (1985) Inhibition of photosynthesis by heavy metals. Photosynth Res 7:31–40

Cooper TF, Gilmour JP, Fabricius KE (2009) Bioindicators of changes in water quality on coral reefs: review and recommendations for monitoring programmes. Coral Reefs 28(3):589–606

Cracknell AP, Newcombe SK, Black AF, Kirby NE (2001) The ABDMAP (algal bloom detection, monitoring and prediction) concerted action. Int J Remote Sens 22:205–247

Dai GH, Wang B, Fu CC, Dong R, Huang J, Deng SB et al (2016) Pharmaceuticals and personal care products (PPCPs) in urban and suburban rivers of Beijing, China: occurrence, source apportionment and potential ecological risk. Environ Sci Process Impacts 18:445–455

Dash NP, Kumar A, Kaushik MS, Abraham G, Singh PK (2017) Agrochemicals influencing nitrogenase, biomass of N_2-fixing cyanobacteria and yield of rice in wetland cultivation. Biocatal Agric Biotechnol 9:28–34

De Lange E (1994) Manual for simple water quality analysis. International Water Tribunal (IWT) Foundation, Amsterdam

Dias E, Oliveira M, Jones-Dias D et al (2015) Assessing the antibiotic susceptibility of freshwater Cyanobacteria spp. Front Microbiol 6:799

Divya TV, Chandwadkar P, Acharya C (2018) NmtA, a novel metallothionein of *Anabaena* sp. strain PCC 7120, imparts protection against cadmium stress but not oxidative stress. Aquat Toxicol 199:152–161

Doerr-MacEwen NA, Haight ME (2006) Expert stakeholders views on the management of human pharmaceuticals in the environment. Environ Manag 38:853–866

Douterelo I, Perona E, Mateo P (2004) Use of cyanobacteria to assess water quality in running waters. Environ Pollut 127:377–384

Draber WT, Tietjen K, Kluth JF, Trebst A (1991) Herbicides in photosynthesis research. Angew Chem 3:1621–1633

Dumas DP, Caldwell SR, Wild JR, Raushel FM (1989) Purification and properties of the phosphotriesterase from *Pseudomonas diminuta*. J Biol Chem 264:19659–19665

Dumas DP, Durst HD, Landis WG, Raushel FM, Wild JR (1990) Inactivation of organophosphorus nerve agents by the phosphotriesterase from *Pseudomonas diminuta*. Arch Biochem Biophys 277:155–159

Dutta S, Bhadury P (2020) Effect of arsenic on exopolysaccharide production in a diazotrophic cyanobacterium. J Appl Phycol 32:2915–2926

Dziga D, Suda M, Bialczyk J, Czaja-Prokop U, Lechowski Z (2007) The alteration of *Microcystis aeruginosa* biomass and dissolved microcystin-LR concentration following exposure to plant-producing phenols. Inc Environ Toxicol 22:341–346

Elersek T, Filipic M (2011) Organophosphorous pesticides - mechanisms of their toxicity. In: Pesticides - the impacts of pesticides exposure. IntechOpen. https://doi.org/10.5772/14020

Fernandez-Piñas F, Mateo P, Bonilla I (1995) Ultrastructural changes induced by selected cadmium concentration in the cyanobacterium *Nostoc* UAM208. J Plant Physiol 147:452–456

Fjerdingstad E (1964) Pollution of streams estimated by benthal phytomicro-organisms I. A saprobic system based on communities of organisms and ecological factors. Int Rev Ges Hydrobiol 49:63–131

Fogg G, Stewart W, Fay P, Walsby A (1973) The blue-green algae. Academic Press, London

Frense D, Muller A, Beckmann D (1998) Detection of environmental pollutants using an optical biosensor with immobilized algae cells. Sensors Actuators B 51:256–260

Fulton M, Key P, Andrews M (1998) The in vitro effects of paraoxon exposure on brain and muscle acetylcholinesterase activity in two estuarine fish species. In: Proceedings, 19th annual meeting, Society of Environmental Toxicology and Chemistry, Charlotte, NC, p 54

Garret RH, Grisham CM (2010) Biochemistry: Cengage learning, 5th edn. Brooks/Cole, Boston

Gaston KJ (2000) Biodiversity: higher taxon richness. Prog Phys Geogr 24:117–127

Gerhardt A (2002) Bioindicator species and their use in biomonitoring. Environmental monitoring I. Encyclopedia of life support systems. UNESCO ed. Eolss Publisher, Oxford

Ghorbani E, Nowruzi B, Nezhadali M et al (2022) Metal removal capability of two cyanobacterial species in autotrophic and mixotrophic mode of nutrition. BMC Microbiol 22:58

Girotti S, Maiolini E, Bolelli L, Ghini S, Ferri E, Barile N, Medvedeva S (2008) Analytical techniques and bioindicators in environmental control: honeybees, mussels, bioluminescent bacteria: rapid immunoassays for pesticide detection. In: Soil chemical pollution, risk assessment, remediation and security, pp 327–347

Grimm NB, Petrone KC (1997) Nitrogen fixation in a desert stream ecosystem. Biogeochemistry 37:33–61

Gupta PK (2018) Toxicity of herbicides. In: Gupta RC (ed) Veterinary toxicology-basic and clinical principals, 3nd edn. Elsevier, USA, pp 553–568. (Chapter 44)

Gutowski A, Foerster J (2009) Benthische Algen ohne Diatomeen und Characeen: Bestimmungshilfe. LANUV, Recklinghausen

Gutowski A, Foerster J, Schaumburg J (2004) Use of benthic algae excluding diatoms and charales for the assessment of the ecological status of running freshwaters: a case history from Germany. Oceanol Hydrobiol Stud 33:3–15

Hans W, Dyble PJ, Moisander PH, Noble RT, Piehler MF et al (2003) Microbial indicators of aquatic ecosystem change: current applications to eutrophication studies. FEMS Microbiol Ecol 46:233–246

Heng LY, Jusoh K, Ling CHM, Idris M (2004) Toxicity of single and combinations of lead and cadmium to the cyanobacteria *Anabaena flos-aquae*. Bull Environ Contam Toxicol 72:373–379

Hollis RP, Killham K, Glover LA (2000) Design and application of a biosensor for monitoring toxicity of compounds to eukaryotes. Appl Environ Microbiol 66:1676–1679

Holt EA, Miller SW (2010) Bioindicators: using organisms to measure environmental impacts. Nat Educ Knowl 3(10):8

Horne AJ, Carnmiggelt C (1975) Algal nitrogen fixation in California streams: seasonal cycles. Freshw Biol 5:461–470

Hosmani S (2014) Freshwater plankton ecology: a review. J Res Manage Technol 3:1–10

https://www.hsph.harvard.edu/ehep/82-2

https://www.unep.org/explore-topics/water/what-we-do/tackling-global-water-pollution

Huang X, Tu Y, Song C et al (2016) Interactions between the antimicrobial agent triclosan and the bloom-forming cyanobacteria *Microcystis aeruginosa*. Aquat Toxicol 172:103–110

Jain A, Singh BN, Singh SP, Singh HB, Singh S (2010) Exploring biodiversity as bioindicators for water pollution. National Conference on Biodiversity, Development and Poverty Alleviation. 2010 May 22, Uttar Pradesh. Uttar Pradesh State Biodiversity Board, Lucknow

Jay AE, Ducruet JM, Duval JC, Pelletier JP (1997) A high sensitivity chlorophyll fluorescence assay for monitoring herbicide inhibition of photosystem II in the chlorophyte *Selenastrum* capricornutum: comparison with effect on cell growth. Arch Hydrobiol 140:273–286

Joanna B (2006) Bioindicators: types, development, and use in ecological assessment and research. Environ Bioind 1:22–39

Kaushik MS, Dash NP, Kumar A, Abraham G, Singh PK (2018) Tolerance of wetland rice field's cyanobacteria to agrochemicals in cultural condition. Biocatal Agric Biotechnol 13:236–243

Kaushik MS, Kumar A, Abraham G et al (2019) Field evaluations of agrochemical toxicity to cyanobacteria in rice field ecosystem: a review. J Appl Phycol 31:471–489

Kelly M (2013) Data rich, information poor? Phytobenthos assessment and the water framework directive. Eur J Phycol 48:437–450

Kennish MJ (1992) Ecology of estuaries: anthropogenic effects. CRC Press, Boca Raton (FL), p 494

Khatri N, Tyagi S (2015) Influences of natural and anthropogenic factors on surface and groundwater quality in rural and urban areas. Front Life Sci 8(1):23–39

Klemm DJ (1990) Macro invertebrate field and laboratory methods for evaluating the biological integrity of surface waters. EPA, Cincinnati (OH)

Kojima Y (1991) Definitions and nomenclature of metallothioneins. Methods Enzymol 205:8–10

Kolkwitz R, Marsson M (1908) O¨ kologie der pflanzlichen Saprobien. Ber Deutschen Botanischen Gesellschaft 26A:505–519

Koma'rek J, Anagnostidis K (1999) Cyanoprocaryota 1. Teil: Chroococcales. In: Sußwasserflora Von mitteleuropa 19/1. Gustav Fischer, Jena-Stuttgart-Lu¨beck-Ulm, p 548

Krastanov A, Alexieieva Z, Yemendzhiev H (2013) Microbial degradation of phenol and phenolic derivatives. Eng Life Sci 13:76–87

Kumar Reddy DH, Lee SM (2012) Water pollution and treatment technologies. J Environ Anal Toxicol 2:e103

Kumar MS, Muralitharan G, Thajuddin N (2009) Screening of a hypersaline cyanobacterium, *Phormidium tenue*, for the degradation of aromatic hydrocarbons: naphthalene and anthracene. Biotechnol Lett 31:1863–1866

Kumar JIN, Patel JG, Kumar RN, Khan SR (2014) Chronic response of three different cyanobacterial species on growth, pigment, and metabolic variations to the high molecular weight polycyclic aromatic hydrocarbon – pyrene. Polycycl Aromat Compd 34:143–160

Kumar S, Kaushik G, Dar MA et al (2018) Microbial degradation of organophosphate pesticides: a review. Pedosphere 28:190–208

Laue P, Bährs H, Chakrabarti S, Steinberg CEW (2014) Natural xenobiotics to prevent cyanobacterial and algal growth in freshwater: contrasting efficacy of tannic acid, gallic acid, and gramine. Chemosphere 104:212–220

Le Page G, Gunnarsson L, Snape J, Tyler CR (2017) Integrating human and environmental health in antibiotic risk assessment: a critical analysis of protection goals, species sensitivity and antimicrobial resistance. Environ Int 109:155–169

Loza V, Berrendero E, Perona E, Mateo P (2013a) Polyphasic characterization of benthic cyanobacterial diversity from biofilms of the Guadarrama River (Spain): morphological, molecular, and ecological approaches. J Phycol 49:282–297

Loza V, Perona E, Mateo P (2013b) Molecular fingerprinting of cyanobacteria from river biofilms as a water quality monitoring tool. Appl Environ Microbiol 79:1459–1472

Loza V, Perona E, Carmona J, Mateo P (2013c) Phenotypic and genotypic characteristics of Phormidium like cyanobacteria inhabiting microbial mats are correlated with the trophic status of running waters. Eur J Phycol 48:235–252

Loza V, Perona E, Mateo P (2014) Specific responses to nitrogen and phosphorus enrichment in cyanobacteria: factors influencing changes in species dominance along eutrophic gradients. Water Res 48:622–631

Lu GH, Piao HT, Gai N, Shao PW, Zheng Y, Jiao XC et al (2018) Pharmaceutical and personal care products in surface waters from the inner city of Beijing, China: influence of hospitals and reclaimed water irrigation. Arch Environ Contam Toxicol 76:255–264

Malik DS, Bharti U (2012) Status of plankton diversity and biological productivity of Sahastradhara stream at Uttarakhand, India. J Appl Nat Sci 4(1):96–103

Markert B (2007) Definitions and principles for bioindication and biomonitoring of trace metals in the environment. J Trace Elem Med Biol 21:77–82

Markert BA, Breure AM, Zechmeister HG (2003) Chapter 1 definitions, strategies and principles for bioindication/ biomonitoring of the environment. Trace Metals Other Contam Environ 6:3–39

Marques JC (2001) Diversity, biodiversity, conservation, and sustainability. Sci World J 1:534–543

Mateo P, Leganés F, Perona E, Loza V, Fernández-Piñas F (2015) Cyanobacteria as bioindicators and bioreporters of environmental analysis in aquatic ecosystems. Biodivers Conserv 24:909–948

Mishra S, Mishra DR (2014) A novel remote sensing algorithm to quantify phycocyanin in cyanobacterial algal blooms. Environ Res Lett 9:114003

Muñoz-Martín MA, Martínez-Rosell A, Perona E, Fernandez-Piñas F, Mateo P (2014) Monitoring bioavailable phosphorus in lotic systems: a polyphasic approach based on cyanobacteria. Sci Total Environ 475:158–168

Nagajyoti PC, Lee KD, Sreekanth TVM (2010) Heavy metals, occurrence and toxicity for plants. A review. Environ Chem Lett 8:199–216

Newby R Jr, Lee LH, Perez JL, Tao X, Chu T (2017) Characterization of zinc stress response in Cyanobacterium *Synechococcus* sp. IU 625. Aquat Toxicol 186:159–117

Nguyen-Ngoc H, Durrieu C, Tran-Minh C (2009) Synchronous-scan fluorescence of algal cells for toxicity assessment of heavy metals and herbicides. Ecotoxicol Environ Saf 72:316–320

Ni LX, Acharya K, Hao XY, Li SY (2013) Antioxidant and metabolism responses to polyphenol stress in cyanobacterium *Microcystis aeruginosa*. J Environ Sci Health, Part B 48:153–161

Nkwoji JA, Igbo JK, Adeleye AO, Obienu JA, Tony-Obiagwu MJ (2010) Implications of bioindicators in ecological health: study of a coastal lagoon, Lagos, Nigeria. Agric Biol J North Am 1(4):683–689

Norris RH, Thoms MC (1999) What is river health? Freshw Biol 41:197–209

Oberholster PJ, Botha A, Ashton PJ (2009) The influence of a toxic cyanobacterial bloom and water hydrology on algal populations and macroinvertebrate abundance in the upper littoral zone of Lake Krugersdrift, South Africa. Ecotoxicology 18(1):34–46

Olafson RW, McCubbin WD, Kay CM (1988) Primary-and secondary-structural analysis of a unique prokaryotic metallothionein from a *Synechococcus* sp. cyanobacterium. Biochem J 251: 691–699

Padhy RN (2001) Monitoring of chemical fertilizers on toxicity of two carbamate insecticides to the cyanobacterium *Anabaena* PCC 7120. Microbios 106:165–175

Padhy RN, Mohapatra K (2001) Toxicity of two carbamate insecticides to the cyanobacterium *Anabaena* PCC 7120 and computations of partial lethal concentrations by the probit method. Microbios 106:81–95

Palmer CM (1969) A composite rating of algae tolerating organic pollution. J Phycol 5:78–82

Patel JG, Kumar JIN, Kumar RN, Khan SR (2013) Chronic toxicity of high molecular weight polynuclear aromatic hydrocarbon - pyrene on freshwater cyanobacterium *Anabaena fertlissima* RAO. Int J Environ 2:175–183

Perona E, Mateo P (2006) Benthic cyanobacterial assemblages as indicators of nutrient enrichment regimes in a Spanish river. Acta Hydrochim Hydrobiol 34:67–72

Perona E, Bonilla I, Mateo P (1998) Epilithic cyanobacterial communities and water quality: an alternative tool for monitoring eutrophication in the Alberche River (Spain). J Appl Phycol 10: 183–191

Phillips DJH, Rainbow PS (1993) Biomonitoring of trace aquatic contaminants. Elsevier Applied Science, New York (NY)

Plafkin JL, Barbour MT, Porter KD, Gross SK, Hughes RM (1989) Rapid assessment protocols for use in streams and rivers: benthic macroinvertebrates and fish. In: Rosenberg DM, Resh VH (eds) Freshwater biomonitoring and benthic macroinvertebrates. Chapman & Hall, New York

Porter SD, Mueller DK, Spahr NE, Munn MD, Dubrovsky NM (2008) Efficacy of algal metrics for assessing nutrient and organic enrichment in flowing waters. Freshw Biol 53:1036–1054

Pradhan A, Bhaumik P, Das S, Mishra M, Khanam S et al (2008) Phytoplankton diversity as indicator of water quality for fish cultivation. Am J Environ Sci 4(4):406–411

Prasad MNV, Strzałka K (1999) Impact of heavy metals on photosynthesis: heavy metal stress in plants. Springer, Berlin

Ramakrishnan B, Megharaj M, Venkateswarlu K, Naidu R, Sethunathan N (2010) The impacts of environmental pollutants on microalgae and cyanobacteria. Crit Rev Environ Sci Technol 40: 699–821

Ramchandra TV, Rishiram R, Karthik B (2006) Zooplanktons as bioindicators: hydro biological investigation in selected Bangalore lakes. Technical report 115

Richardson LL, Stanic D, May A, Brownell A, Gantar M, Campagna SR (2014) Ecology and physiology of the pathogenic cyanobacterium *Roseofilum reptotaenium*. Life 4:968–987

Rodriguez V, Aguirre de Ca'rcer D, Loza V, Perona E, Mateo P (2007) A molecular fingerprint technique to detect pollution-related changes in river cyanobacterial diversity. J Environ Qual 36:464–468

Rodriguez-Mozaz S, Chamorro S, Marti E, Huerta B, Gros M, Sànchez-Melsió A et al (2015) Occurrence of antibiotics and antibiotic resistance genes in hospital and urban wastewaters and their impact on the receiving river. Water Res 69:234–242

Rott E, Schneider S (2014) A comparison of ecological optima of soft-bodied benthic algae in Norwegian and Austrian rivers and consequences for river monitoring in Europe. Sci Total Environ 475:180–186

Shashirekha S, Uma L, Subramanian G (1997) Phenol degradation by the marine cyanobacterium *Phormidium valderianum* BDU 30501. J Ind Microbiol Biotechnol 19:130–133

Sierra MV, Go'mez N (2007) Structural characteristics and oxygen consumption of the epipelic biofilm in three lowland streams exposed to different land uses. Water Air Soil Pollut 186:115–127

Singh DP, Khattar JIS (2016) Impact of insecticides on cyanobacteria. Seaweed Res Utiln 38:165–180

Singh S, Veram E, Niveshika TB, Mishra AK (2016) Exopolysaccharide production in *Anabaena* sp. PCC 7120 under different $CaCl_2$ regimes. physiol Mol Biol Plants 22:557–566

Singh AK, Singh PP, Tripathi V (2018) Distribution of cyanobacteria and their interactions with pesticides in paddy field: a comprehensive review. J Environ Manag 224:361–375

Sla'decek V (1973) System of water quality from the biological point of view. Ergeb Limnol (Germany, FR) 7:1–18

Soltani N, Khodaei K, Alnajar N, Shahsavari A, Ashja Ardalan A (2012) Cyanobacterial community patterns as water quality bioindicators. Iran J Fish Sci 11:876–891

Stancheva R, Fetscher AE, Sheath RG (2012) A novel quantification method for stream-inhabiting, nondiatom benthic algae, and its application in bioassessment. Hydrobiologia 684:225–239

Stancheva R, Sheath RG, Read BA et al (2013) Nitrogen-fixing cyanobacteria (free-living and diatom endosymbionts): their use in southern California stream bioassessment. Hydrobiologia 720:111–127

Su L, Jia W, Hou C, Lei Y (2011) Microbial biosensors: a review. Biosens Bioelectron 26:1788–1799

Sundaram S, Soumya KK (2011) Study of physiological and biochemical alterations in cyanobacterium under organic stress. Am J Plant Physiol 6:1–16

Sundaram S, Soumya KK, Ramgopal PJK, Rahman A (2011) Impact of organic stress on growth, photosynthetic and physiological response of some cyanobacterial isolates. J Environ Sci Technol 4:264–283

Surosz W, Palinska KA (2004) Effects of heavy-metal stress on cyanobacterium *Anabaena flos-aquae*. Arch Environ Contam Toxicol 48:40–48

Takatera K, Watanabe T (1992) Application of high performance liquid chromatography/inductively coupled plasma mass spectrometry to the speciation of cadmium-binding metallothionein-like protein in a cyanobacterium. Anal Sci 8:469–474

Teta R, Della Sala G, Glukhov E, Gerwick L, Gerwick WH, Mangoni A, Costantino V (2015) Combined LC-MS/MS and molecular networking approach reveals new cyanotoxins from the 2014 cyanobacterial bloom in green Lake, Seattle. Environ Sci Technol 49(24):14301–14310

Teta R, Romano V, Sala GD, Picchio S et al (2017) Cyanobacteria as indicators of water quality in Campania coasts, Italy: a monitoring strategy combining remote/proximal sensing and in situ data. Environ Res Lett 12:024001

Tiwari B, Chakraborty S, Srivastava AK, Mishra AK (2017a) Biodegradation and rapid removal of methyl parathion by the paddy field cyanobacterium *Fischerella* sp. Algal Res 25:285–296

Tiwari B, Singh S, Chakraborty S, Verma E, Mishra AK (2017b) Sequential role of biosorption and biodegradation in rapid removal degradation and utilization of methyl parathion as a phosphate source by a new cyanobacterial isolate *Scytonema* sp. BHUS-5. Int J Phytorem 19(10):884–893

Tiwari B, Verma E, Chakraborty S, Srivastava AK, Mishra AK (2018) Tolerance strategies in cyanobacterium *Fischerella* sp. under pesticide stress and possible role of a carbohydrate-binding protein in the metabolism of methyl parathion (MP). Int Biodeterior Biodegradation 127:217–226

Turner JS, Robinson NJ (1995) Cyanobacterial metallothioneins: biochemistry and molecular genetics. J Ind Microbiol 14:119–125

Underwood DLA, Shapiro AM (1999) Evidence for division of labor in the social caterpillar *Eucheira socialis* (Lepidoptera: Pieridae). Behav Ecol Sociobiol 46(4):228–236

Vaishampayan A (1985) Mutagenic activity of alachlor, butachlor and carbaryl to a nitrogen-xing cyanobacterium Nostoc muscorum. J Agric Sci 104:571–576

Voet D, Voet JG, Pratt CW (2012) Principles of biochemistry, 4th edn. Wiley, New York

Walley WJ, Grbovic J, Dzˇeroski S (2001) A reappraisal of saprobic values and indicator weights based on Slovenian river quality data. Water Res 35:4285–4292

Walsh GE (1978) Toxic effects of pollutants on plankton. In: Butler GC (ed) Principles of ecotoxicology. Wiley, New York (NY), pp 257–274

Whitton BA (2013) Use of benthic algae and bryophytes for monitoring rivers. J Ecol Environ 36: 95–100

Whitton BA, Mateo P (2012) Rivulariaceae. In: Whitton B (ed) Ecology of cyanobacteria II. Their diversity in space and time. Springer, London, pp 561–591

Whitton BA, Neal C (2011) Organic phosphate in UK rivers and its relevance to algal and bryophyte surveys. Ann Limnol 47:3–10

Whitton BA, Potts M (2000) Introduction to the cyanobacteria. In: Ecology of cyanobacteria: their diversity in time and space. Kluwer, Dordrecht, pp 1–10

WHO (1999) Toxic cyanobacteria in water: a guide to their public health consequences, monitoring and management eds I Chorus and J Bartram (E and FN Spon)

Wong LS, Surif S, Lee YH (2008) Toxicity biosensor for the evaluation of cadmium toxicity based on photosynthetic behavior of cyanobacteria *Anabaena torulosa*. Asian J Biochem 3:162–168

Wong LS, Lee YH, Salmijah S (2012) The fluorometric response of cyanobacteria to short exposure of heavy metal. Adv Environ Biol 6:103–108

Wong LS, Lee YH, Surif S (2013) Whole cell biosensor using *Anabaena torulosa* with optical transduction for environmental toxicity evaluation. J Sens 2013:8

Xu J, Zhang Y, Zhou C et al (2014) Distribution, sources and composition of antibiotics in sediment, overlying water and pore water from Taihu Lake, China. Sci Total Environ 497-498:267–273

Yang L, He JT, Su SH, Cui YF, Huang DL, Wang GC (2017) Occurrence, distribution, and attenuation of pharmaceuticals and personal care products in the riverside groundwater of the Beiyun River of Beijing, China. Environ Sci Pollut Res 24:1–14

Zaghloul A, Saber M, Gadow S, Awad F (2020) Biological indicators for pollution detection in terrestrial and aquatic ecosystems. Bull Natl Res Cent 44:127

Zannatul F, Muktadir AKM (2009) A review: potentiality of zooplankton as bioindicator. Am J Appl Sci 6(10):1815–1819

Zhang Y, Lu D, Ju T et al (2013) Biodegradation of phenol using *Bacillus cereus* WJ1 and evaluation of degradation efficiency based on a grapheme-modified electrode. Int J Electochem Sci 8:504–519

Zhang X, Liu L, Zhang S et al (2016) Biodegradation of dimethyl phthalate by freshwater unicellular cyanobacteria. Biomed Res Int 2016:1–8

Zhang M, Wang X, Tao J et al (2018) PAHs would alter cyanobacterial blooms by affecting the microcystin production and physiological characteristics of *Microcystis aeruginosa*. Ecotoxicol Environ Saf 157:134–142

Zhang P, Zhou H, Li K et al (2018b) Occurrence of pharmaceuticals and personal care products, and their associated environmental risks in a large shallow lake in North China. Environ Geochem Health 40:1525–1539

Zhong X, Zhu Y, Wang Y, Zhao Q, Huang H (2021) Effects of three antibiotics on growth and antioxidant response of *Chlorella pyrenoidosa* and *Anabaena cylindrical*. Ecotoxicol Environ Safety 211:111954

Degradation of Xenobiotics by Cyanobacteria

10

Júlia Ronzella Ottoni, Caroline da Costa Silva Gonçalves, Keith Dayane Leite Lira, Suzan Pantarotto de Vasconcellos, Luis Fernando Romanholo Ferreira, and Michel Rodrigo Zambrano Passarini

Abstract

Known as one of the first inhabitants of the Earth, cyanobacteria contribute significantly to the biogeochemical cycles of carbon and nitrogen. These organisms produce a variety of bioactive compounds, which may have applications in diverse biotechnological sectors including bioremediation process. Due to the ability to thrive under harsh and extreme conditions, cyanobacteria can metabolize a wide variety of toxic polluting compounds found in the environment. In this context, this chapter focuses on gathering information about the use of cyanobacteria as biotechnological sources of enzymes and/or secondary metabolites with prominent potential to aid the treatment of xenobiotic and recalcitrant environmental pollutants, as well as elucidating the results of the processes involved in the bioremediation of these substances.

Keywords

Environmental polluting compounds · Dyes · Hydrocarbons · Drugs · Pesticides · Heavy metals

J. R. Ottoni · C. da Costa Silva Gonçalves · M. R. Z. Passarini (✉)
Laboratory of Environmental Biotechnology, Federal University of Latin-American Integration (UNILA), Foz do Iguaçu, Parana, Brazil
e-mail: michel.passarini@unila.edu.br

K. D. L. Lira · S. P. de Vasconcellos
Laboratory of Health and Environment – LABMSMA, Department of Pharmaceutical Sciences, Federal University of São Paulo (UNIFESP), Diadema, São Paulo, Brazil

L. F. R. Ferreira
Institute of Research and Technology, University of Tiradentes – UNIT, Aracaju, Sergipe, Brazil

B. Neilan et al. (eds.), *Cyanobacterial Biotechnology in the 21st Century*,
https://doi.org/10.1007/978-981-99-0181-4_10

10.1 Introduction

Cyanobacteria are primitive photosynthetic prokaryotes that supposedly appeared on the planet during the Precambrian period. They possibly constitute the group of the first photosynthetic microorganisms in a period of 2–3 billion years and were fundamental for the evolution of higher organisms (Dubey et al. 2011). They are present in virtually all the habitats on Earth, being found even in extreme environments such as deserts, polar sites, and hypersaline lakes. The extensive evolutionary history, together with the need to adapt to different environments, may be associated with the considerable and complex nature of their secondary metabolites (Nandagopal et al. 2021). Due to biological characteristics such as resistance to desiccation, persistence in extreme environments, autotrophic nutrition, and the ability to fix nitrogen, these organisms have advantages over other forms of life (Zinicovscaia and Cepoi 2016).

More than 1100 natural products from 39 cyanobacterial genera are already known, and of these, 731 are active against eukaryotic and prokaryotic cells or act as enzymes inhibitors. These sets of compounds include alkaloids, depsipeptides, lipopeptides, macrolides/lactones, peptides, terpenes, polysaccharides, lipids, polyketides, and others. The chemical diversity observed is the result of several biosynthetic pathways and may be related to the broad spectrum of biological activities of these natural products originated from cyanobacteria (Singh et al. 2011; Dittmann et al. 2015). The complex toxins (cyanotoxins) biosynthesized by these organisms can negatively affect human and environmental health; but on the other hand, due to their allelochemical nature, these toxins can be used in agriculture as pesticides (Singh et al. 2017). Besides, cyanobacterial cells can produce secondary metabolites with a broad diversity of biological activities including antimicrobial, immunosuppressant, anticancer, anticoagulant, anti-inflammatory, antiprotozoal, antituberculosis, antiviral, and antitumor, as well as enzymes that can be applied in several sectors of biotechnology (Khalifa et al. 2021).

Due to the natural ability to adapt to different habitats, including aquatic and heavily polluted environments, and inexpensive growth and maintenance, cyanobacteria are promising agents to the bioremediation process (El-Bestawy 2008). These organisms are effective on the transformation and removal of heavy metals, besides other organic environmental pollutants, such as pesticides, crude oil, naphthalene, naphthalene, phenanthrene, phenol, catechol, and xenobiotic (Ahmad 2021). In this way, cyanobacteria present a potential role in the treatment of wastewater and industrial effluents, such as from oil refinery, dye, paper mill, and pharmaceutical industries (El-Sheekh et al. 2014).

However, in recent years, population growth has led to increased demand for food and industrialization, actions that have contributed to environmental pollution raise due to the large amount of toxic polluting compounds that are inappropriately released into the environment, including dyes, hydrocarbons, drugs, pesticides, and heavy metals, mainly harming water resources (Fioravante et al. 2010; Rodrigues et al. 2016; Yu et al. 2020; Bernal et al. 2021; Zinicovscaia and Cepoi 2016).

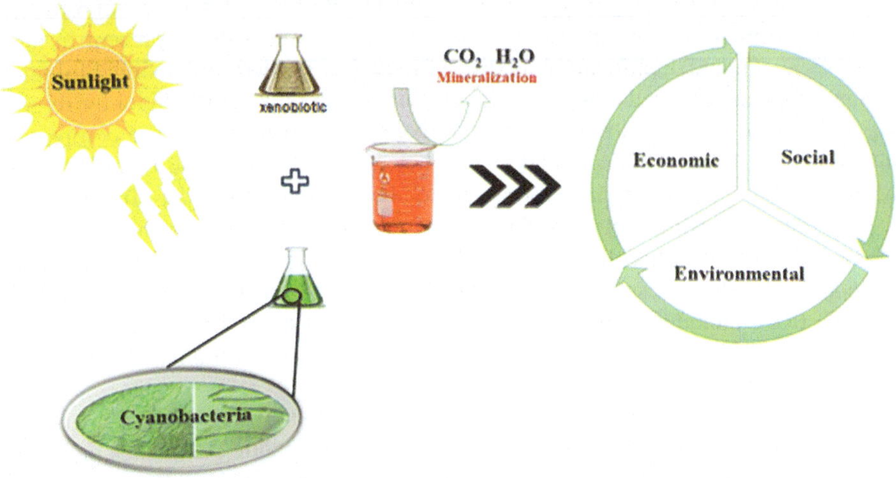

Fig. 10.1 Scheme of xenobiotic biodegradation by cyanobacteria

Persistent organic pollutants have a low water solubility, accumulate in the environment, and have carcinogenic, teratogenic, and neurotoxic potential to human cells, causing disturbances in the physical and chemical properties of water and interfering with the self-purification process (Zinicovscaia and Cepoi 2016).

Bioremediation may be a promising technique, as it consists of using organisms to eliminate environmental contaminants by transforming pollutants into less toxic compounds, even when they are completely mineralized (Yu et al. 2020) (Fig. 10.1).

Cyanobacteria may be able to metabolize a variety of toxic compounds, including persistent organic pollutants, transforming them into less toxic compounds or into nontoxic compounds that are utilized as nutrients. Another ability of these organisms is the capacity to use ammonium ions as a source of nitrogen, an important process in domestic and agricultural wastewater purification (Zinicovscaia and Cepoi 2016). In this context, this chapter addresses the main information on the bioremediation processes of environmental polluting compounds, including dyes, hydrocarbons, drugs, pesticides, and heavy metals, using cyanobacteria as an alternative source in the treatment of these pollutant substances.

10.2 Dye Biodegradation by Cyanobacteria

Synthetic dyes are recalcitrant compounds capable of resisting conventional biological treatments. They are compounds found as the main waste in the effluent of the textile dyeing industry. The discharge of residual dye effluent, due to chronic effects such as mutagenic damage and carcinogenicity to biota, is of particular concern to aquatic ecosystems (Dellamatrice et al. 2017). About 15% of dyes used in fabric dyeing operations in textile industries does not settle on fabric fibers and is lost in wastewater. Consequently, colored effluents are formed, which can represent a serious problem of water pollution due to the toxic components present in these dyes. Conventional treatment of textile effluents is carried out using biological systems such as activated sludge; however, this process may not efficiently remove the dyes due to the recalcitrant nature and diverse composition of these compounds (O'Neill et al. 1999; Ralanapongleko and Phetsom 2014; Dellamatrice et al. 2017). Alternatively, cyanobacteria strains have been used as cost-effective and ecologically promising tools in the remediation of colored textile effluents. Strains of *Westiellopsis* sp., *Lyngbya* sp., *Chlorella* sp., *Oscillatoria* sp., *Synechococcus* sp., *Nodularia* sp., *Nostoc* sp., and *Cyanothece* sp. have been reported to be able to degrade dyes, including remazol dyes (Dellamatrice et al. 2017; Vyayakumar and Manoharau 2012; Henciya et al. 2013; Jaysudha and Sampathkumar 2014; Dubey et al. 2011).

Microorganisms such as bacteria, fungi, and cyanobacteria are capable of bioadsorbing or biodegrading the effluent dye using enzymatic complexes including azoreductases, peroxidases, laccases, and various oxidative and reducing enzymes synthesized by these microorganisms, which play a vital role in biodegradation and bioconversion of synthetic dyes (Shetty and Krishnakumar 2020). There are reports on the ability to synthesize enzymes such as laccases and azoreductases by members of the *Cyanophyceae* and *Chlorophyceae* classes of cyanobacteria (Acuner and Dilek 2004; Palanisami et al. 2010). The presence of functional groups such as NH, C=O, C=C, and C–H on the surface of the algal biomass facilitates chemical reactions such as aliphatic and aromatic substitution, which occurs between dyes and biomass (Chia et al. 2014; Shetty and Krishnakumar 2020).

Afreen et al. (2017) isolated a new extracellular laccase enzyme produced by the *Spirulina platensis* strain CFTRI which was able to discolor 96% of the reactive Blue 4 anthraquinone dye in 4 h. The authors concluded that cyanobacterial laccase can be used to aid wastewater treatment. The decolorization potential and efficiency of different dyes by cyanobacteria were evaluated by Abd Ellatif et al. (2021). Seven isolates were tested, i.e., *Nostoc muscorum*, *Nostoc humifusum*, *Spirulina platensis*, *Anabaena oryzae*, *Wollea saccata*, *Oscillatoria* sp., and *Chlorella vulgaris*, and the authors found that the *Chlorella vulgaris* isolate was responsible for the highest percentages of discoloration of brazilwood, orange G, and naphthol green B dyes, with a reduction of 99.5%, 99.5%, and 98.5%, respectively. The best discoloration percentages of crystal violet and malachite green dyes were achieved by *A. oryzae* (97.4%) and *W. saccata* (93.3%). The authors also performed enzymatic assays to analyze the activities of the laccase and lignin peroxidase enzymes and observed that the isolates of *C. vulgaris, A. oryzae*, and *W. saccata* were efficient in the degradation of lignin. *C. vulgaris* was the largest producer of laccase and lignin peroxidase in the assays with the dyes brazilwood, orange G, and naphthol green B, while *A. oryzae* and *W. saccata* showed the highest production of laccase with the dyes crystal violet and green malachite. The results show the potential of *C. vulgaris*, *A. oryzae*, and *W. saccata* as possible candidates for bioremediation and preprocessing to remove dyes from textile effluents.

Cyanobacteria proved to be organisms with a great potential for bioremediation of environments containing dyes, which shows the importance of further studies of exploration and improvement of this capacity for environmental decontamination purposes.

10.3 Pharmaceutical Biotransformation by Cyanobacteria

Pharmaceuticals and their metabolites have emerged as a new class of environmental pollutants, displaying harmful effects on natural flora and fauna. These compounds have already been identified in ground and surfaces waters, wastewater treatment stations, landfills, industrial effluents, and sludge. Although the potential long-term effects of these environmental pollutants on human health are not completely understood, it has been noticed that they can generate biological toxicity, induce resistance to antibiotics in pathogenic bacteria, produce genotoxic effect, and compromise the equilibrium of aquatic communities. Furthermore, current wastewater treatment plants still are not able to affect these pharmaceutical compounds (Grignet et al. 2022; Patel et al. 2019).

Cyanobacteria are especially sensitive to antimicrobials that affect the growth, pigment synthesis, and influence the production and release of cyanotoxins (Ye et al. 2017). Through the study of the cyanobacteria *Phormidium valderianum* BDU 30501, Prabaharan (1994) was able to notice their ability to evolve resistance to antimicrobials. The cyanobacteria were able to grow in the presence of higher concentrations of ampicillin and penicillin. A higher growth occurred in the presence of ampicillin, suggesting that this compound may be used as an additional nitrogen input by these cyanobacteria, according to Pan et al. (2021).

Wu and coauthors (2020) analyzed the effects of multiple exposures of *Microcystis aeruginosa* to the antibiotic erythromycin. This is one of the most commonly cyanobacteria identified in eutrophic freshwater, and many of its strains are known to produce hepatotoxins, called microcystins, which pose a serious threat to environmental and human health (Zhou et al. 2021). Erythromycin inhibits *M. aeruginosa*'s cell growth by altering photosynthesis and inducing ROS-mediated membrane lipid peroxidation. However, after repeated exposure to the antibiotic, an increase in the growth tax of *M. aeruginosa* cells and in the production and release of microcystins was observed.

The antioxidant response of *M. aeruginosa* to spiramycin and amoxicillin (two typical antibiotic contaminants) was analyzed by Liu and coauthors (2012). The presence of amoxicillin stimulated the activities of three antioxidant enzymes (peroxidase (POD), catalase (CAT), and superoxide dismutase (SOD)). The exposure of *M. aeruginosa* to test concentrations (500 ng/L) of streptomycin, however, inhibited the activities of POD and CAT. Spiramycin-treated *M. aeruginosa* also showed a higher amount of malondialdehyde, a nonenzymatic antioxidant, indicating a higher toxicity of this antibiotic in comparison with amoxicillin. Besides, after 7 days of exposure, partial degradation of spiramycin (12.5%–32.9%) and amoxicillin (30.5%–33.6%) was also observed. Pan and collaborators (2021) have analyzed this cyanobacteria ability to degrade tetracycline, another widely used antibiotic, in high concentrations, against which *M. aeruginosa* exhibited strong resistance, being able to remove 98% of tetracycline after 24 hours. This antibiotic, through the cyanobacteria, was able to inhibit the release of microcystin-LR.

The potential of the cyanobacteria *Spirulina platensis* in the removal of carbamazepine (CBZ), an anticonvulsant medication, was analyzed by Wang and coauthors (2020). At a relatively low concentration of CBZ (1 mg/L), an increase in the *S. platensis* biomass was observed, although in a higher concentration, the pharmaceutical (2.5–100 mg/L) negatively affects this cyanobacteria's growth. However, the inhibition effect of CBZ on *S. platensis* decreased with the cultivation time. On the other hand, its resistance and the removal of CBZ increased with the cultivation time. Moreover, the content of chlorophylls, carotenoids, carbohydrates, and lipids in *S. platensis* and the SOD and CAT activities were significantly increased on the CBZ presence, indicating the existence of a protective mechanism.

Therefore, cyanobacteria are fascinating organisms that present a vast array of biotechnological applications, including a strong potential for bioremediation of pharmaceutical-contaminated wastewater. These fascinating organisms are widely distributed throughout aquatic environments, presenting major effects on the water quality and functioning of these ecosystems, which suggests that more applications for them may be found through the development of researches in the area.

10.4 Heavy Metal Biotransformation by Cyanobacteria

The group of heavy metals, mainly copper (Cu), iron (Fe), chromium (Cr), lead (Pb), and aluminum (Al), is considered as dangerous pollutants for health and environments (Al-Amin et al. 2021). These heavy metals can reach the environment through untreated industrial effluents, harming the environment and the food chain at various trophic levels (Joshi 2014). Microbial communities can interact with heavy metals in a variety of ways, both intracellularly and extracellularly, by the mobilization, transformation, precipitation, immobilization, accumulation, and absorption of metals and by the formation of complexes (Iram et al. 2012).

Studies have been carried out using microbial cells from the cyanobacterial group for the removal and/or treatment of environments impacted with heavy metals. Zinicovscaia et al. (2015) carried out a study with a culture of *Spirulina platensis* to remove heavy metals in effluents from a machinery company. Results showed the absorption of zinc and iron (after 5 min of contact) by the microbial biomass. The authors concluded that bioaccumulation can be an expensive process for cells as it requires the maintenance of vital conditions for microorganisms. Gaur and Dhankhar (2009) evaluated zinc removal by free and immobilized *Anabaena variabilis*. The authors observed that cyanobacteria were able to absorb the metal for up to 90 min after microbial contact with ion, after which the zinc concentration remained constant. On the other hand, with an increase in zinc concentration, there was a reduction in microbial adsorption, which was justified by the possible saturation of the binding sites with zinc ions. At different concentrations of chromium ions, Jayashree et al.

(2012) observed that live and dead cells of *Oscillatoria* sp. can absorb this metal. The authors observed that, in the treatment of cells with heat (musts), the results showed that there was a greater amount of metal absorbed in comparison with the live cells of *Oscillatoria* sp., thus concluding that cyanobacteria can absorb metals in a non-metabolic way. Regarding copper ions, it is a very toxic metal for cyanobacteria, whose mechanisms to reduce the toxicity of this metal are the immobilization of intracellular copper with metallothioneins or phytochelatins (Zinicovscaia and Cepoi 2016).

The production of exopolysaccharides (EPS) by cyanobacteria has been considered promising for the removal of metal ions due to their ability to act as biological chelating agents (De Philippis et al. 2011). Freire-Nordi et al. (2005) carried out a study to evaluate the characteristics of EPS produced by *Anabaena spiroides* for the chelation of the heavy metals copper, lead, and mercury. Other cyanobacterial strains, such as *Gloeothece* sp. (for copper and lead remediation), *Oscillatoria willei* (silver ion remediation), and *Chroococcus paris* (copper, zinc, and cadmium), have also shown promising results in the metal removal process by the production of EPS (De Philippis et al. 2011). So, it is possible to affirm that cyanobacteria strains can be a promising and sustainable alternative for the removal of heavy metals present in the ecosystems due to anthropogenic activities.

10.5 Pesticide Biotransformation by Cyanobacteria

The agricultural sector is essential for the development of society, as several million tons of pesticides are applied each year to improve crop productivity (Deviram et al. 2020). Pesticides are important classes of persistent organic pollutants defined as a substance or mixture of substances used for direct or indirect remediation against the action of insects, bacteria, and other forms of animal or plant life harmful to agriculture (Nicolopoulou-Stamati et al. 2016). Xenobiotics have organic fragments in their structure, such as chloroaromatics (chloroanilines, chloroacetamides, and triazines), and these substances represent the most widely used classes of pesticides worldwide, being extremely persistent in the environment and harmful to human health (Nicolopoulou-Stamati et al. 2016; Sharma et al. 2019).

Due to the adverse effects of their use, pesticides are classified as carcinogens, since they can induce or produce the formation of cancer, as well as their residues can accumulate in adipose tissues causing acute and chronic disorders in human health (Kalyabina et al. 2021). In the ecological scope, the effects caused by the use of pesticides affect mainly the waters and soils, which reach these environments via the air, through the evaporation of substances, and by leaching the pollutant from the soil into the water (Pérez-Lucas et al. 2019; Damalas and Eleftherohorinos 2011). In addition, commercialized foods may contain toxic residues from the production, processing, storage, and packaging processes (Rather et al. 2017; Santos et al. 2019; Jesus et al. 2015).

Conventional remediation processes for these organic pollutants are expensive and inefficient and contribute to the worsening of environmental problems by adopting systems such as waste incineration, which result in the production of dioxins (Baldissarelli et al. 2019). In this context, many countries have restricted the use of these substances, establishing guidelines and concentration limits and investing in research aimed at the development of efficient methods of decontamination and appropriate disposal of these persistent environmental pollutants, such as biodegradation. Microorganisms have a dose-response relationship to exposure to xenobiotics (Tsatsakis et al. 2018), since they can acquire new catabolic genes necessary for the degradation of the target compound from other microorganisms, through conjugative or transformational events (Reineke 1986; Chaudhry and Chapalamadugu 1991).

Among microorganisms, cyanobacteria constitute a group of prokaryotic and photoautotrophic organisms that exhibit oxygenated photosynthesis and are susceptible to ecophysiological adaptations. This functional plasticity assists in the resistance of cyanobacteria to different stress conditions, and among their abilities is the ease of fixation of atmospheric nitrogen and the ability to produce efficient protective chemicals against oxidative and radical stressors (Deviram et al. 2020; Kabra et al. 2014; Lau et al. 2015).

Forlani et al. (2008) evaluated the resistance of the herbicide glyphosate in six strains of cyanobacteria (*Anabaena* sp., *Arthrospira fusiformis*, *Leptolyngbya boryana*, *Microcystis aeruginosa*, *Nostoc punctiforme*, and *Spirulina platensis*). All strains were significantly tolerant, and cell proliferation was not affected by the herbicide (Forlani et al. 2008). Recently, Żyszka-Haberecht et al. (2019) reported the tolerance of cyanobacteria to the high concentrations of pesticide and attributed this tolerance to the ability of these organisms to develop a wide range of sensory mechanisms and adaptive systems (Żyszka-Haberecht et al. 2019).

Cyanobacteria have biotechnological potential for remediation of environmental pollutants, either through biodegradation or bioaccumulation. Cyanobacteria such as *Anacystis nidulans*, *Synechococcus elongatus*, *Nostoc* sp., and *Microcystis aeruginosa* are species commonly used in the degradation of organophosphate and organochlorine pesticides (Singh et al. 2016). To date, the mechanisms of chemical biodegradation by cyanobacteria have not been elucidated.

Hamed et al. (2021) investigated the degradation of different concentrations of the organophosphate pesticide pyridafenthion (PY) by the cyanobacteria *Anabaena laxa* and *Nostoc muscorum* for 7 days. The authors observed that *Nostoc muscorum* showed a greater bioaccumulation and degradation of PY capacities, and the pesticide degradation resulted in the compound 6-hydroxy-2-phenylpyridazin-3(2H)-one, considered an environmentally safe product. Tiwari et al. (2017) studied the adsorption and degradation of the organophosphate pesticide methyl parathion (MP) as a phosphate source by rice paddy cyanobacteria *Fischerella* sp. Fourier transform infrared spectra showed efficient MP removal (~80%) due to pesticide adsorption on the surface of *Fischerella* sp. from the interaction of the surface hydroxyl groups (–OH) of the cyanobacterial cell. Furthermore, the formation of p-nitrophenol was observed, suggesting that there is a synergistic process of biosorption followed by simultaneous biodegradation of MP.

Among the classes of pesticides, the use of pyrethroids is preferable since the biodegradation of these substances by microorganisms generates less or nontoxic metabolites. In addition, pyrethroids have a low toxicity to mammals and birds, except for fish, which are vulnerable to these compounds (Di Giulio and Hinton 2008; Bano et al. 2021). Bano et al. (2021) evaluated the response of the filamentous cyanobacteria *Pseudanabaena* sp. and *Leptolyngbya* sp. isolated from the marine environment in the removal and degradation of the pyrethroids cypermethrin and deltamethrin. *Pseudanabaena* sp. was able to remove 98.0% of cypermethrin and 99.0% for deltamethrin, while the cyanobacteria *Leptolyngbya* sp. showed a removal efficiency of 82.0% for cypermethrin and 97.8% for deltamethrin. In the biodegradation process of these pyrethroids, 3-phenoxy-benzaldehyde and 1-(2-methyl-cyclopropyl)-ethanone were identified as products through the technique of gas chromatography coupled to mass spectrum.

Avila et al. (2022) evaluated the degradation of three pesticides commonly found in fresh water (acetamiprid, bentazone, and propanil) in laboratory-scale reactors employing four cyanobacteria: *Phormidium* sp., *Oscillatoria* sp., *Nostoc* sp., and *Tolypothrix* sp. It was observed that propanil and acetamiprid were completely removed after 7 days, while bentazone was not removed. The pesticides acetamiprid and propanil generated four and two products respectively in the biodegradation process. Acetamiprid products increased over reaction time. On the other hand, propanil products were not detected on a pilot scale, suggesting complete mineralization of propanil or residual formation at concentrations below the detection limit of the analytical method (Avila et al. 2022).

Martínez-Aguilar et al. (2018) evaluated the degradation of organophosphate pesticides (malathion and its transformation product malaoxon) widely used in agricultural crops on the biomass of the cyanobacteria *Chroococcus* sp. cultured in BG-11 medium at a temperature of 28 ± 2 °C and pH 6. Degradation rates of 69 and 65% were observed for malathion and malaoxon, respectively.

The findings presented here elucidate the potential of cyanobacteria to remove pesticides present in the environment or transform these compounds into either less toxic substances or substances with no toxicity, thus contributing to the reduction of pesticide pollution.

10.6 Hydrocarbons

Cyanobacteria have been described as efficient hydrocarbon biodegraders and constitute a strong strategy for the treatment of hydrocarbon-polluted environments, employing bioremediation techniques. They can be found in places where oil spills and leaks occur, such as during the Gulf War in Kuwait, when an oil spill occurred and where a bloom of cyanobacteria was observed, showing that algae can use hydrocarbons as a source of energy/carbon (Atyah and Al-Mayaly 2018). The main advantage of cyanobacteria when compared to other types of microorganisms, with regard to their ability to biodegrade hydrocarbons, is related to their efficiency as bioaccumulators of organic and inorganic materials. These microorganisms may have a moldable metabolism, with a high photosynthetic and absorption capacity in aquatic environments, increasing their potential to promote an effective remediation system (Touliabah et al. 2022). Cyanobacteria respond to the presence of oil according to the extension and to the time of leakage and may differ in time and space (Xie et al. 2018). In this sense, studies have evidenced the capability of cyanobacteria to decontaminate hydrocarbon-polluted areas, mitigating environmental impacts in sensitive ecosystems.

Hydrocarbons are substances widely disseminated and are associated with many episodes of pollution worldwide, involving not only aquatic ecosystems but also terrestrial ones. The main pollution sources of hydrocarbons are related to industry processes, irregular disposal of wastewater, oil spills, and accidents involving petroleum refineries and transportation procedures (Hassanshahian et al. 2020). About 35 million barrels of crude oil are managed and transported worldwide annually, elucidating the potential risks associated with contamination of aquatic and terrestrial ecosystems, therefore showing the importance of efficient, ecological, and cost-effective remediation techniques (Atyah and Al-Mayaly 2018). Hydrocarbons such as polycyclic aromatic compounds are normally highly toxic and carcinogenic, highlighting the importance of emergency actions towards the cleanup of hydrocarbon-affected areas (Mojiri et al. 2019; Lawal 2017; Ibraheem 2010).

The adoption of phytoremediation technique, using cyanobacteria for the decontamination of environments polluted with hydrocarbons, can be quite advantageous, since when consuming hydrocarbons as source of carbon and energy, cyanobacteria transform these compounds into less toxic or nontoxic products through the mineralization of the substrates (Touliabah et al. 2022).

Some species of cyanobacteria have already been described as potential hydrocarbon degraders, including *Phormidium* sp., *Nostoc* sp., *Anabaena* sp., *Aphanothece conferta*, and *Synechocystis aquatilis* (Ibraheem 2010). *Synechococcus* and *Prochlorococcus* are the most abundant species in marine environments with the ability to biotransform hydrocarbons (Lea-Smith et al. 2015). Atyah and Al-Mayaly (2018) also described the capacity of *Anabaena* sp. to degrade crude oil from Al-Dora refinery. The authors also reported that the species *Oscillatoria salina*, *Plectonema terebrans*, *Aphanocapsa* sp., and *Synechococcus* sp. can, singly, degrade up to 89% of crude oil hydrocarbon content to nontoxic products. These species can also act as microbial mats, which improve their biodegradation abilities in aquatic ecosystems, thus providing an efficient strategy to successfully metabolize hydrocarbons.

Xie et al. (2018) performed assays to track taxonomic profile fluctuations in microcosms of estuarine sediments containing different oil concentrations. The authors verified that cyanobacterial bloom occurred only on the surface of the sediment containing a low oil concentration and that there was little growth of these organisms in microcosms with a higher oil concentration. The main groups found in this study were *Leptolyngbya*, *Oscillatoria*, *Arthrospira*, *Geitlerinema*, and *Cyanothece*.

Biodegradation of hydrocarbons by cyanobacteria can occur either individually, with isolated strains, or with the application of a consortium of cyanobacteria. The mechanism is based on the action of enzymes produced by cyanobacteria, which can be extra- or intracellular and which interact with hydrocarbons, leading to the formation of less complex by-products when compared to the substrates (Touliabah et al. 2022). The main enzymes associated to hydrocarbon degradation include azo-dye reductase (Touliabah et al. 2022), laccase, and polyphenol oxidase (Shashirekha et al. 1997).

Studies on the biodegradation of hydrocarbons are directed to the main pollutants, generally aliphatic and aromatic compounds. Crude oil hydrocarbons, specifically, are of great relevance, due to their potential to promote environmental contamination of high proportions, including dispersion of saturate and aromatic compounds, asphaltenes, and resins (Hassanshahian et al. 2020). Polychlorinated biphenyls, polychlorinated dibenzo-p-dioxins, dibenzofurans, and dibenzothiophenes are also described as hydrocarbon contaminants metabolizable by cyanobacterial species, not only isolatedly but also as microbial mats (Touliabah et al. 2022; El-Sheekh and Hamouda 2014; McGenity et al. 2012; Gupta et al. 2012; Ibraheem 2010).

Filamentous cyanobacteria such as *Phormidium autumnale* have been also reported as hydrocarbon degrader (Palinska and Marquardt 2008), and *Anabaena flos-aquae* has shown the ability to degrade phthalate. Strains of *Microcystis aeruginosa* were found to be diethyl phthalate (DEP), di-n-butyl phthalate (DBP), and dimethyl phthalate (DMP) biodegraders (Babu and Wu 2010), and dimethyl phthalate has been degraded by cyanobacteria strains of *Synechocystis*, *Cyanothece*, and *Synechococcus* (Zhang et al. 2016).

Here, it was possible to bring some evidence of the effectiveness of cyanobacteria as a high-level biocatalyst to employ in bioremediation processes looking for hydrocarbon decontamination. The algae characteristics of cyanobacteria allow them to feed on various sources of nutrients, under diverse and uncommon conditions. Different approaches have been described in literature, where the treatment can occur in situ or ex situ, and the cyanobacterial species can work alone or in consortium with other species.

References

Abd Ellatif S, El-Sheekh MM, Senousy HH (2021) Role of microalgal ligninolytic enzymes in industrial dye decolorization. Int J Phytoremediation 23(1):41–52

Acuner E, Dilek FB (2004) Treatment of tectilon yellow 2G by *Chlorella vulgaris*. Process Biochem 39(5):623–631. https://doi.org/10.1016/S0032-9592(03)00138-9

Afreen S, Shamsi TN, Baig MA, Ahmad N et al (2017) A novel multicopper oxidase (laccase) from cyanobacteria: purification, characterization with potential in the decolorization of anthraquinonic dye. PLoS One 12(4):e0175144

Ahmad IZ (2021) The usage of cyanobacteria in wastewater treatment: prospects and limitations. Lett Appl Microbiol

Al-Amin A, Parvin F, Chakraborty J, Kim YI (2021) Cyanobacteria mediated heavy metal removal: a review on mechanism, biosynthesis, and removal capability. Environ Technol Rev 10(1): 44–57

Atyah BS, Al-Mayaly I (2018) Biodegradation of crude oil by *Anabaena variabilis* isolated from Al-Dora refinery IOSR. J Pharm Biol Sci 13:51

Avila R, García-Vara M, López-García E, Postigo C, de Alda ML, Vicent T, Blánquez P (2022) Evaluation of an outdoor pilot-scale tubular photobioreactor for removal of selected pesticides from water. Sci Total Environ 804:150040. https://doi.org/10.1016/j.scitotenv.2021.150040

Babu B, Wu JT (2010) Biodegradation of phthalate esters by cyanobacteria. J Phycil 46:1106–1113

Baldissarelli DP, Vargas GDLP, Korf EP, Galon L, Kaufmann C, Santos JB (2019) Remediation of soils contaminated by pesticides using physicochemical processes: a brief review. Planta Daninha 37

Bano S, Nadir M, Ahmed A, Rasool SG, Siddiqui PJA, Rasheed M (2021) Removal efficiency of marine filamentous cyanobacteria for pyrethroids and their effects on the biochemical parameters and growth. Algal Res 60:102546. https://doi.org/10.1016/j.algal.2021.102546

Bernal SPF, Lira MMA, Jean-Baptiste J, Garcia PE, Batista E, Ottoni JR, Passarini MRZ (2021) Biotechnological potential of microorganisms from textile effluent: isolation, enzymatic activity and dye discoloration. An Acad Bras Cienc 93:e20191581. https://doi.org/10.1590/0001-3765202120191581

Chaudhry GR, Chapalamadugu S (1991) Biodegradation of halogenated organic compounds. Microbiol Rev 55(1):59–79

Chia MA, Odoh OA, Ladan Z (2014) The indigo blue dye decolorization potential of immobilized *Scenedesmus quadricauda*. Water Air Soil Pollut 225(4):1920. https://doi.org/10.1007/s11270-014-1920-2

Damalas CA, Eleftherohorinos IG (2011) Pesticide exposure, safety issues, and risk assessment indicators. Int J Environ Res Public Health 8(5):1402–1419

De Philippis R, Colica G, Micheletti E (2011) Exopolysaccharide-producing cyanobacteria in heavy metal removal from water: molecular basis and practical applicability of the biosorption process. Appl Microbiol Biotechnol 92(4):697–708

Dellamatrice PM, Silva-Stenico ME, Moraes LABD, Fiore MF, Monteiro RTR (2017) Degradation of textile dyes by cyanobacteria. Braz J Microbiol 48:25–31

Deviram G, Mathimani T, Anto S, Ahamed TS, Ananth DA, Pugazhendhi A (2020) Applications of microalgal and cyanobacterial biomass on a way to safe, cleaner and a sustainable environment. J Clean Prod 253:119770. https://doi.org/10.1016/j.jclepro.2019.119770

Di Giulio RT, Hinton DE (2008) The toxicology of fishes. CRC Press. https://doi.org/10.1201/9780203647295

Dittmann E, Gugger M, Sivonen K, Fewer DP (2015) Natural product biosynthetic diversity and comparative genomics of the cyanobacteria. Trends Microbiol 32(10):642–652

Dubey SK, Dubey J, Mehra S, Tiwari P, Bishwas AJ (2011) Potential use of cyanobacterial species in bioremediation of industrial effluents. Afr J Biotechnol 10(7):1125–1132

El-Bestawy E (2008) Treatment of mixed domestic–industrial wastewater using cyanobacteria. J Ind Microbiol Biotechnol 35(11):1503–1516

El-Sheekh MM, Hamouda RA (2014) Biodegradation of crude oil by some cyanobacteria under heterotrophic conditions. Desalin Water Treat 52(7–9):1448–1454. https://doi.org/10.1080/19443994.2013.794008

El-Sheekh MM, El-Shouny WA, Osman ME, El-Gammal EW (2014) Treatment of sewage and industrial wastewater effluents by the cyanobacteria *Nostoc muscorum* and *Anabaena subcylinderica*. J Water Chem Technol 36(4):190–197

Fioravante IA, Barbosa FAR, Augusti R, Magalhães SMS (2010) Removal of methyl parathion by cyanobacteria Microcystis novacekii under culture conditions. J Environ Monit 12(6): 1302–1306

Forlani G, Pavan M, Gramek M, Kafarski P, Lipok J (2008) Biochemical bases for a widespread tolerance of cyanobacteria to the phosphonate herbicide glyphosate. Plant Cell Physiol 49(3): 443–456

Freire-Nordi CS, Vieira AAH, Nascimento OR (2005) The metal binding capacity of *Anabaena spiroides* extracellular polysaccharide: an EPR study. Process Biochem 40:2215–2224

Gaur N, Dhankhar R (2009) Removal of Zn^{+2} ions from aqueous solution using *Anabaena variabilis*: equilibrium and kinetic studies. Int J Environ Res 3:605–616
Grignet RS, Maria GA, Barros MGA, Panatta AAS, Bernal SPF, Ottoni JR, Passarini MRZ, Gonçalves CCS (2022) Medicines as an emergent contaminant: the review of microbial biodegration potential. Folia Microbiol. https://doi.org/10.1007/s12223-021-00941-6
Gupta RK, Choudhary KK, Kumar M, Negi A, Rai H (2012) Bioremediation and cyanobacteria an overview. BioNano Frontiers 9:190–196
Hamed SM, Hozzein WN, Selim S, Mohamed HS, AbdElgawad H (2021) Dissipation of pyridaphenthion by cyanobacteria: insights into cellular degradation, detoxification and metabolic regulation. J Hazard Mater 402:123787
Hassanshahian M, Amirinejad N, Behzadi MA (2020) Crude oil pollution and biodegradation at the Persian Gulf: a comprehensive and review study. J Environ Health Sci Eng 18(2):1415–1435
Henciya S, Shankar M, Maliga P (2013) Decolorization of textile dye effluent by marine cyanobacterium *Linghya* sp. BDU 9001 with coir pith. Int J Environ Sci 3:1909–1918
Ibraheem IBM (2010) Biodegradability of hydrocarbons by cyanobacteria. J Phycol 46(4):818–824
Iram S, Parveen K, Usman J, Nasir K, Akhtar N, Arouj S, Ahmad I (2012) Heavy metal tolerance of filamentous fungal strains isolated from soil irrigated with industrial wastewater. Biologija 58(3)
Jayashree S, Thangaraju N, Gnanadoss JJ (2012) Toxic effects of chromium on the aquatic cyanobacterium *Oscillatoria* sp. and removal of chromium by biosorption. J Exp Sci 3:28–34
Jaysudha S, Sampathkumar P (2014) Nutrient removal from tannery effluent by free and immobilized cells of marine microalgae Chlorella salina. Int J Environ Biol 4:21–26
Jesus RA, Santos LFS, Navickiene S, Mesquita ME (2015) Evaluation of metal-organic framework as low-cost adsorbent material in the determination of pesticide residues in soursop exotic fruit (*Annona muricata*) by liquid chromatography. Food Anal Methods 8(2):446–451
Joshi BH (2014) Evaluation and characterization of heavy metal resistant fungi for their prospects in bioremediation. J Environ Res Dev 8(4):876–882
Kabra AN, Ji MK, Choi J, Kim JR, Govindwar SP, Jeon BH (2014) Toxicity of atrazine and its bioaccumulation and biodegradation in a green microalga, *Chlamydomonas mexicana*. Environ Sci Pollut Res 21(21):12270–12278
Kalyabina VP, Esimbekova EN, Kopylova KV, Kratasyuk VA (2021) Pesticides: formulants, distribution pathways and effects on human health–a review. Toxicol Rep 8:1179–1192
Khalifa SA, Shedid ES, Saied EM, Jassbi AR, Jamebozorgi FH, Rateb ME, Du M, Bdel-Daim MM, Kai G-Y, Al-Hammady MAM, Xiao J, Guo Z, El-Seedi HR (2021) Cyanobacteria—from the oceans to the potential biotechnological and biomedical applications. Mar Drugs 19(5):241
Lau NS, Matsui M, Abdullah AAA (2015) Cyanobacteria: photoautotrophic microbial factories for the sustainable synthesis of industrial products. BioMed research international
Lawal AT (2017) Polycyclic aromatic hydrocarbons. A review. Cogent Environ Sci 3(1):1339841
Lea-Smith DJ, Biller SJ, Davey MP, Cotton CA, Sepulveda BMP, Turchyn AV, Scanlan DJ, Smith AG, Chisholm SW, Howe CJ (2015) Contribution of cyanobacterial alkane production to the ocean hydrocarbon cycle. Proc Natl Acad Sci 112(44):13591–13596
Liu Y, Guan Y, Gao B, Yue Q (2012) Antioxidant responses and degradation of two antibiotic contaminants in Microcystis aeruginosa. Ecotoxicol Environ Saf 86:23–30
Martínez-Aguilar K, Pérez-Legaspi IA, Ramírez-Fuentes E, Trujillo-Tapia MN, Alfredo Ortega-Clemente L (2018) Growth, photosynthesis and removal responses of the cyanobacteria *Chroococcus* sp. to malathion and malaoxon. J Environ Sci Health B 53(12):771–776. https://doi.org/10.1080/03601234.2018.1505070
McGenity TJ, Folwell BD, McKew BA, Sanni GO (2012) Marine crude-oil biodegradation: a central role for interspecies interactions. Aquatic Biosyst 8(1):1–19
Mojiri A, Zhou JL, Ohashi A, Ozaki N, Kindaichi T (2019) Comprehensive review of polycyclic aromatic hydrocarbons in water sources, their effects and treatments. Sci Total Environ 696:133971

Nandagopal P, Steven NA, Chan L-W, Rahmat Z, Jamaluddin H, Noh NIM (2021) Bioactive metabolites produced by cyanobacteria for growth adaptation and their pharmacological properties. Biology 10(10):1061

Nicolopoulou-Stamati P, Maipas S, Kotampasi C, Stamatis P, Hens L (2016) Chemical pesticides and human health: the urgent need for a new concept in agriculture. Front Public Heal 4. https://doi.org/10.3389/fpubh.2016.00148

O'Neill C, Hawkes FR, Hawkes DL, Lourenco ND, Pinheiro HM, Delle W (1999) Colour in textile effluents-sources, measurement, discharge contents and stimulation: a review. J Chem Technol Biotechnol 74(1009–1018):7

Palanisami S, Saha SK, Lakshmanan U (2010) Laccase and polyphenol oxidase activities of marine cyanobacteria: a study with poly R-478 decolorization. World J Microbiol Biotechnol 26(1):63. https://doi.org/10.1007/s11274-009-0143-y

Palinska KA, Marquardt J (2008) Genotypic and phenotypic analysis of strains assigned to the widespread cyanobacterial morphospecies *Phormidium autumnale* (Oscillatoriales). Arch Microbiol 189(4):325–335

Pan M, Lyu T, Malamoros V, Angelidaki I, Cooper M, Pan G (2021) Mitigating antibiotic pollution using cyanobacteria: removal efficiency, pathways and metabolism. Water Res 190:116735

Patel M, Kumar R, Kishor K, Mlsna T, Pittman CU, Mohan D (2019) Pharmaceuticals of emerging concern in aquatic systems: chemistry, occurrence, effects, and removal methods. Chem Rev 119:3510–3673

Pérez-Lucas G, Vela N, El Aatik A, Navarro S (2019) Environmental risk of groundwater pollution by pesticide leaching through the soil profile. Pesticides-use and misuse and their impact in the environment, 1–28

Prabaharan A, Sumathi M, Subramanian G (1994) Ability to use ampicillin as a nitrogen source by the marine cyanobacterium *Phormidium valderianum* BDU 30501. Curr Microbiol 28:315–320

Ralanapongleko K, Phetsom J (2014) Decolorization by conventional treatment is inefficient. Int J Chem Eng Appl 5:26–30

Rather IA, Koh WY, Paek WK, Lim J (2017) The sources of chemical contaminants in food and their health implications. Front Pharmacol 8:830

Reineke W (1986) Construction of bacterial strains with novel degradative capabilities for chloroaromatics. J Basic Microbiol 26(9):551–567

Rodrigues GN, Alvarenga N, Vacondio B, Vasconcellos SP, Passarini MR, Seleghim MH, Porto AL (2016) Biotransformation of methyl parathion by marine-derived fungi isolated from ascidian *Didemnum ligulum*. Biocatal Agric Biotechnol 7:24–30

Santos LFS, de Jesus RA, Costa JAS, Gouveia LGT, de Mesquita ME, Navickiene S (2019) Evaluation of MCM-41 and MCM-48 mesoporous materials as sorbents in matrix solid phase dispersion method for the determination of pesticides in soursop fruit (*Annona muricata*). Inorg Chem Commun 101:45–51

Sharma A, Kumar V, Shahzad B, Tanveer M, Sidhu GPS, Handa N, Kohli SK, Yadav P, Bali AS, Parihar RD, Dar OI, Singh K, Jasrotia S, Bakshi P, Ramakrishnan M, Kumar S, Bhardwaj R, Thukral AK (2019) Worldwide pesticide usage and its impacts on ecosystem. SN Appl Sci 1: 1446. https://doi.org/10.1007/s42452-019-1485-1

Shashirekha S, Uma L, Subramanian G (1997) Phenol degradation by the marine cyanobacterium *Phormidium valderianum* BDU 30501. J Ind Microbiol Biotechnol 19(2):130–133

Shetty K, Krishnakumar G (2020) Algal and cyanobacterial biomass as potential dye biodecolorizing material: a review. Biotechnol Lett 42(12):2467–2488

Singh RK, Tiwari SP, Rai AK, Mohapatra TM (2011) Cyanobacteria: an emerging source for drug discovery. J Antibiot 64:401–412

Singh JS, Kumar A, Rai AN, Singh DP (2016) Cyanobacteria: a precious bio-resource in agriculture, ecosystem, and environmental sustainability. Front Microbiol 7:529

Singh R, Parihar P, Singh M, Bajguz A, Kumar J, Singh S, Singh VP, Prasad SM (2017) Uncovering potential applications of cyanobacteria and algal metabolites in biology, agriculture and medicine: current status and future prospects. Front Microbiol 8:1–37

Tiwari B, Chakraborty S, Srivastava AK, Mishra AK (2017) Biodegradation and rapid removal of methyl parathion by the paddy field cyanobacterium *Fischerella* sp. Algal Res 25:285–296. https://doi.org/10.1016/j.algal.2017.05.024

Touliabah HES, El-Sheekh MM, Ismail MM, El-Kassas H (2022) A review of microalgae - and cyanobacteria - based biodegradation of organic pollutants. Molecules 27(3):1141

Tsatsakis AM, Vassilopoulou L, Kovatsi L, Tsitsimpikou C, Karamanou M, Leon G, Liesivuori J, Hayes AW, Spandidos DA (2018) The dose response principle from philosophy to modern toxicology: the impact of ancient philosophy and medicine in modern toxicology science. Toxicol Rep 5:1107–1113

Vyayakumar S, Manoharau C (2012) Treatment of dye industry effluent using free and immobilized cyanobacteria. J Bioremed Biodegr 3(1–6):13

Wang Q, Liu W, Li X, Wang R, Zhai J (2020) Carbamazepine toxicity and its co-metabolic removal by the cyanobacteria Spirulina platensis. Sci Total Environ 706:135686

Wu Y, Wan L, Zhang W, Ding H, Yang W (2020) Resistance of cyanobacteria *Microcystis aeruginosa* to erythromycin with multiple exposure. Chemosphere 249:126147

Xie C, Sun K, Zhang K, Su Y, Lu Z (2018) Cyanobacterial blooms in oil-contaminated subtidal sediments revealed by integrated approaches. Environ Microbiol Rep 10(4):444–452. https://doi.org/10.1111/1758-2229.12660

Ye J, Du Y, Wang L, Qian J, Chen J, Wu Q, Hu X (2017) Toxin release of cyanobacterium *Microcystis aeruginosa* after exposure to typical tetracycline antibiotic contaminants. Toxins 9: 53–76

Yu X, Sui Q, Lyu S, Zhao W, Liu J, Cai Z, Yu G, Barcelo D (2020) Municipal solid waste landfills: an underestimated source of pharmaceutical and personal care products in the water environment. Environ Sci Technol 54(16):9757–9976. https://doi.org/10.1021/acs.est.0c00565

Zhang X, Liu L, Zhang S, Pan Y, Li J, Pan H, Xu S, Luo F (2016) Biodegradation of dimethyl phthalate by freshwater unicellular cyanobacteria. BioMed Research International

Zhou CC, Chen H, Zhao H, Wang Q (2021) Microcystin biosynthesis and toxic effects. Algal Res 55:102277

Zinicovscaia I, Cepoi L (eds) (2016) Cyanobacteria for bioremediation of wastewaters. Springer International Publishing

Zinicovscaia I, Duca G, Cepoi L, Chiriac T, Rudi L, Mitina T, Frontasyeva MV, Pavlov S, Gundorina SF (2015) Biotechnology of metal removal from industrial wastewater: zinc case study. Clean Soil Air Water 42:1–6

Żyszka-Haberecht B, Niemczyk E, Lipok J (2019) Metabolic relation of cyanobacteria to aromatic compounds. Appl Microbiol Biotechnol 103(3):1167–1178

Impact of Pesticides on Cyanobacteria in Aquatic Ecosystems

11

Gabriela Sosa Benegas, Cecilio Correa-Perez, and Sergio Mendez-Gaona

Abstract

The use of pesticides in agricultural land worldwide will increase to satisfy the future food demand with a high risk of pesticide pollution into water systems. The route of pesticides to water may be through direct application or indirect when it moves mainly from the soil by runoff. However, several characteristics make it challenging to predict the resultant effect of aquatic pesticide pollution on cyanobacteria, primarily the occurrence of mixtures and the strong influence of environmental factors such as temperature and precipitations. Here, we examined the links and effects of pesticides on cyanobacteria in aquatic systems; with particular attention to warmer waters in conjunction with pesticide risk and precipitations, we propose as a case study the four important lakes in South America. Results indicate that the lakes with less warming conditions presented a better state of bloom conditions in regions with a high risk of pesticide pollution, warm temperatures, and significant precipitations; it is evidenced that high temperatures may be the main factor affecting herbicide toxicity to cyanobacteria and thus can be a key factor to explain and determine which lake

Supplementary Information The online version contains supplementary material available at https://doi.org/10.1007/978-981-99-0181-4_11.

G. S. Benegas (✉)
Latin American Institute for the Science of Life and Nature, Federal University of Latin American Integration (UNILA), Foz do Iguaçu, Brazil
e-mail: gabriela.sosa@uc.edu.py

C. Correa-Perez
CEMIT Multidisciplinary and Technological Center of Studies and Research of the University of Asuncion (UNA), San Lorenzo, Paraguay

S. Mendez-Gaona
Polytechnic Faculty, University of Asuncion (UNA), San Lorenzo, Paraguay

B. Neilan et al. (eds.), *Cyanobacterial Biotechnology in the 21st Century*,
https://doi.org/10.1007/978-981-99-0181-4_11

is prone to the worst harmful cyanobacteria bloom conditions. Finally, cyanobacteria may be favored, and/or they are sensitive to pesticides, and there is a trend that pesticide stress stimulates cyanobacterial growth.

Keywords

Pesticide pollution · Aquatic system · Cyanobacteria · Warm temperatures · Precipitations · Herbicide · Fungicide · Insecticide · South America

11.1 Introduction

By 2050, the demand for food production is estimated to increase by 70%. As the availability of additional agricultural land is limited, the expansion will happen at the expense of natural habitats containing wildlife and natural enemies of crop pests (Popp et al. 2013; FAO 2017). Given these limitations, all the efforts now are focused on accomplishing the challenge of growing more food on less land, with less water, and using less energy with sustainability.

In a pesticide-free scenario, it is estimated that at least 70% of crop yields will be lost to pests (Oerke 2006). Thus, the food demand can only be accomplished through increased agricultural productivity, achieved with pesticides, biotechnology, and sustainable agriculture (FAO 2017). Pesticides are widely used in agriculture and have considerable benefits by enhancing crop yield. On the other hand, they present adverse effects on nontarget species (OECD 2006; FAO 2017). Every year, globally, three billion kilograms of pesticides is used. However, only 1% of the total is effectively used against target organisms. The remaining large amounts reach nontarget plants and the environment's compartments like the atmosphere, soil, surface waters, and groundwater (Cordell et al. 2009; Oerke 2006; Tang et al. 2021).

Another critical threat to water quality worldwide is the harmful cyanobacterial blooms (HCBs). These phenomena have increased in intensity, frequency, and space scale throughout the same period as the accelerated pesticide usage (Codd et al. 1999; Sivonen and Jones 1999; Harris and Smith 2016). In the same way, HCBs and pesticides are enhanced by fertilizers, water temperature, and runoff. This evidence proposes that synergistic interaction between pesticide pollution and cyanobacteria may increase the frequency and nuisance of HCBs.

This chapter examines the links and effects of pesticides on cyanobacteria in warm aquatic systems. As cyanobacteria are photosynthetic organisms that can be affected by herbicides, these are the major market segment globally (Lürling and Roessink 2006; Harris and Smith 2016; Sharma et al. 2019; Chorus and Welker 2021). In the same way that Harris and Smith (2016) did, we divide our analysis into two broad groups of active ingredients: pesticides that affect target photosynthetic organisms (herbicides) and those that do not and present effects on cyanobacteria (insecticides and fungicides). The impact of pesticides with a high frequency of occurrence as the herbicide atrazine, the insecticide imidacloprid, and the fungicide carbendazim, is highlighted.

11.2 Pesticides in Aquatic Environments

Pesticides are substances or a mixture of substances employed in agricultural and aquaculture practices to control pests, weeds, and diseases in plants (EPA). These substances include hundreds of active ingredients which differ in their chemical property, environmental behavior, and ecotoxicity (Chow et al. 2020). Pesticides negatively affect biodiversity and ecosystem function in aquatic environments. These substances are widely used in modern agriculture and aquaculture. They move from soil into water or are applied directly; they are ubiquitous (Tang et al. 2021). Consequently, water bodies within 40% of the land surface worldwide are at risk of pesticide runoff (Ippolito et al. 2014; Sharma et al. 2019).

As we can see in Fig. 11.1, there are various pesticide classifications from different sources (Akashe et al. 2018). Pesticides are classified based on the pests they kill, their chemical composition, mode of entry, mode of action, the risk for human health, persistence, etc. Based on the use and the global market classification, pesticides include a wide range of herbicides, insecticides, fungicides, rodenticides, nematicides, algaecides, miticides, and acaricides (OECD 2006; WHO 2019). The herbicide segment in the pesticide market on a global basis is the largest, with nearly 50% of the total share (Sharma et al. 2019).

Several factors and characteristics differentiate aquatic pesticide pollution effects from other pollution types, making them especially challenging; among them are the occurrence of mixtures and the difficulty to predict the resultant effect on aquatic organisms. Various studies evidenced that globally aquatic systems are contaminated with complex combinations of pesticides, posing risks to aquatic life (DeLorenzo et al. 2001; Stehle and Schulz 2015; Norman et al. 2020). According to Norman et al. (2020), in a study of 203 current-use pesticides in surface waters, herbicides were the most frequently detected, with frequencies ranging from 24 to 56%, with the herbicide atrazine, the insecticide imidacloprid, and the fungicide carbendazim being the most detected in all sample types. This exposure can involve a mixture of toxic substances. As a result, the resultant effect on susceptible individuals can be unpredictable and different from the exposures to each pollutant if applied separately (Metcalf and Codd 2020).

The route of pesticides to the water compartment may be either direct through the direct application or indirect when it moves mainly from the soil of pesticide-treated areas or by runoff (Fig. 11.2). When released in aquatic systems, pesticides may cause sublethal or lethal effects in sensitive species. Thus, a cascade of consumer-resource interactions takes place; the reduced food source for predators or a release from the competition can occur and indirectly leads to an increase in abundance of tolerant species such as cyanobacteria (Bérard et al. 1999; DeLorenzo et al. 2001; Fleeger et al. 2003). In these environments, indirect effects can be more important

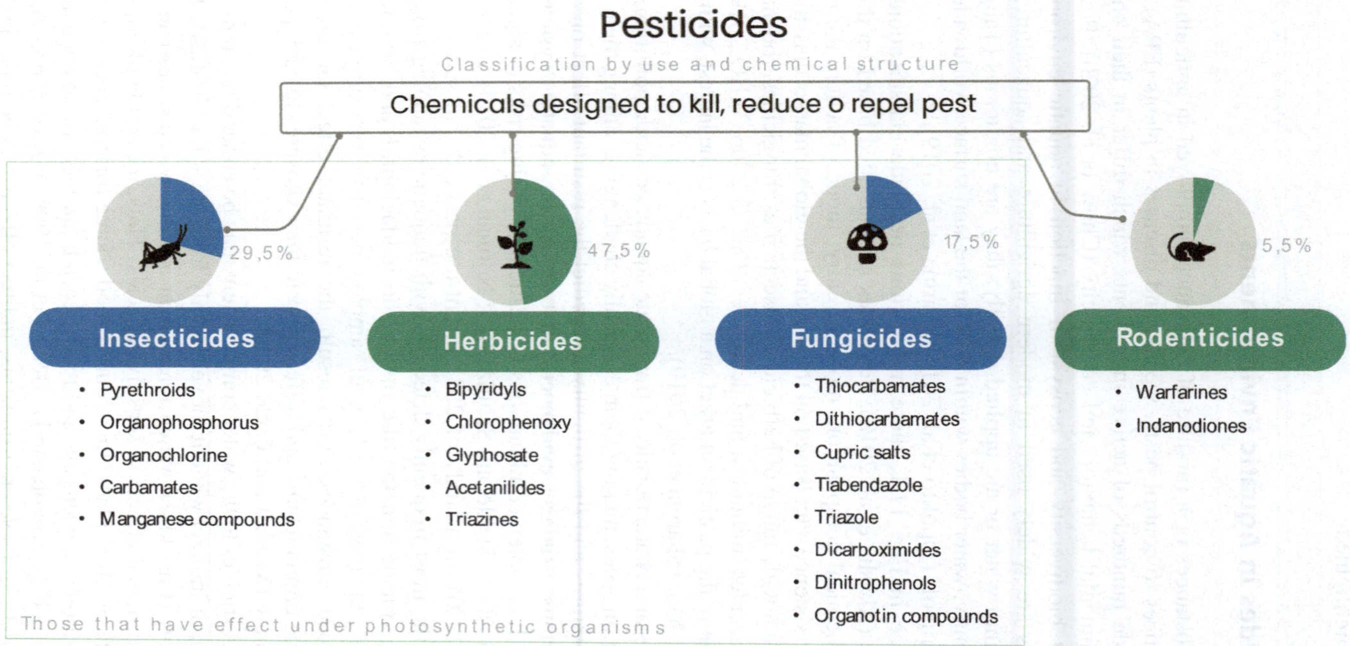

Fig. 11.1 Schematic representation of pesticides by use and chemical structure considering the WHO (2019) classification. The green square includes those classes that affect photosynthetic organisms and cyanobacteria according to Harris and Smith (2016); percentages in circular graphs represent the world usage of subclass reported by Sharma et al. (2019)

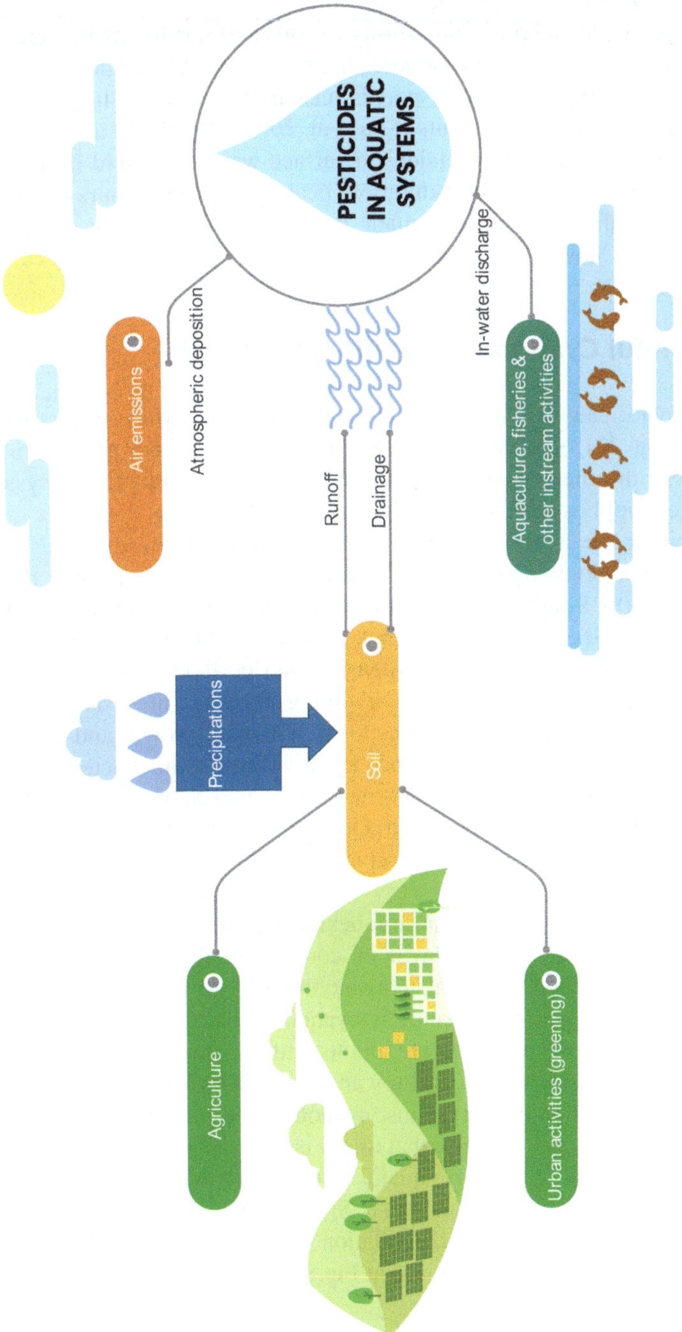

Fig. 11.2 The route of pesticides to the water compartment

than direct effects, making nontarget organisms better indicators of pesticide stress to natural communities (Lampert et al. 1989).

There are pesticides with a solid affinity to soils and sediments that enter surface waters and are mobilized some time after their application (Delgado-Moreno et al. 2011; Chow et al. 2020). This transport is mishandled by draught (Kreuger 1998) and enhanced by precipitation events (Leu et al. 2004; Doppler et al. 2012), and as a consequence, the presence of pesticides in surface waters is found in high variable concentration ranges, from short-duration peaks (i.e., few hours) and frequent presence, making the exposure of aquatic organisms, acute and/or chronic (Stehle and Schulz 2015; Chow et al. 2020).

11.3 Harmful Cyanobacterial Blooms (HCBs)

HCBs are a major cause of water quality degradation in surface waters worldwide (Paerl and Otten 2013). In addition, many cyanobacteria taxa produce a suite of potent toxins with other metabolites that negatively affect aquatic ecosystems and drinking water supplies (Codd 2000; Graham et al. 2010).

Cyanobacteria are a group of photosynthetic prokaryotes that are diverse in morphology, physiology, and development, with a low level of cellular differentiation (Vermaas 2001; Singh et al. 2016). They are ubiquitous and grow in all sorts of aquatic and terrestrial environments. When exposed to stressors like nutrient starvation, salinity, temperature, pH, light intensity, and pesticides, various species can survive under adverse conditions by evolving a range of fascinating, unique biomolecules and secondary metabolites to support their growth and survival (Stal 2007; Sharma et al. 2019). The abundance and toxicity of cyanobacteria are strongly influenced by eutrophication by the oversupply of two nutrients, phosphorus and nitrogen, to surface waters (Paerl and Otten 2013); environmental factors also influence the nuisance of HCBs with changes in food web structure and global warming (Kosten et al. 2012).

After pesticide use, cyanobacteria frequently appear in large masses, and these applications may suppress the growth of competing phytoplankton, control the population of grazers, and/or stimulate cyanobacterial growth (Padhy 1985; Harris and Smith 2016). On the other hand, as cyanobacteria share many physiological features with higher plants, they are photosynthetic microorganism targets for herbicides in aquatic ecosystems (Whitton and Potts 2000; Ma et al. 2010). Evidence varies, showing that in aquatic systems, cyanobacteria may be favored and/or are sensitive to pesticides; the question that arises is thus: as a side effect of the use of pesticides, will HCBs turn our aquatic ecosystems blue-green? (Lürling and Roessink 2006; Elliott 2011; Harris and Smith 2016). A schematic of possible general interactions based on the compilation of experimental results summarized in the supplementary material is shown in Fig. 11.3.

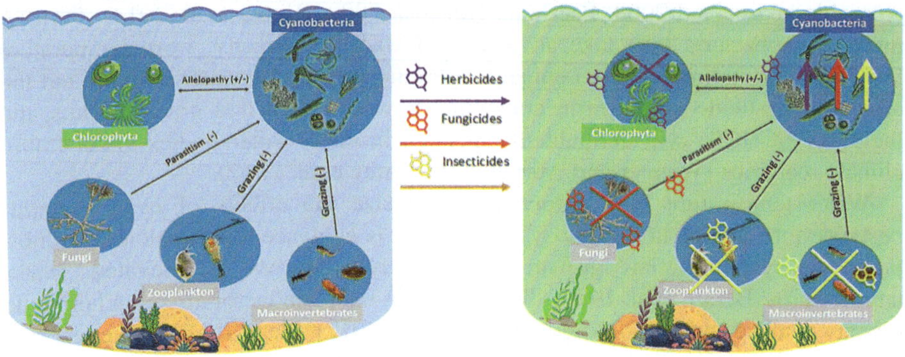

Fig. 11.3 Schematic overview of the main biotic factors modulating cyanobacteria's relative abundance in the presence of pesticides. Biotic factors mainly include Chlorophyta, fungi, zooplankton, and macroinvertebrates. The purple font indicates herbicides, the red font indicates fungicides, and the yellow font indicates insecticides. The arrows represent the general interactions. The plus sign indicates potentiated effect, and the minus sign indicates weakened effect. Figure based on Copyright ©2019, Tao Lu et al., used under Creative Commons CC BY license and adapted considering the results of Tables 11.S1, 11.S2, and 11.S3

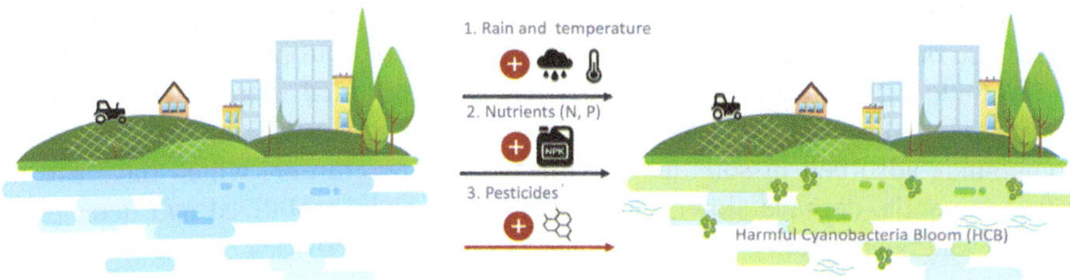

Fig. 11.4 Potential effects of the increase of temperature, rainfalls, nutrients, and pesticides on cyanobacteria abundance

11.3.1 How Will Warmer Climatic Conditions Influence Pesticide Effects in Cyanobacteria?

Environmental factors strongly influence pesticide runoff, toxicity, and HCB frequency and nuisance; this shift toward an HCB is illustrated in Fig. 11.4 (Bérard et al. 1999; Ippolito et al. 2014; Harris and Smith 2016; Gomes and Juneau 2017). At the same time, cyanobacterial species havel species-specific responses to complex interactions that affect the toxicity of pesticides to cyanobacteria, interactions between light, temperature, and pesticides impossibilities the consensus of a generalist model of how climate parameters affect the toxicity of pesticides (Gomes et al. 2017).

Based on the extensive research conducted about the evolutionary history, ecophysiology, and dynamics of cyanobacteria, it is suggested that they will thrive and probably get favored under conditions predicted for global climate change in eutrophic lakes (Paul 2008; Paerl and Huisman 2009; Rigosi et al. 2014). Furthermore, cyanobacteria may take advantage of warming directly, from temperature increases, and indirectly, from enhanced stratification of the water column and the effects of pesticide pollution in competing phytoplankton and grazers. Thus, the incidence of HCBs will undoubtedly increase in many water bodies under future climate scenarios (Wagner and Adrian 2009; Carey et al. 2012).

Water temperature may reduce or emphasize the effects of cyanobacteria-pesticide interactions in aquatic systems. There is evidence that herbicide's inhibition of cyanobacterial growth significantly reduces at elevated water temperatures (Bérard et al. 1999). Temperature is the main factor affecting herbicide toxicity to algae and cyanobacteria (Gomes et al. 2017). Many studies have reported an increase in toxicity of pesticides at warmer temperatures, but it is important to contrast that with the highest rate of pesticide degradation and consequent shorter exposure (Hooper et al. 2013). The probable synergism between temperature and HCBs is far to be established, and it is likely to be more related to nutrient and trophic conditions of the aquatic ecosystem (Rigosi et al. 2014; Op de Beeck et al. 2017).

The rain washes off phosphorus and pesticides from croplands into the streams, rivers, and lakes. As a result, the water flows into the aquatic system and, when it reaches the water system, fuels the growth of cyanobacteria, turning the color of water into bright green (Fig. 11.4) (Chorus 2001). In (sub)tropical areas, the blooms of the toxic cyanobacteria *Microcystis* have been reported more often during the rainy season (Daam et al. 2009). Rain events also can mobilize persistent active compounds adsorbed in soils and sediments, like pyrethroids which have a solid affinity to soils, degrade slowly, and have shown to be mobilized and detected in surface waters after rain events (Delgado-Moreno et al. 2011).

11.4 Risk of Pesticide Pollution Effects in Warm Waters and High Precipitation: Example of South America

Locations of large-scale agriculture in Asia, Latin America, Africa, Australia, and the Pacific Basin reported the most intense impacts of HCBs and cyanotoxins (Huisman et al. 2018; Brêda-Alves et al. 2021). On a global scale, algal blooms are getting more widespread and intense. The worsening of bloom conditions in lakes around the globe was revealed in a recent long-term observational study of 71 lakes worldwide; the authors observed an increase in phytoplankton bloom intensity in 68% of the lakes studied, and only those that experienced the least warming were able to sustain improvements in bloom conditions (8%). It is clear that aquatic systems worldwide in the last 30 years followed a trend toward more intense blooms; in some lakes, the water temperature may explain the improvement in some less warmed lakes, but the authors did not find consistent patterns among the lakes where blooms have gotten worse when considering fertilizer use, rainfall, or temperature (Ho et al. 2019). This suggests that climate influences lake recovery but does not significantly explain the deterioration. However, none of the studies includes the pesticide use and pollution risk with resultant water deterioration by HCBs.

Currently, it is pointed out that an alarming 74.8% of the global agricultural land (approximately 28.8 million km^2) is at risk of pesticide pollution, and in Latin America and the Caribbean, 76% (6.2 million km^2) is at a high risk (Tang et al. 2021). Additionally, in this region, repercussions of pesticide dispersion in the aquatic environment remain largely unknown due to the high monitoring cost that leads to a lack of comprehensive quantification of active ingredient use and residue (Zamora-Sequeira et al. 2019; Norman et al. 2020). It is also noteworthy that the levels of pesticide use in Latin American countries are underestimated due to failures in registration, data availability, and laxer protection measures in the countries (Echeverría-Sáenz et al. 2021).

South America has a diversified environmental landscape, comprehending different climates (Basso et al. 2001). The continent hosts regions with significant precipitations (Fig. 11.5c): in the Amazonas basin, which is the zone with the highest rainfalls with annual accumulations up to 4000 mm, and in the La Plata River basin, which can present accumulated precipitations in the order of 2000 mm per year. Important basins such as the Amazon, the Parana Plata, and the Orinoco carry into the Atlantic Ocean up to 30% of the world's freshwater (Basso et al. 2001; Marengo 2009). On the other hand, in temperature terms (Fig. 11.5b), tropical regions of South America present warmer temperatures in the order of 24 °C. In the central area of the continent, cooler temperatures rounding 14 °C are observed. In the Andine high-plane and south region of the continent, cool temperatures are observed between 1 °C and 3 °C (WMO 2021).

The southeast of the continent, including the south of Brazil, Uruguay, Paraguay, and Argentina, is characterized by large-scale agriculture and livestock raising and increasing urbanization and intensification of economic activities. In the last decade, these countries have emerged to become among the worldwide top nations of agricultural production, mainly soybean, maize, and sugarcane (Buytaert and Breuer 2013). As warm climates allow year-round cultivation, the application of agrochemicals is made with short or absence of pulses, so pesticides become persistent water pollutants with high frequencies of detection (Hernando et al. 2006; Echeverría-Sáenz et al. 2021). For example, the fungicide pyrimethanil and the herbicide diuron have a detection frequency of almost 90% in the water samples from the Madre de Dios River basin in Costa Rica, in the same way. In the Guapuré River in Brazil, the compounds 2,4-D, atrazine, simazine, carbendazim, tebuconazole, and imidacloprid were detected between 75 and 100% (de Castro Lima et al. 2020). Literature shows a common pattern in pesticide detection; according to Norman et al.(2020), the main compounds detected in streams are herbicides, fungicides, and insecticides. This is confirmed in streams in Brazil, Paraguay, Argentina, and Chile by recent studies (Climent et al. 2019; Corcoran et al. 2020; Becker et al. 2021).

The effect of the rainfall associated with runoff and pesticide pollution and warmer waters can increase the amount of pesticide that enters a watershed and the toxicity on nontarget organisms (Michalak et al. 2013; Ippolito et al. 2014; Tang et al. 2021). Summarizing these factors, we can hypothesize that the conversion of risk of pesticide pollution, precipitation, warm temperature, fertilizers use, and the

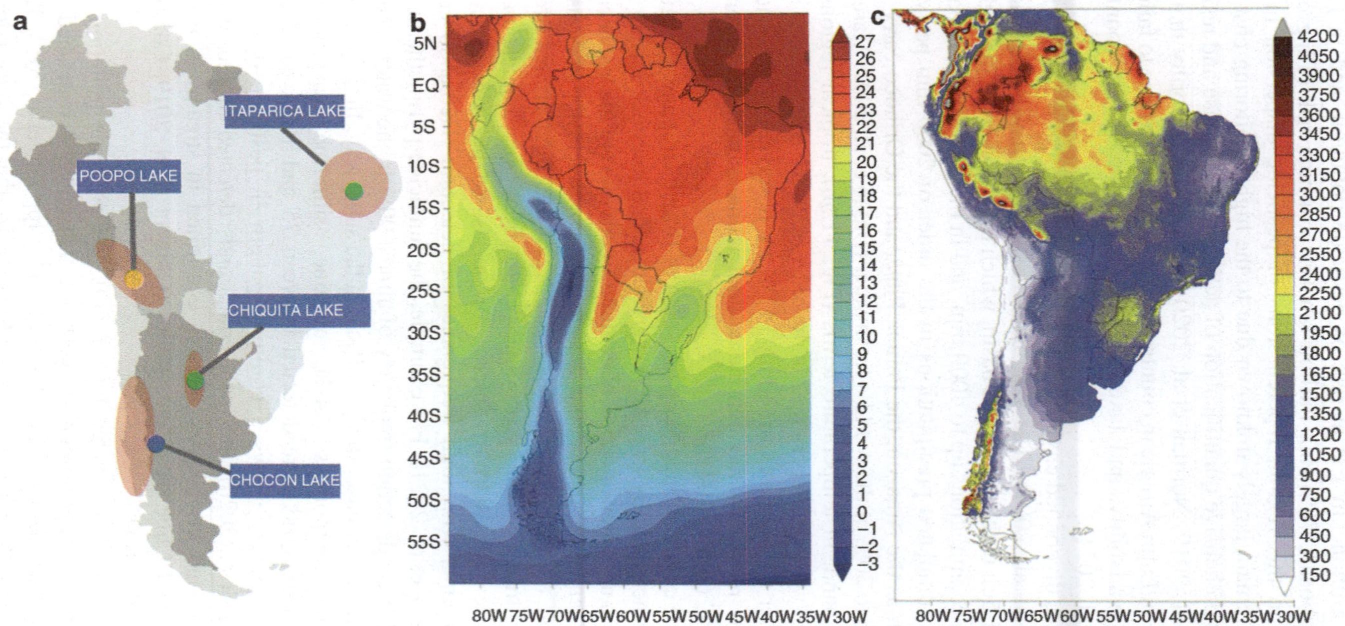

Fig. 11.5 Schematic maps of South America. (**a**) Map of risk of HCBs in South America that resulted from summarizing the data of pesticide pollution risk map proposed by Tang et al. (2021), the map of monitored lakes with blooming conditions of Ho et al. (2019), and the climatic conditions of temperature and rainfall data for the region. The blue dot represents HCB condition improving; the orange dot represents moderate improvement; and the green dots represent a deterioration of the HCB conditions of the lakes according to Ho et al. (2019). The red zones in (**a**) represent the regions at a high risk of pesticide pollution. (**b**) Map of average temperatures reported by Kalnay et al. (1996) in the period 1981–2010; the color scale is in °C. (**c**) Map of annual accumulated precipitation from 1981 to 2010 in CHIRPS (2022) database; the color scale is in millimeters

occurrence of HCBs can generate a map of regions of concern highly susceptible to HCB events. As a case study, we propose a map risk of HCBs of South America (Fig. 11.5a); this illustrates the importance of climate zone as an environmental factor affecting the sensitivity of the aquatic system to pesticides.

Recent and important studies make available georeferenced data of the pesticide pollution risk worldwide (Tang et al. 2021); long-term trends in phytoplankton blooms for 71 lakes, 4 of them located in South America (Ho et al. 2019); and temperature and precipitations already available in several database platforms (Kalnay et al. 1996; Funk et al. 2015; CHIRPS 2022). It can be assumed that as cyanobacteria are more characteristic of warm aquatic systems (Butterwick et al. 2005), they are responsible for the major part of the phytoplankton blooms observed around the globe (Sivonen and Jones 2003).

Exploring and analyzing the available data in the context of HCBs, in Fig. 11.5a, we summarize trends for the four lakes in South America. The red balloons represent the zones with a high risk of pesticide pollution, the color dots represent the state of deterioration or improvement of the bloom conditions, and blue means improvement, orange moderate deterioration, and green deterioration. Figure 11.5b, c represent climate conditions, temperature from 1990 to 2019, and accumulated precipitations from 1981 to 2010, respectively.

Two lakes presented a significant deterioration, one in the northeast zone of Brazil, Lake Itaparica, with a median temperature of 23.7 °C. In a tropical zone without annual variation, the total annual precipitation is 530.7 mm, with the rainiest period in December to April (Supplementary Fig. S.1a). The second lake, Lake Chiquita in Argentina, with an annual temperature of 20.8 °C, has a clear thermal amplitude during the year with hot summers and cold winters. The total annual precipitation is 819.4 mm, with the rainiest period from November to March (Supplementary Fig. S.1b).

Of the four lakes, Lake Chocon was the only one with improvement in bloom conditions despite the high pesticide pollution risk. In the lake, an annual median temperature of 6.9 °C is observed, with a variation in summer and winter months, and the total annual precipitation is 119.7 mm, with rainy periods in November and February (Supplementary Fig. S.1c). Lake Poopo presented a moderate deterioration, located in the Andine high plane with a median temperature of 4.1 °C, without considerable thermic amplitude in a year period; the total annual precipitation is 379.4 mm, showing the main rainfalls during December and March (Supplementary Fig. S.1d).

Taken together, as Ho et al. (2019) pointed out, the lakes with less warming condition presented a better state considering the bloom conditions, in regions with a high risk of pesticide pollution, warm temperatures, and important precipitations; it is evidenced that high temperatures may be the main factor affecting herbicide toxicity to cyanobacteria (Gomes and Juneau 2017) and, thus, can be a key factor to explain and determine which lake is prone to the worst HCB conditions.

11.5 Impact of Pesticides on Cyanobacteria

Cyanobacteria may be favored in the presence of pesticides because they (1) have a higher tolerance to pesticides; (2) can use nutrients bound within pesticides to stimulate their growth (Harris and Smith 2016); (3) pesticide affects the top-down control of cyanobacterial competing organisms such as phytoplankton, zooplankton, and macroinvertebrates (Bérard et al. 1999; Fleeger et al. 2003; Norman et al. 2020); and (4) can degrade active ingredients into nontoxic forms and use them as a source of nutrients to support growth (Brêda-Alves et al. 2021). In natural systems, heterotrophic bacteria, zooplankton, fungi, other phytoplankton, and microbial community composition play an essential role in the response of cyanobacteria in the presence of pesticides (Harris and Smith 2016). There are also a few reports of strains of cyanobacteria resistant to fungicides, insecticides, and atrazine (Whitton 2000).

The literature shows ambiguous responses of cyanobacteria to pesticides (Supplementary tables and Fig. 11.6). There are several contradictory studies, and it is hypothesized that this can be due to the differences between field and laboratory conditions (Harris and Smith 2016). In laboratory experiments, responses to questions about a particular pesticide and cyanobacteria are obtained where the growing conditions are controlled, mostly with single-species tests in actively growing cells. In contrast, in natural waters, cyanobacteria tests are multispecies, cells differ in age, and spores are exposed to pesticides in a system under the influence of environmental factors, introducing elements that lead to a variation of the observed responses (Padhy 1985; Bérard et al. 1999; Harris and Smith 2016).

In this analysis, we considered 111 experiments from year 2000 to date. Summarizing the revised experiments, herbicides favor cyanobacteria (30 positive effect studies vs. 8 with negative effect); responses to fungicides and insecticides that

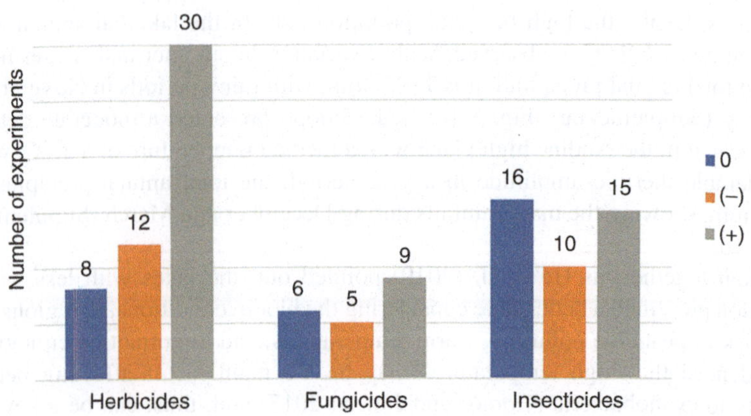

Fig. 11.6 Histograms showing the response of cyanobacteria in single-species, mesocosm, and microcosm experiments in the presence of pesticides. The color bars refer to positive (gray), neutral (blue), and negative (orange) categories. The table listing the experiments are available in Tables 11.S1, 11.S2, and 11.S3

act mainly indirectly on cyanobacteria are varied. However, as cyanobacteria strains resistant to pesticides are reported, the category no effect (0) may indicate that there's some evidence that cyanobacteria thrive on adverse pesticide effects.

11.6 Herbicides

Herbicides affect photosynthetic organisms through several modes of action (Jha and Mishra 2005; Harris and Smith 2016). The main ways that can affect cyanobacteria are (1) amino acid pathway inhibitors (e.g., glyphosate); (2) photosynthesis inhibitors (e.g., triazines); (3) growth regulators (e.g., 2-4D); (4) cell membrane disruptors (e.g., paraquat and diquat); and (5) nitrogen-fixing inhibitors.

The influence of herbicides on cyanobacteria has been investigated extensively (Padhy 1985; DeLorenzo et al. 2001; Gomes et al. 2017), though most have been restricted to laboratory cultures. In older literature, cyanobacteria are thought to be quite sensitive to herbicides due to their photosynthetic activity, the feature of similarity with plants, and the site of action of many herbicides (Whitton 2000). Actually, evidence shows that cyanobacteria photoregulation processes decrease in the presence of atrazine (Deblois et al. 2013), and other results demonstrate that exposure to common and widely applied herbicides, like metribuzin, at very low concentrations indirectly affects invertebrate populations favoring cyanobacterial system dominance (Brock et al. 2004; Lürling and Roessink 2006; Harris and Smith 2016).

The majority of the photosynthesis-inhibiting herbicides can be categorized into one of the two groups: those that impede photosystem I (PSI) and those that inhibit photosystem II (PSII) (Tucci et al. 2019). These inhibiting herbicides may affect the phytoplankton community structure and provide cyanobacteria an opportunity for dominance, promoting HCBs (Lürling and Roessink 2006). Consequently, the increased levels of cyanotoxins and herbicides can trigger cascades of adverse toxicological effects on nontargeted organisms, altering the food chain (Brêda-Alves et al. 2021).

PSI inhibitors can accept electrons from PSI-forming radicals, and their high reactivity can destroy living cells (Al-Khatib 2022). Examples of PSI inhibitors are paraquat, diquat, and bipyridilium (Tucci et al. 2019). Among this type of herbicide, paraquat is the most used in the environment and is very persistent because of its strong affinity to soil particles, with a half-life of up to 20 years (Huang et al. 2019). Photosystem II is responsible for forming oxygen from water (Lodish et al. 2007). PSII inhibitors block the binding site of the primary electron acceptor, a protein called plastoquinone, and thus they act precluding the electron transfer to the intermediate (Metz et al. 1986). Examples of PSII inhibitors are triazines, ureas, triazinones, phenyl carbamates, benzonitriles, and benzothiadiazoles (Tucci et al. 2019).

The deleterious effects on photosynthesis of some herbicides are associated with cell oxidative bursts due to ROS accumulation and, therefore, the lipid peroxidation that leads to chloroplast's cell membrane destruction (Jursinic and Stemler 1983;

Gomes and Juneau 2017). In cyanobacteria, this ROS-induced stress is scavenged by antioxidant defense systems, including redox homeostasis of glutamine (GSH) (Babele et al. 2019). Glutathionylation is a central mechanism in cyanobacteria to thrive on oxidative stress caused by herbicides (Chardonnet et al. 2015). On the other hand, chloroplast mutations may lead to loss of herbicide binding and inhibition, e.g., diuron-resistant strains have been discovered in phytoplankton species (Faust et al. 1995; Babele et al. 2019), and dissipation of endosulfan, a potent PSII inhibitor (Prasad et al. 2011), was observed in laboratory conditions, indicating biodegradation of this pesticide by cyanobacteria (Jha and Mishra 2005).

In addition to the direct toxicological effect of pesticides in surface waters, some studies have shown that organophosphorus pesticides like glyphosate, which contains phosphorous, may increase eutrophication (Vera et al. 2010) and stimulate cyanobacterial growth and nuisance at environmentally relevant concentrations (Sun et al. 2013). Some species of cyanobacteria have the competitive advantage over eukaryotic algae; they can use the phosphorus bound within glyphosate molecule to support their growth, e.g., in an experiment, *M. aeruginosa* presented an increased specific growth rate, and additionally, glyphosate stimulated the production of tryptophane-like substances, fulvic acid-like substances, and the cyanotoxin microcystin-leucine-arginine (Wang et al. 2021). However, algae and cyanobacteria are considered tolerant to glyphosate (Wang et al. 2016). Like other pesticides, their toxicity in environmental compartments is modulated by climatic conditions, mainly by temperature, even though studies on algae and cyanobacteria involving glyphosate and temperature are still scarce (Pesce et al. 2009; Gomes and Juneau 2017).

11.6.1 Atrazine

Atrazine (ATZ) is an s-triazine herbicide used primarily to control broadleaf plants and grassy weeds, inhibiting photosynthesis by blocking electron transport reaction of photosystem II and inducing damage in the photosynthetic apparatus (Forney and Davis 1981; Vencill et al. 2002). However, ATZ possesses a high potential to contaminate groundwater, so it was removed from the list of approved products in the European Union (Ackerman 2007); despite this, ATZ is extensively used and routinely detected in surface waters because of the runoff from agricultural fields (DeLorenzo et al. 2001).

The most toxic metabolite of atrazine is deethylatrazine (DEA). This compound is nearly as harmful as atrazine and significantly reduces chlorophyll content, phototrophic carbon assimilation, biovolume, and dissolved oxygen levels of estuarine microbial communities at 50 mg/L (Downing et al. 2004). DEA is rapidly transported in surface waters, and there it is persistent during 45–65 days (Pereira and Hostettler 1993; DeLorenzo et al. 2001).

The effects of atrazine in aquatic environments were recently reviewed by Castro et al. (2021) and Pereira de Albuquerque et al. (2020). In those analyses, the authors remarked a scarce number of studies in aquatic organisms compared to the number of studies with terrestrial organisms (Pereira De Albuquerque et al. 2020; Castro et al. 2021).

In aquatic systems, ATZ is algistatic, and it can directly influence the ecophysiology and dynamics of phytoplankton populations (Bérard et al. 1999). Atrazine and metribuzin in environmentally relevant concentrations may favor cyanobacteria growth at the expense of other phytoplankton species (Lürling and Roessink 2006). ATZ in concentrations as low as 0.1 μ/L may cause significant effects on algal community structure (Pannard et al. 2009). Pannard et al. (2009) pointed out that in conditions of high nutrient supply, cyanobacteria were more tolerant to ATZ than other phytoplankton taxa. This evidenced that ATZ becomes more toxic to phytoplankton and may favor cyanobacteria under nutrient-enriched conditions (DeLorenzo et al. 2001).

11.7 Non-herbicide Pesticides

11.7.1 Insecticides

Stehle and Schulz (2015), analyzing agricultural insecticide in surface waters worldwide, observed that insecticides were rarely detected, but they exceeded water quality thresholds when detected (Stehle and Schulz 2015).

In natural habitats, predators grazing on cyanobacteria control populations and after insecticide application, cyanobacteria may appear in large masses (Padhy 1985). Pesticide affects grazer populations and indirectly beneficiates cyanobacteria. According to Stehle and Schulz (2015), in the presence of pesticides, the richness of macroinvertebrate families was reduced by approximately 30%. Other examples are the studies made by Daam et al. (2008), who found a significant decrease in cladocerans after repeated applications of the organophosphorus insecticide chlorpyrifos (Daam et al. 2008), and Cerbin et al. (2010), who reported the synergistic adverse effect of the toxin-forming cyanobacteria *Microcystis aeruginosa* and the insecticide carbaryl on the life cycle and offspring of the zooplankton *Daphnia pulicaria* (Cerbin et al. 2010).

Pesticides bioaccumulated in prey species may cause toxic effects in the consumers or continue to biomagnify through the food web (DeLorenzo et al. 2001; Timbrell 2019). For example, lipophilic active compounds like endosulfan bioaccumulate in phytoplankton consumed in higher concentrations by grazers as zooplankton and macroinvertebrates (DeLorenzo et al. 2001).

At higher temperatures, surface water viscosity decreases, promoting the sinking of larger, nonmotile phytoplankton (e.g., diatoms), giving cyanobacteria a further advantage (Paerl and Huisman 2009). Several cyanobacteria have modulated buoyancy through polysaccharide synthesis (Reynolds et al. 1987). Due to devoid mechanisms, insecticides may have a pronounced effect on cyanobacteria, reducing polysaccharide production (Jha and Mishra 2005); this facilitates migration to the water surface to exploit light and nutrients (Zhang et al. 2019). In warm conditions and insecticide pollution, cyanobacteria may have an important advantage based on this evidence.

Cyanobacteria respond to short-term exposure to the organophosphate insecticide methyl-parathion, ensuring its survival through the induction of antioxidative enzymes and proteins that may have a role in degrading the pesticide (Tiwari et al. 2018). For example, in experiments performed by Ibrahim et al. (2014), the cyanobacteria *Nostoc muscorum* can tolerate different concentrations of malathion and be responsible for 91% of the pesticide biodegradation.

11.7.1.1 Imidacloprid

Neonicotinoids are broad-spectrum systemic insecticides, are the fastest-growing class of insecticides worldwide, and are now registered for use in over 120 different countries (Jeschke and Nauen 2008). They act selectively on the insect's central nervous system in the postsynaptic nicotinic acetylcholine receptors (Bai et al. 1991). In the environment, they are persistent, have a high runoff potential, and are toxic to a wide range of invertebrates (De Perre Dimitri et al. 2021).

Imidacloprid is by far the most widely studied and detected neonicotinoid in surface waters, with concentrations in surface waters worldwide ranging between 0.001 and 320 mg/L (Morrissey et al. 2015; Norman et al. 2020). Its aquatic concentrations are reduced by the high levels of UV light and, to a minor extent, by sorption to sediment (Lu et al. 2015).

Dimitri et al. (2021) observed that in warm conditions, imidacloprid-induced population declines on larger zooplankton species (*Diaptomus* sp. and Ostracoda) and shifted the phytoplankton community to the dominance of graze-resistant gelatinous cyanobacteria. Imidacloprid affects the top-down control that macroinvertebrate and zooplankton taxa impose on phytoplankton (Fleeger et al. 2003).

HCBs can indirectly reduce the photolysis of imidacloprid by shading and thus decrease its degradation; in laboratory experiments with imidacloprid blooms, as Dimitri et al. (2021) observed. However, at present, there are no studies that have investigated the combined effect of imidacloprid and cyanobacteria (De Perre Dimitri et al. 2021).

11.7.2 Fungicides

Widely used agricultural fungicides might promote harmful algal blooms by inhibiting cyanobacterial natural antagonists. Commonly used fungicides tebuconazole and azoxystrobin in two experiments induced harmful algae blooms by suppressing fungal pathogens that may otherwise control their growth (Ortiz-Cañavate et al. 2019; Lu et al. 2019).

11.7.2.1 Carbendazim

To date, few studies provide evidence of the effect of carbendazim on cyanobacteria. In a study of Kannaiyan (1996) where symbiotic forms of cyanobacterial cultures were exposed to carbendazim, it was observed that ammonia excretion was enhanced, glutamine synthetase activity was markedly inhibited, and the pesticide treatment stimulated growth and nitrogenase activity of the cyanobacterial cultures.

In a microcosm experiment performed by Cuppen et al. (2000), carbendazim proved very persistent in the water layer with values that varied between 6 and 25 weeks; in the same experiment, the macroinvertebrate community was seriously affected by carbendazim application (Cuppen et al. 2000). Daam et al. (2008) observed the death of sensitive invertebrates and noted the increased levels of chlorophyll-*a* and indicators of increasing phytoplankton in the aquatic system, as a consequence of the overall decreased grazing pressure after carbendazim exposure (Daam et al. 2009). These few pieces of evidence may indicate an indirect effect of carbendazim on phytoplankton and cyanobacteria, where the decrease of the grazing pressure and competition may shift the aquatic system to a possible cyanobacterial dominance (Fig. 11.3).

11.8 Concluding Remarks

This chapter examined the links and effects of pesticides on cyanobacteria in aquatic systems; with particular attention to warmer waters in conjunction with pesticide risks and precipitations, we propose as a case study four important lakes in South America. Results may indicate, as Ho et al. (2019) pointed out, that the lakes with less warming condition presented better state considering the bloom conditions, in regions with a high risk of pesticide pollution, warm temperatures, and important precipitations; it is evidenced that the high temperatures may be the main factor affecting herbicide toxicity to cyanobacteria (Gomes and Juneau 2017) and, thus, can be a key factor to explain and determine which lake is prone to the worst harmful cyanobacteria bloom conditions.

The analysis was divided into two broad groups of active ingredients: pesticides that affect target photosynthetic organisms (herbicides) and those that do not and present effects on cyanobacteria (insecticides and fungicides). Focusing in the effects of the pesticides with a high frequency of occurrence according to Norman et al. (2020) thought herbicide atrazine, the insecticide imidacloprid, and the fungicide carbendazim.

The route of pesticides to the water compartment may be either direct, through the direct application, or indirect, when it moves mainly from the soil of pesticide-treated areas or by runoff. When released in aquatic systems, pesticides can initiate a cascade of consumer-resource interactions, increasing the abundance of tolerant species such as cyanobacteria (Bérard et al. 1999; DeLorenzo et al. 2001; Fleeger et al. 2003). Thus, pesticides can cause direct or indirect toxicity to microalgae and cyanobacteria, inducing changes through ecological relations, so evaluations of the effects of any given pesticide on cyanobacteria should not be limited to its toxicity to a single species, and the studies' approach may focus on indirect effects that can be more important than direct effects, making nontarget organisms better indicators of pesticide stress to natural communities (Lampert et al. 1989).

After pesticide use, cyanobacteria frequently appear in large masses, and these applications may suppress the growth of competing phytoplankton, control the population of grazers, and/or stimulate cyanobacterial growth (Padhy 1985; Harris and Smith 2016). On the other hand, cyanobacteria are targets for herbicides in aquatic ecosystems (Whitton and Potts 2000; Ma et al. 2010). Evidence varies, showing that cyanobacteria may be favored and/or they are sensitive to pesticides. Literature shows ambiguous responses of cyanobacteria to pesticides. There are several contradictory studies, and it is hypothesized that this can be due to the differences between field and laboratory conditions (Harris and Smith 2016). Summarizing the experiments revised, herbicides are prone to favor cyanobacteria (30 positive effect studies vs. 8 with negative effect), and responses to fungicides and insecticides that act mainly indirectly on cyanobacteria are varied.

Environmental factors strongly influence pesticide effects on cyanobacteria (Bérard et al. 1999; Ippolito et al. 2014; Harris and Smith 2016; Gomes and Juneau 2017). At the same time, different cyanobacterial species present specific responses to complex environmental interactions such as light, temperature, and pesticides. This scenario, make not possible the consensus of a generalist model of how climate parameters affect the toxicity of pesticides to cyanobacteria (Gomes et al. 2017).

To sum up, we are aware that important topics might be overlooked; there is a well-established active field of studies taking place that take into account global scenarios and trends in active ingredients, considering the information and data in the literature that was covered; and we can conclude that besides the variable effects of pesticides in cyanobacteria and HCBs, there is a tendency that future climatic, agricultural, and environmental scenarios will certainly favor cyanobacteria blooms and threaten water quality (Paerl and Huisman 2009; Kosten et al. 2012; Echeverría-Sáenz et al. 2021).

References

Ackerman F (2007) The economics of atrazine. Int J Occup Environ Health 13:437–445. https://doi.org/10.1179/OEH.2007.13.4.437

Akashe MM, Pawade UV, Nikam AV (2018) Classification of pesticides: a review. J Res Ayurveda Pharm 9:144. https://doi.org/10.7897/2277-4343.094131

Al-Khatib K (2022) Photosystem I (PSI) electron diverter - herbicide symptoms. https://herbicidesymptoms.ipm.ucanr.edu/MOA/Photosystem_I_PSI_Electron_Diverter/. Accessed 13 Mar 2022

Babele PK, Kumar J, Chaturvedi V (2019) Proteomic De-regulation in cyanobacteria in response to abiotic stresses. Front Microbiol 10:1315. https://doi.org/10.3389/FMICB.2019.01315/BIBTEX

Bai D, Lummis SCR, Leicht W et al (1991) Actions of imidacloprid and a related nitromethylene on cholinergic receptors of an identified insect motor neurone. Pestic Sci 33:197–204. https://doi.org/10.1002/PS.2780330208

Basso E, Compagnucci R, Fearnside P, et al (2001) Latin America. In: Latin in America: IPCC, Intergovernmental Panel on Climate Change. Nobre, C, pp 699–707

Becker RW, Araújo DS, Sirtori C et al (2021) Pesticides in surface water from Brazil and Paraguay cross-border region: screening using LC-QTOF MS and correlation with land use and occupation through multivariate analysis. Microchem J 168:106502. https://doi.org/10.1016/J.MICROC.2021.106502

Bérard A, Leboulanger C, Pelte T (1999) Tolerance of Oscillatoria limnetica Lemmermann to atrazine in natural phytoplankton populations and in pure culture: influence of season and temperature. Arch Environ Contam Toxicol 37:472–479. https://doi.org/10.1007/s002449900541

Brêda-Alves F, de Oliveira Fernandes V, Chia MA (2021) Understanding the environmental roles of herbicides on cyanobacteria, cyanotoxins, and cyanoHABs. Aquat Ecol 552(55):347–361. https://doi.org/10.1007/S10452-021-09849-2

Brock TCM, Crum SJH, Deneer JW et al (2004) Comparing aquatic risk assessment methods for the photosynthesis-inhibiting herbicides metribuzin and metamitron. Environ Pollut 130:403–426. https://doi.org/10.1016/J.ENVPOL.2003.12.022

Butterwick C, Heaney SI, Talling JF (2005) Diversity in the influence of temperature on the growth rates of freshwater algae, and its ecological relevance. Freshw Biol 50:291–300. https://doi.org/10.1111/J.1365-2427.2004.01317.X

Buytaert W, Breuer L (2013) Water resources in South America: sources and supply, pollutants and perspectives, 359

Carey CC, Ibelings BW, Hoffmann EP et al (2012) Eco-physiological adaptations that favour freshwater cyanobacteria in a changing climate. Water Res 46:1394–1407. https://doi.org/10.1016/J.WATRES.2011.12.016

Castro MS, Barbosa FG, Guimarães PS et al (2021) A scientometric analysis of ecotoxicological studies with the herbicide atrazine and microalgae and cyanobacteria as test organisms. Environ Sci Pollut Res 28:25196–25206. https://doi.org/10.1007/S11356-020-12213-W

Cerbin S, Kraak MHS, de Voogt P et al (2010) Combined and single effects of pesticide carbaryl and toxic Microcystis aeruginosa on the life history of Daphnia pulicaria. Hydrobiologia 643:129–138. https://doi.org/10.1007/S10750-010-0130-1/FIGURES/3

Chardonnet S, Sakr S, Cassier-Chauvat C et al (2015) First proteomic study of S-glutathionylation in cyanobacteria. J Proteome Res 14:59–71. https://doi.org/10.1021/PR500625A

CHIRPS (2022) CHIRPS DAILY: climate hazards group InfraRed precipitation with station data (version 2.0 final) | Earth Engine Data Catalog | Google Developers. https://developers.google.com/earth-engine/datasets/catalog/UCSBCHG_CHIRPS_DAILY. Accessed 14 Mar 2022

Chorus I (2001) Cyanotoxins: occurrence, causes, consequences. Springer, Berlin

Chorus I, Welker M (2021) Toxic cyanobacteria in water, 2nd edn. CRC Press, London

Chow R, Scheidegger R, Doppler T et al (2020) A review of long-term pesticide monitoring studies to assess surface water quality trends. Water Res X 9:100064

Climent MJ, Herrero-Hernández E, Sánchez-Martín MJ et al (2019) Residues of pesticides and some metabolites in dissolved and particulate phase in surface stream water of Cachapoal River basin, Central Chile. Environ Pollut 251:90–101. https://doi.org/10.1016/J.ENVPOL.2019.04.117

Codd GA (2000) Cyanobacterial toxins, the perception of water quality, and the prioritisation of eutrophication control. Ecol Eng 16:51–60. https://doi.org/10.1016/S0925-8574(00)00089-6

Codd G, Bell S, Kaya K et al (1999) Cyanobacterial toxins, exposure routes and human health. Eur J Phycol 34:405–415. https://doi.org/10.1080/09670269910001736462

Corcoran S, Metcalfe CD, Sultana T et al (2020) Pesticides in surface waters in Argentina monitored using polar organic chemical integrative samplers. Bull Environ Contam Toxicol 104:21–26. https://doi.org/10.1007/S00128-019-02758-Z

Cordell D, Drangert J-O, White S (2009) The story of phosphorus: global food security and food for thought. https://doi.org/10.1016/j.gloenvcha.2008.10.009

Cuppen JGM, Van Den Brink PJ, Camps E et al (2000) Impact of the fungicide carbendazim in freshwater microcosms. I. Water quality, breakdown of particulate organic matter and responses of macroinvertebrates. Aquat Toxicol 48:233–250. https://doi.org/10.1016/S0166-445X(99)00036-3

Daam MA, Van Den Brink PJ, Nogueira AJA (2008) Impact of single and repeated applications of the insecticide chlorpyrifos on tropical freshwater plankton communities. Ecotoxicology 17: 756–771. https://doi.org/10.1007/S10646-008-0227-8

Daam MA, Rodrigues AMF, Van den Brink PJ, Nogueira AJA (2009) Ecological effects of the herbicide linuron in tropical freshwater microcosms. Ecotoxicol Environ Saf 72:410–423. https://doi.org/10.1016/J.ECOENV.2008.07.009

de Castro Lima JAM, Labanowski J, Bastos MC et al (2020) "Modern agriculture" transfers many pesticides to watercourses: a case study of a representative rural catchment of southern Brazil. Environ Sci Pollut Res 2710(27):10581–10598. https://doi.org/10.1007/S11356-019-06550-8

De Perre Dimitri V, Yao K-S, Li D et al (2021) Imidacloprid treatments induces cyanobacteria blooms in freshwater communities under sub-tropical conditions. Aquat Toxicol 240:105992. https://doi.org/10.1016/j.aquatox.2021.105992

Deblois CP, Dufresne K, Juneau P (2013) Response to variable light intensity in photoacclimated algae and cyanobacteria exposed to atrazine. Aquat Toxicol 126:77–84. https://doi.org/10.1016/J.AQUATOX.2012.09.005

Delgado-Moreno L, Lin K, Veiga-Nascimento R, Gan J (2011) Occurrence and toxicity of three classes of insecticides in water and sediment in two Southern California coastal watersheds. J Agric Food Chem 59:9448–9456. https://doi.org/10.1021/jf202049s

DeLorenzo ME, Scott GI, Ross PE (2001) Toxicity of pesticides to aquatic microorganisms: a review. Environ Toxicol Chem 20:84–98. https://doi.org/10.1002/ETC.5620200108

Doppler T, Camenzuli L, Hirzel G et al (2012) Spatial variability of herbicide mobilisation and transport at catchment scale: insights from a field experiment. Hydrol Earth Syst Sci 16:1947–1967. https://doi.org/10.5194/HESS-16-1947-2012

Downing HF, Delorenzo ME, Fulton MH et al (2004) Effects of the agricultural pesticides atrazine, chlorothalonil, and endosulfan on South Florida microbial assemblages. Ecotoxicology 13:245–260. https://doi.org/10.1023/B:ECTX.0000023569.46544.9F

Echeverría-Sáenz S, Spínola-Parallada M, Soto AC (2021) Pesticides burden in Neotropical Rivers: Costa Rica as a case study. Molecules 26. https://doi.org/10.3390/MOLECULES26237235

Elliott JA (2011) Is the future blue-green? A review of the current model predictions of how climate change could affect pelagic freshwater cyanobacteria. https://doi.org/10.1016/j.watres.2011.12.018

FAO (2017) The future of food and agriculture

Faust M, Altenburger R, Boedeker W, Grimme LH (1995) Algal toxicity of binary combinations of pesticides. Bull Environ Contam Toxicol

Fleeger JW, Carman KR, Nisbet RM (2003) Indirect effects of contaminants in aquatic ecosystems. Sci Total Environ 317:207–233. https://doi.org/10.1016/S0048-9697(03)00141-4

Forney DR, Davis DE (1981) Effects of low concentrations of herbicides on submersed aquatic plants. Weed Sci 29:677–685. https://doi.org/10.1017/S0043174500040261

Funk C, Peterson P, Landsfeld M et al (2015) The climate hazards infrared precipitation with stations—a new environmental record for monitoring extremes. Sci Data 21(2):1–21. https://doi.org/10.1038/sdata.2015.66

Gomes MP, Juneau P (2017) Temperature and light modulation of herbicide toxicity on algal and cyanobacterial physiology. Front Environ Sci 5:50. https://doi.org/10.3389/FENVS.2017.00050/BIBTEX

Gomes MP, Garcia QS, Barreto LC et al (2017) Allelopathy: an overview from micro-to macroscopic organisms, from cells to environments, and the perspectives in a climate-changing world. Biol 72:113–129

Graham JL, Loftin KA, Meyer MT, Ziegler AC (2010) Cyanotoxin mixtures and taste-and-odor compounds in cyanobacterial blooms from the Midwestern United States. Environ Sci Technol 44:7361–7368. https://doi.org/10.1021/ES1008938

Harris TD, Smith VH (2016) Inland waters do persistent organic pollutants stimulate cyanobacterial blooms? View supplementary material 6:124. https://doi.org/10.5268/IW-6.2.887

Hernando MD, Mezcua M, Fernández-Alba AR, Barceló D (2006) Environmental risk assessment of pharmaceutical residues in wastewater effluents, surface waters and sediments. Talanta 69: 334–342. https://doi.org/10.1016/J.TALANTA.2005.09.037

Ho JC, Michalak AM, Pahlevan N (2019) Widespread global increase in intense lake phytoplankton blooms since the 1980s. Nat 5747780(574):667–670. https://doi.org/10.1038/s41586-019-1648-7

Hooper MJ, Ankley GT, Cristol DA et al (2013) Interactions between chemical and climate stressors: a role for mechanistic toxicology in assessing climate change risks. Environ Toxicol Chem 32:32–48. https://doi.org/10.1002/ETC.2043

Huang Y, Zhan H, Bhatt P, Chen S (2019) Paraquat degradation from contaminated environments: current achievements and perspectives. Front Microbiol 10. https://doi.org/10.3389/FMICB.2019.01754

Huisman J, Codd GA, Paerl HW et al (2018) Cyanobacterial blooms. Nat Rev Microbiol 16:471–483. https://doi.org/10.1038/s41579-018-0040-1

Ibrahim WM, Karam MA, El-Shahat RM, Adway AA (2014) Biodegradation and utilization of organophosphorus pesticide malathion by cyanobacteria. Biomed Res Int 2014:1. https://doi.org/10.1155/2014/392682

Ippolito A, Kattwinkel M, Rasmussen JJ et al (2014) Modeling global distribution of agricultural insecticides in surface waters. https://doi.org/10.1016/j.envpol.2014.12.016

Jeschke P, Nauen R (2008) Neonicotinoids—from zero to hero in insecticide chemistry. Pest Manag Sci 64:1084–1098. https://doi.org/10.1002/PS.1631

Jha MN, Mishra SK (2005) Biological responses of cyanobacteria to insecticides and their insecticide degrading potential. Bull Environ Contam Toxicol 75:374–381. https://doi.org/10.1007/S00128-005-0764-2

Jursinic P, Stemler A (1983) Changes in [14C]Atrazine binding associated with the oxidation-reduction state of the secondary quinone acceptor of photosystem II. Plant Physiol 73:703. https://doi.org/10.1104/PP.73.3.703

Kalnay E, Kanamitsu M, Kistler R et al (1996) The NCEP/NCAR 40-year reanalysis project. Bull Am Meteorol Soc 77:437–472. https://doi.org/10.1175/1520-0477(1996)077<0437:TNYRP>2.0.CO;2

Kannaiyan (1996) Effect of the systemic fungicide, Bavistin on the nitrogen status of cyanobacteria under immobilized state in polyurethane foam. S Afr J Bot 62:127–132

Kosten S, Huszar VLM, Bécares E et al (2012) Warmer climates boost cyanobacterial dominance in shallow lakes. Glob Chang Biol 18:118–126. https://doi.org/10.1111/j.1365-2486.2011.02488.x

Kreuger J (1998) Pesticides in stream water within an agricultural catchment in southern Sweden, 1990-1996. Sci Total Environ 216:227–251. https://doi.org/10.1016/S0048-9697(98)00155-7

Lampert W, Fleckner W, Pott E et al (1989) Herbicide effects on planktonic systems of different complexity. Hydrobiol 1881(188):415–424. https://doi.org/10.1007/BF00027809

Leu C, Singer H, Stamm C et al (2004) Simultaneous assessment of sources, processes, and factors influencing herbicide losses to surface waters in a small agricultural catchment. Environ Sci Technol 38:3827–3834. https://doi.org/10.1021/ES0499602

Lodish H, Berk A, Kaiser CA et al (2007) Energetica celular. In: Freeman WH (ed) Molecular cell biology, 5th edn. W.H. Freeman, New York, pp 331–347

Lu Z, Challis JK, Wong CS (2015) Quantum yields for direct photolysis of neonicotinoid insecticides in water: implications for exposure to nontarget aquatic organisms. Environ Sci Technol Lett 2:188–192. https://doi.org/10.1021/ACS.ESTLETT.5B00136/SUPPL_FILE/EZ5B00136_SI_001.PDF

Lu T, Zhang Q, Lavoie M et al (2019) The fungicide azoxystrobin promotes freshwater cyanobacterial dominance through altering competition. Microbiome 7:128. https://doi.org/10.1186/s40168-019-0744-0

Lürling M, Roessink I (2006) On the way to cyanobacterial blooms: impact of the herbicide metribuzin on the competition between a green alga (Scenedesmus) and a cyanobacterium (Microcystis). https://doi.org/10.1016/j.chemosphere.2006.01.073

Ma J, Tong S, Wang P, Chen J (2010) Toxicity of seven herbicides to the three cyanobacteria Anabaena flos-aquae, Microcystis flos-aquae and Mirocystis aeruginosa. Int J Environ Res 4: 347–352. https://doi.org/10.22059/IJER.2010.27

Marengo JA (2009) Long-term trends and cycles in the hydrometeorology of the Amazon basin since the late 1920s. Hydrol Process 23:3236–3244. https://doi.org/10.1002/HYP.7396

Metcalf JS, Codd GA (2020) Co-occurrence of cyanobacteria and cyanotoxins with other environmental health hazards: impacts and implications. Toxins 12:629. https://doi.org/10.3390/TOXINS12100629

Metz JG, Pakrasi HB, Seibert M, Arntzer CJ (1986) Evidence for a dual function of the herbicide-binding D1 protein in photosystem II. FEBS Lett 205:269–274. https://doi.org/10.1016/0014-5793(86)80911-5

Michalak AM, Anderson EJ, Beletsky D et al (2013) Record-setting algal bloom in Lake Erie caused by agricultural and meteorological trends consistent with expected future conditions. Proc Natl Acad Sci U S A 110:6448–6452. https://doi.org/10.1073/pnas.1216006110

Morrissey CA, Mineau P, Devries JH et al (2015) Neonicotinoid contamination of global surface waters and associated risk to aquatic invertebrates: a review. Environ Int 74:291–303. https://doi.org/10.1016/J.ENVINT.2014.10.024

Norman JE, Mahler BJ, Nowell LH et al (2020) Daily stream samples reveal highly complex pesticide occurrence and potential toxicity to aquatic life. Sci Total Environ 715:136795. https://doi.org/10.1016/j.scitotenv.2020.136795

OECD (2006) OECD guidance for country data review reports on microbial pest control products and their microbial pest control agents

Oerke EC (2006) Crop losses to pests. J Agric Sci 144:31–43

Op de Beeck L, Verheyen J, Olsen K, Stoks R (2017) Negative effects of pesticides under global warming can be counteracted by a higher degradation rate and thermal adaptation. J Appl Ecol 54:1847–1855. https://doi.org/10.1111/1365-2664.12919

Ortiz-Cañavate BK, Wolinska J, Agha R (2019) Fungicides at environmentally relevant concentrations can promote the proliferation of toxic bloom-forming cyanobacteria by inhibiting natural fungal parasite epidemics. Chemosphere 229:18–21. https://doi.org/10.1016/J.CHEMOSPHERE.2019.04.203

Padhy RN (1985) Cyanobacteria and pesticides. Residue Rev 95:1–44. https://doi.org/10.1007/978-1-4612-5132-3_1

Paerl HW, Huisman J (2009) Climate change: a catalyst for global expansion of harmful cyanobacterial blooms. Environ Microbiol Rep 1:27–37. https://doi.org/10.1111/J.1758-2229.2008.00004.X

Paerl HW, Otten TG (2013) Harmful cyanobacterial blooms: causes, consequences, and controls. Microb Ecol 65:995–1010. https://doi.org/10.1007/s00248-012-0159-y

Pannard A, Le Rouzic B, Binet F (2009) Response of phytoplankton community to low-dose atrazine exposure combined with phosphorus fluctuations. Arch Environ Contam Toxicol 57: 50–59. https://doi.org/10.1007/S00244-008-9245-Z

Paul VJ (2008) Global warming and cyanobacterial harmful algal blooms. Adv Exp Med Biol 619: 239–257. https://doi.org/10.1007/978-0-387-75865-7_11

Pereira De Albuquerque F, Luiz De Oliveira J, Moschini-Carlos V, Fernandes Fraceto L (2020) An overview of the potential impacts of atrazine in aquatic environments: perspectives for tailored solutions based on nanotechnology. Sci Total Environ. https://doi.org/10.1016/j.scitotenv.2019.134868

Pereira WE, Hostettler FD (1993) Nonpoint source contamination of the Mississippi river and its tributaries by herbicides. Environ Sci Technol 27:1542–1552. https://doi.org/10.1021/ES00045A008

Pesce S, Batisson I, Bardot C et al (2009) Response of spring and summer riverine microbial communities following glyphosate exposure. Ecotoxicol Environ Saf 72:1905–1912. https://doi.org/10.1016/J.ECOENV.2009.07.004

Popp J, Pető K, Nagy J (2013) Pesticide productivity and food security. A review. Agron Sustain Dev 33:243–255

Prasad SM, Zeeshan M, Kumar D (2011) Toxicity of endosulfan on growth, photosynthesis, and nitrogenase activity in two species of Nostoc (Nostoc muscorum and Nostoc calcicola). Toxicol Environ Chem 93:513–525. https://doi.org/10.1080/02772248.2010.542157

Reynolds CS, Oliver RL, Walsby AE (1987) Cyanobacterial dominance: the role of buoyancy regulation in dynamic lake environments cyanobacterial dominance: the role of buoyancy regulation in dynamic lake environments. New Zeal J Mar Freshw Res 21:379–390. https://doi.org/10.1080/00288330.1987.9516234

Rigosi A, Carey CC, Ibelings BW, Brookes JD (2014) The interaction between climate warming and eutrophication to promote cyanobacteria is dependent on trophic state and varies among taxa. Limnol Oceanogr 59:99–114. https://doi.org/10.4319/LO.2014.59.1.0099

Sharma A, Kumar V, Shahzad B et al (2019) Worldwide pesticide usage and its impacts on ecosystem. SN Appl Sci 1:1–16

Singh DP, Khattar JIS, Kaur G, Singh Y (2016) Toxicological impact of herbicides on cyanobacteria. Annu Res Rev Biol 9:1–39. https://doi.org/10.9734/ARRB/2016/22614

Sivonen K, Jones G (1999) Cyanobacterial toxins. E & FN Spon, London

Sivonen K, Jones G (2003) Algae and cyanobacteria in fresh water. In: Guidelines for safe recreational water environments, vol 1: coastal and fresh waters, pp 136–158

Stal LJ (2007) Cyanobacteria, pp 659–680. https://doi.org/10.1007/978-1-4020-6112-7_36

Stehle S, Schulz R (2015) Agricultural insecticides threaten surface waters at the global scale. Proc Natl Acad Sci 112:5750–5755. https://doi.org/10.1073/PNAS.1500232112

Sun K, Liu W, Liu L et al (2013) Ecological risks assessment of organophosphorus pesticides on bloom of Microcystis wesenbergii. Int Biodeterior Biodegrad 77:98–105. https://doi.org/10.1016/J.IBIOD.2012.11.010

Tang FHM, Lenzen M, McBratney A, Maggi F (2021) Risk of pesticide pollution at the global scale. Nat Geosci 144(14):206–210. https://doi.org/10.1038/s41561-021-00712-5

Timbrell J (2019) Introduction to toxicology, 3rd edn. CRC Press, Bristol

Tiwari B, Verma E, Chakraborty S et al (2018) Tolerance strategies in cyanobacterium Fischerella sp. under pesticide stress and possible role of a carbohydrate-binding protein in the metabolism of methyl parathion (MP). Int Biodeterior Biodegradation 127:217–226. https://doi.org/10.1016/J.IBIOD.2017.11.025

Tucci M, Bombelli P, Howe CJ et al (2019) A storable mediatorless electrochemical biosensor for herbicide detection. Microorganisms 7:630. https://doi.org/10.3390/MICROORGANISMS7120630

Vencill WK, William K, Armbrust K, Weed Science Society of America, Weed Science Society of America, Herbicide Handbook Committee (2002) Herbicide handbook, 8th edn. Weed Science Society of America, Lawrence KS

Vera MS, Lagomarsino L, Sylvester M et al (2010) New evidences of roundup (glyphosate formulation) impact on the periphyton community and the water quality of freshwater ecosystems. Ecotoxicology 19:710–721. https://doi.org/10.1007/S10646-009-0446-7

Vermaas WF (2001) Photosynthesis and respiration in Cyanobacteria eLS. https://doi.org/10.1038/NPG.ELS.0001670

Wagner C, Adrian R (2009) Cyanobacteria dominance: quantifying the effects of climate change. Limnol Oceanogr 54:2460–2468. https://doi.org/10.4319/LO.2009.54.6_PART_2.2460

Wang C, Lin X, Li L, Lin S (2016) Differential growth responses of marine phytoplankton to herbicide glyphosate. PLoS One 11:e0151633. https://doi.org/10.1371/JOURNAL.PONE.0151633

Wang W, Jiang M, Sheng Y (2021) Glyphosate accelerates the proliferation of Microcystis aeruginosa, a dominant species in cyanobacterial blooms. Environ Toxicol Chem 40:342–351. https://doi.org/10.1002/ETC.4942

Whitton BA (2000) Soils and rice-fields. In: The ecology of cyanobacteria. Springer, Dordrecht, pp 233–255

Whitton BA, Potts M (2000) Introduction to the cyanobacteria. Kluwer Academic, Dordrecht

WHO (2019) WHO | the WHO recommended classification of pesticides by hazard. World Health Organization, Rome

WMO (2021) State of the climate in Latin America and the Caribbean

Zamora-Sequeira R, Starbird-Pérez R, Rojas-Carillo O, Vargas-Villalobos S (2019) What are the main sensor methods for quantifying pesticides in agricultural activities? A review molecules 24. https://doi.org/10.3390/MOLECULES24142659

Zhang M, Lu T, Paerl HW et al (2019) Feedback regulation between aquatic microorganisms and the bloom-forming cyanobacterium Microcystis aeruginosa. Appl Environ Microbiol 85. https://doi.org/10.1128/AEM.01362-19

International Environmental Standards for the Regulation of Freshwater Cyanobacterial Blooms and Their Biotoxins

12

Fernando Cobo, Sandra Barca, Rufino Vieira-Lanero,
and M. Carmen Cobo

Abstract

Among the harmful algal blooms (HABs), cyanobacterial blooms (CyanoHABs) are one of the most studied and well known, as they constitute a worldwide environmental problem, with serious repercussions on human and animal health. Cyanobacteria can produce a wide variety of substances with different biological activities, many identified as potent toxins. Thus, the need to control and create regulations on this issue is crucial to ensure human health. In this article we provide a short and condensed general review of toxin concentration levels and exposure scenarios which are included in the legislation, standards, and guidelines that are in effect in different countries. Most national regulations are aimed to ensure the safety of potable water supply, and, as far as cyanotoxins are concerned, they usually include the minimum concentrations above which a health hazard is presumed. However, a very limited spectrum of toxins is considered. In addition, the global picture shows clear regional inequalities in terms of indicator variables, action levels, and existing knowledge about the affected ecosystems, the distribution of cyanobacteria, and their toxins. A general drawback lies in the fact that the regulations do not usually recommend specific measurement analytics, which hampers the intercalibration and application of standards.

F. Cobo (✉) · S. Barca · R. Vieira-Lanero
Laboratorio de Hidrobioloxía, Departamento de Zooloxía, Xenética e Antropoloxía Física, Universidade de Santiago de Compostela, Santiago de Compostela, Spain
e-mail: fernando.cobo@usc.es

M. C. Cobo
Department of Biological Sciences, University of Alabama, Tuscaloosa, AL, USA

B. Neilan et al. (eds.), *Cyanobacterial Biotechnology in the 21st Century*,
https://doi.org/10.1007/978-981-99-0181-4_12

Keywords

Cyanobacteria · Cyanotoxins · Environmental legislation

12.1 Introduction

Cyanobacteria show two opposite sides: while blooms can cause major problems for water quality, they are of great scientific and industrial interest, since they are the main primary producers in aquatic food webs, and an excellent source of compounds of potential commercial value (e.g., fertilizers, biofuels, etc.) (Cobo 2015). Thus, the study of cyanobacteria and their blooms is a broad discipline with different branches, many of them directly related to human welfare and therefore with the need for legislation and the establishment of rules and standards.

Harmful algal blooms (HABs) refers to the rapid growth and accumulation of different water organisms (under favorable nutrients, climate, and water conditions) that reach high concentrations and cause significant adverse impacts on the ecosystem (Paerl and Otten 2013; Huisman et al. 2018; Chorus and Welker 2021; Kapsalis and Kalavrouziotis 2021; Sha et al. 2021). Among them, dinoflagellate blooms, diatom blooms, and specially cyanobacteria blooms (CyanoHABs) have attracted researchers and environmental agencies' attention (Sha et al. 2021). Cyanobacterial blooms are the most studied and well known, as they constitute a major economic and ecological problem, with serious repercussions on human and animal health and with losses associated with increasing costs of drinking water treatments and the decrease of recreational water values (Cobo 2015; Svirčev et al. 2019; Kapsalis and Kalavrouziotis 2021). Cyanobacteria can produce a wide variety of bioactive metabolites, some of them identified as potent toxins (cyanotoxins) (Bláha et al. 2009; Paerl and Otten 2013; Svirčev et al. 2019), including the most potent toxins known (Humpage 2008). In many cases, toxins are secondary metabolites in the formation of photopigments, while some other toxins, such as microcystins, exhibit attributes of intracellular nitrogenous compounds (Orr and Jones 1998). Such is their potential danger that some national and international treaties and organizations, as the Chemical Weapons Convention or the "Australia Group," which is an unofficial forum of countries seeking to ensure that exports do not contribute to the development of chemical or biological weapons, have begun to consider the potential of saxitoxins and microcystins as agents of bioterrorism or as possible chemical weapons (Metcalf and Codd 2009).

There are numerous known cases of lethal poisoning of animals and livestock by drinking water contaminated with cyanobacteria (Francis 1878; Codd et al. 1989; Mcbarron and May 1966; Svirčev et al. 2019; Huisman et al. 2018; Yu 1995). Poisoning of human populations has also occurred worldwide (Sellner et al. 2003; Huisman et al. 2018; Svirčev et al. 2019). The first records in scientific literature are of poisonings in Australia, England, China, and South Africa in the late nineteenth and the early twentieth centuries (Sellner et al. 2003; Huisman et al. 2018). However, there are episodes reported in medieval European manuscripts that could be attributed to blooms and cyanotoxins. Despite this earliest knowledge on the existence of cyanobacterial blooms, it was not until the 1980s when they began to be recognized as a potential health hazard (Codd et al. 1989). Moreover, it was not until cases of poisoning such as that of 1996 in Caruaru (Brazil) (Dunn 1996) or the great human mortality caused in China by hepatic carcinomas (caused by the ingestion of water contaminated with *Microcystis* sp.) (Yu 1995) when the problem received global public attention. Consequently, the World Health Organization (WHO) included in the addendum to Volume 2 of the "Guidelines for Drinking Water Quality (1998)" the first provisional reference value for a cyanotoxin: microcystin-LR. This led to the consideration of cyanotoxins as contaminants subject to limitation and legal regulation in a growing number of countries (Chorus and Welker 2021).

The proliferation of cyanobacteria in lakes and water reservoirs is a worldwide problem whose increased frequency and intensity have been attributed to anthropogenic changes, more specifically nutrient over-enrichment of waters by urban, agricultural, and industrial development and river regulation (Xia et al. 2016). And this upward trend is likely to continue. Although the possibility that the higher numbers are due to a better monitoring effort has been bring up, several studies have highlighted the effect of climate change in the increase of CyanoHABs (e.g., Jeppesen et al. 2009; Paerl and Huisman 2009; Gkelis et al. 2014; Moe et al. 2016; Xia et al. 2016; Cottingham et al. 2021; Wang et al. 2021). Many of the lakes and reservoirs in which these events occur are intended for drinking water or recreation activities. Therefore, with CyanoHABs being a major problem in terms of human health, and given their global increase, the need to control and create regulations on this issue is imperative.

In this article we present a brief and condensed general review on the concentration levels of cyanotoxins in drinking and bathing waters as provided for in the regulations, standards, and guidelines in force in different countries and regions, based on the available literature. For more detailed information, the reader is referred to the original publications. Regulations concerning the risk of exposure to cyanotoxins in food will not be discussed here.

12.2 Materials and Methods

A systematic review of scientific articles has been carried out by consulting more than 20 databases with access from the library service of the University of Santiago de Compostela (Spain) (Aquatic Sciences & Fisheries Abstracts (ASFA), Annual Reviews, BioMed Central, Nature, Ovid, Agris, American Chemical Society (ACS), Directory of Open Access Journals (DOAJ), Oxford Journals, ProQuest Dissertations & Theses, ScienceDirect, SCOPUS, Taylor & Francis, Wiley Online Library, Web of Science (WOS), etc.), without restriction of the date or type of study, in English, French, Spanish, and Portuguese. Gray literature was included by manual Internet search. Review was focus on abstracts and then, when necessary, full articles, and the main sources were assessed.

The following compilation works are essential because of their relevance to the topic under discussion here: Chorus (2005, 2012), Chorus and Welker (2021), Burch (2008), and Ibelings et al. (2014). These works include the recommendations of the World Health Organization (WHO) and the European Union, as well as the contributions of several scientists and water managers on standards for the control of risks posed by exposure to toxic cyanobacteria in about 20 countries on 5 continents. The monograph edited by Hudnell et al. (2012) is also relevant and more specifically Chapter 36 (Burch 2008), which includes effective doses, guidelines, and regulations in different countries. Besides, the publication of Hudnell (2010) summarizes the situation regarding the regulation of freshwater harmful algae blooms in the USA, and Graham (2021) addresses the occurrence of cyanobacteria in this country over the past 20 years.

Other literature sources are very valuable for obtaining an accurate picture of the scope and capacity of regulations to reduce or control potential risks, for example: Niamien-Ebrottie et al. (2015), which reviews the occurrence of cyanobacterial blooms and cyanotoxins worldwide; Svirčev et al. (2019), which includes a global overview of cyanobacterial distribution and poisoning events; Dietrich and Hoeger (2005) and Farrer et al. (2015), which elaborate a critical review of the proposed guideline values for some cyanotoxins; or Merel et al. (2013), which provides a general review of knowledge on the occurrence and management of harmful freshwater cyanobacterial blooms and a compilation of some relevant literature.

In addition, we considered personal observations, specialist communications, and a multitude of articles of local scope that provide information on special aspects in different countries.

12.3 Results

12.3.1 Cyanotoxin in Drinking Waters

In view of the points set out in the introduction, the regulation of cyanobacteria and their potential blooms is essential to ensure human health in terms of drinking water consumption. However, the regions or countries for which regulation exists are not globally uniform (Fig. 12.1) and do not include all potential toxins (Fig. 12.1 and Table 12.1). Cyanotoxins vary widely both in terms of their chemical nature (cyclic peptides, alkaloids, lipopolysaccharides, etc.) and the effects and symptoms they produce (irritant toxins, cytotoxins, dermatotoxins, neurotoxins, and hepatotoxins) (Pérez et al. 2008) for most of which toxicological data are insufficient to determine concentration limits. In this section we will focus on microcystins, because of their importance and as they are the most considered in the regulations, and other toxins of interest for which data is available.

12.3.1.1 Microcystins

Microcystins are the most common cyanotoxins, and they are considered the main pollutants in freshwater in terms of risk for human health (Chorus and Bartram

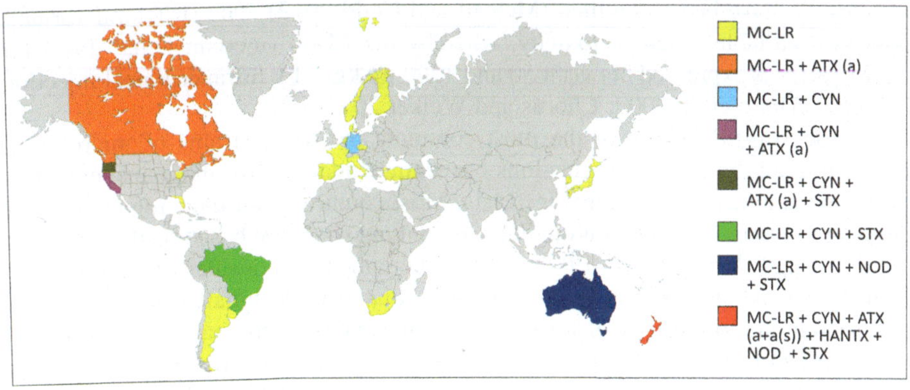

Fig. 12.1 Countries and US states where regulations or recommendations exist for the management of cyanotoxins in drinking water [*ATX (a)* anatoxin-a; *ATX (a(s))* anatoxin-a(s); *CYN* cylindrospermopsins; *HANTX* homoanatoxin-a; *MC-LR* microcystin-LR or equivalents; *NOD* nodularin; *STX* saxitoxins]

Table 12.1 Summary of toxicological characteristics of cyanotoxins and examples of guideline limit values for drinking water. (*LD50* mean lethal dose; *ip* intraperitoneal)

Toxins	LD_{50} (ip mouse) of pure toxin (μg/Kg)	Limit values for drinking water
Protein phosphatase blockers		
Microcystins (≈90 congeners)	50–1200	1 μg/L microcystin-LR (WHO)
Nodularins (≈9 congeners)	30–50	1 μg/L (New Zealand)
Neurotoxins		
Anatoxin-a	–	3–6 μg/L (in different regions)
Anatoxin-a(s)	20	1 μg/L (New Zealand)
Homoanatoxin-a	–	2 μg/L (New Zealand)
Saxitoxins (≈30 congeners)	10	3 μg/L (proposed by Fitzgerald et al. 1999)
β-N-Methylamino-L-alanine (BMAA)	–	Not proposed
Cytotoxins		
Cylindrospermopsin (5 congeners)	2100 in 1 day	15 μg/L (1 μg/L has been proposed)
Dermatotoxins	300	Not proposed
Lipopolysaccharides (LPS)	70,000	Not proposed

1999). These cyanotoxins are accumulated in the cells in the form of intracellular metabolites and are mostly release during cell lysis (Rohrlack and Hyenstrand 2007). Microcystins tolerate a wide range of chemical conditions of the water in which they are dissolved, so they are resistant, for example, to enzymatic hydrolysis and boiling, and are not destroyed by the most used oxidizing treatments in drinking water treatment plants (Douma et al. 2010). To date, more than 90 chemical variants of microcystin have been identified (Metcalf and Codd 2012). The structural variants possess a variable range of toxicity, whereby the LD50 (median lethal dose) for microcystins is expressed between 50 and 1200 μg kg^{-1} by intraperitoneal injection in mouse (Hotto et al. 2007; Chorus and Welker 2021).

Microcystin-LR is one of the most common and toxic strains of microcystin included in all the existing regulations or recommendations for the management of cyanotoxins in drinking water (Fig. 12.1). Most countries that have regulations for defining cyanotoxin concentration limits in drinking water are based on the reference value of 1 μg l^{-1} microcystin-LR (or what is the same they consider a tolerable daily intake of 0. 04 μg kg^{-1} body weight d^{-1}), that was designated at the time as provisional due to the lack of toxicological information. This value was proposed by the WHO to be measured at the outlet of drinking water treatment plants (tap water), and it was adopted by the legislation of many countries (Brazil, Uruguay, Ohio, the Czech Republic, Norway, Spain, France, Japan, Korea, and South Africa). In some cases (e.g., Australia, Canada, Florida, and Oregon), minor national or regional adaptations are made based on their own statistical data on quantity of water consumed or average body weight, leading to variations between limits ranging from 1.3 to 8 μg l^{-1}. Usually, regulations use the term "microcystin-LR" (e.g., Canada, the Czech Republic, or Singapore), but it is also common to find "microcystin-LR equivalents" (sum of all variants) (e.g., Australia, France, Finland, or Turkey).

The Directive 98/83/EC of the European Union on the quality of water intended for human consumption calls on member countries to "take the necessary measures to ensure that water does not contain concentrations of microorganisms, parasites or harmful substances which may be a danger to human health, and that it complies with minimum microbiological and chemical standards." However, it does not explicitly mention cyanotoxins. For this reason, Poland, which had adopted the guideline value of 1 μg l^{-1} for microcystin-LR in 2002, withdrew it in 2012 in the transposition of the Directive into its national legislation (Mankiewicz-Boczek et al. 2005).

For water masses "intended to produce drinking water" (e.g., reservoirs or lakes), the WHO establishes three levels according to cell biovolume: "surveillance level" (0.02 mm^3 l^{-1}), "alert level 1" (0.2 mm^3 l^{-1}), and "alert level 2" (10 mm^3 l^{-1}). In general, microcystin analysis is imperative when the "alert level 1" is reached in the water intake (prior to water treatment) (e.g., France, Spain, the Czech Republic, Finland, and Turkey). When there are monitoring programs, the national legislations do not expressly include a limit (Cuba, Denmark, Germany, Greece, Hungary, and the Netherlands). Nevertheless, the WHO recommendation is usually followed if a phytoplankton bloom situation (determined by cell quantification) could develop into risk to human health (e.g., Hungary, Germany, Finland, or Italy), in which case the microcystin-LR concentration of 1 μg l^{-1} is considered as the start of a "dangerous" situation. In cases such as in Paraguay, Chile, Argentina, China, etc., where legislation does not provide for toxin control in drinking water, some drinking water treatment plants or research laboratories routinely perform cell counts and microcystin analysis. When it comes to livestock water, only the Australian and

New Zealand legislations include a permitted concentration of microcystins (11,500 cells ml^{-1} of *Microcystis aeruginosa* or 2.3 μg l^{-1} of microcystin-LR) (see Burch 2008 and references therein).

12.3.1.2 Other Cyanotoxins

Because cyanotoxins are very diverse both in their chemical nature and in the effects and symptoms produced, the toxicological data available are insufficient to determine concentration limits.

Nodularins have a high liver toxicity (Dawson 1998). To date there are known nine different nodularins, and their LD50 is between 30 and 50 μg kg^{-1} (by intraperitoneal injection in mice) (Chorus et al. 2000, Chorus and Welker 2021). Only New Zealand has proposed a provisional maximum concentration of 1 μg l^{-1}, while the Australian legislation includes the monitoring of the cell density of *Nodularia spumigena* (a toxin-producing species) and has set a health alert level of 40,000 cells ml^{-1} or a biovolume of 9.1 mm^3 l^{-1}.

Cylindrospermopsin in an alkaloid has cytotoxic and hepatotoxic activity (Mazur-Marzec 2006). This cyanotoxin, of which five natural analogues are known, is typical of tropical areas, although it is also present in Europe and particularly in Spain (Quesada et al. 2006). One of the major risks of this toxin is that a high proportion is extracellular, and during a bloom more than 60% of the toxin may be dissolved in water (Mazur-Marzec 2006). The toxicological studies conducted by SHAW et al. (2000) suggest a maximum acceptable limit of 15 μg l^{-1} for drinking water. However, Humpage and Falconer (2003) propose a limit of 1 μg l^{-1} based on chronic oral toxicity in mice. Both limits are included in the legislation of Brazil (guideline value of 15 μg l^{-1}) and Australia, New Zealand, and Oregon (health alert level or provisional maximum concentration value of 1 μg l^{-1}). In addition, Australia sets a *Cylindrospermopsis raciborskii* cell density alert level of 15,000–20,000 cells ml^{-1} (biovolume 0.6–0.8 mm^3 l^{-1}).

Anatoxin-a, homoanatoxin-a, and anatoxin-a(s) have been associated with rapid virulence toxicosis in wild and domestic animals (Metcalf and Codd 2012). The LD50 for anatoxin-a(s) is 20 μg kg^{-1} by intraperitoneal injection in mouse (Mazur-Marzec 2006). The occurrence of this type of toxin worldwide is low, and no cases of serious human poisonings have been documented (Merel et al. 2013). Nevertheless, a reference level of 3 μg l^{-1} has been suggested (van Appeldoorn et al. 2007) in Oregon, while in Canada, there is a provisional maximum limit of 3.7 μg l^{-1} for anatoxin-a and in New Zealand of 6 μg l^{-1} for anatoxin-a, 1 μg l^{-1} for anatoxin-a(s), and 2 μg l^{-1} for homoanatoxin-a (Giddings et al. 2012; Wood et al. 2012).

The saxitoxin group can produce approximately 30 analogs (Llewellyn 2009), some of which are found only in freshwater cyanobacteria. The toxicity of this group of alkaloids is variable; the most potent saxitoxin exhibits an LD50 of 10 μg kg^{-1} by intraperitoneal injection in mice (Molica and Azevedo 2009). Although the WHO considers that there are still insufficient data to establish a maximum acceptable concentration for saxitoxins in drinking water (Chorus and Bartram 1999), Fitzgerald et al. (1999) proposed 3 μg l^{-1} saxitoxin equivalent in water for public supply, a limit that is included in the regulations of Australia ($\simeq$20,000 cells ml^{-1} or at a biovolume of 5 mm3 l^{-1} of *Anabaena circinalis*), New Zealand, Brazil, and Ohio (Chorus 2012).

β-N-Methylamino-L-alanine (BMAA) is a nonprotein amino acid that has been recently found in many groups of cyanobacteria worldwide (Cox et al. 2005). This cyanotoxin has been associated with neurological disease (Cox et al. 2005; Spencer et al. 2016; Baker et al. 2018). The toxic effects of BMAA as well as the methodologies for its detection (Baker et al. 2018; Rutkowska et al. 2019) have attracted the attention of researchers. Nevertheless, no toxicological data such as LD50 or NOAEL (no observable adverse effect level) is yet available, so a guideline level for BMAA has not yet been proposed (Merel et al. 2013).

The mean lethal dose for dermatotoxins is around 300 μg kg^{-1} per intraperitoneal injection in mice (Yadav et al. 2011). Nevertheless, there are no guideline levels for them nor for lipopolysaccharides (LPS), although their LD50 value is 70,000 μg kg^{-1} per intraperitoneal injection in mice (Vergara Larrayad et al. 2005), and they are considered a health risk due to the high concentrations that cyanobacteria easily reach, resulting in the high levels of these compounds (Codd 2000).

12.3.2 Cyanotoxins in Bathing Water

Regulations on cyanobacteria for bathing waters are even less globally uniform than for drinking waters (Fig. 12.2)—what can be explained by the complexity of establishing base levels. Cyanotoxin production can be considered directly correlated to the increase in biomass and is maximal when growth conditions are optimal (Roset et al. 2001). In addition, the accumulation of cells and foams, and their arrangement in the body of water, which can vary with wind direction and

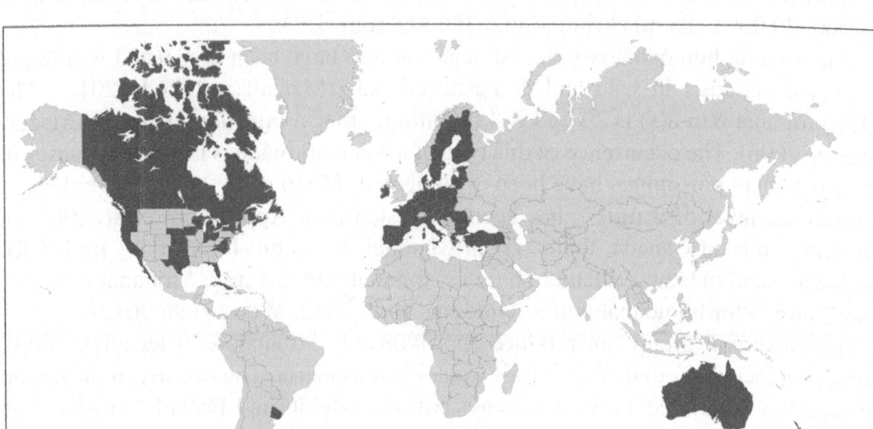

Fig. 12.2 Countries and US states where regulations or recommendations exist for the management of cyanotoxins in recreational waters

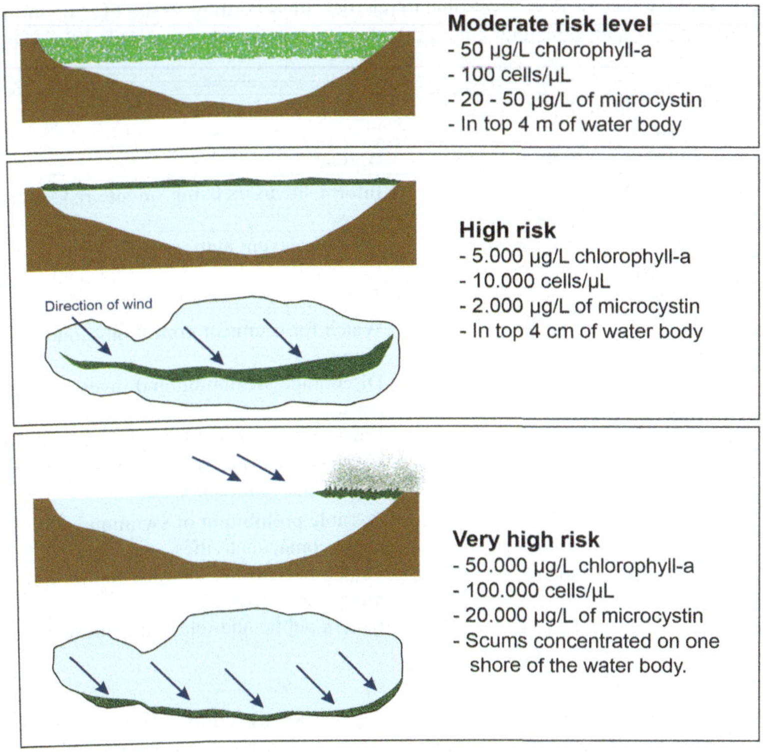

Fig. 12.3 Diagram of foam formation in blooms and its relation to emergency levels. (Modified from Falconer et al. 1999)

intensity, is also an important factor (Fig. 12.3). Thus, the WHO considered that for recreational waters, a single reference value for cyanotoxins is not appropriate (WHO 2003). Instead, a series of reference values of number of cells, or their equivalent in biovolume or chlorophyll-a values, related to the increased likelihood of adverse health effect, and three guide levels for bathing water (Table 12.2) were defined: (1) "Level 1" (2 mm 3 l^{-1}; 20,000 cells/ml^{-1} or 10 µg l^{-1} of chlorophyll-a) implies "relatively low probability of adverse health effects." At this level if toxin-producing cyanobacteria species are present, dissolved microcystin concentrations of up to 2–4 µg l^{-1} (even 10 µg l^{-1} depending on the species responsible for the bloom) can be reached. (2) "Level 2" (10 mm^3 l^{-1}; 100,000 cells/ml^{-1} or 50 µg l^{-1} of chlorophyll-a) implies "moderate likelihood of adverse health effects," with amounts up to 20 µg l^{-1} of microcystins. When this level is exceeded, a foam of cyanobacteria may occur on the surface of the water body (Fig. 12.3). (3) "Level 3" (>1 mg/L) implies scum formation and a "high risk," as inhalation of cyanobacteria by aerosols produced by foam is an important route of intoxication in the case of water sports activities (Falconer 1996; Plaas and Paerl 2020).

Table 12.2 Bathing waters. Risk levels due to microcystin according to the likelihood of causing adverse health effects and guidelines for safe bathing water management (Modified from WHO 2003)

Guideline levels in bathing waters			
Risk level	WHO (2003)	Grupta (1998)	Actions
Relatively low	2 mm^3/L 10 μg/L chl-a 20,000 cel./ml	2–4 μg/L dissolved microcystins	Inform site users using on-site risk advisory signs. Inform relevant authorities
Moderate	10 mm^3/L 50 μg/L chl-a 100,000 cel./ml	10–20 μg/L dissolved microcystins	Watch for scums or conditions conducive to scums Discourage swimming and investigate the hazard Inform site users using on-site risk advisory signs, post hazard warnings on social media, etc. Inform relevant authorities
High	>1 mg/L Scum formation	>20 μg/L dissolved microcystins	Possible prohibition of swimming and other water contact activities Temporary closure of water bodies for recreational activities Inform public and relevant authorities

Besides, for recreational waters Grupta (1998) indicates three levels of risk: (1) "low risk with allergenic effects" (2–4 μg l^{-1} microcystin-LR), (2) "medium risk with hepatotoxicity by accidental ingestion" (4–20 μg l^{-1} microcystin-LR), and (3) "high risk" (>20 μg l^{-1} microcystin-LR) when foams appear indicating a high number of cyanobacterial cells. These levels can be correlated with the ones established by the WHO (Table 12.2).

The European Parliament and Council Directive 2006/7/EC concerning the management of bathing water quality and the laws of the member countries transposing it into their national legal framework state that: "When the bathing water profile shows a propensity to cyanobacteria, appropriate monitoring shall be carried out to enable their identification and reported to the health authority, which shall assess the health risks." Following these recommendations, many countries apply a two- or three-level alert framework with increasing severity. The lowest level "vigilance" does not result in any action and simply leads authorities to continue or intensify monitoring efforts. At the next level "alert," the public is warned through signs at the site, via the Internet, emergency hotlines, etc. Warning signs can inform the public about increased risks of skin irritation and gastrointestinal illnesses. The highest level "action" is based on one of these conditions: (1) the presence of persistent cyanobacterial blooms or surface foams, (2) exceeding cyanobacterial biomass thresholds (counts, biovolume, or pigment concentration), or (3) the existence of some medical report of toxic effects attributed to cyanobacteria in the area. At this third action level, the public is normally advised to refrain from recreational activities involving contact with water (particularly with risk of ingestion), and water bodies may be temporarily closed for recreational activities such as swimming.

In general, national guidelines for recreational waters merge different combinations of microcystin-LR levels and cell numbers, biovolume, foam formation, or pigment concentration for the definition of alert levels, within more or less sophisticated operational frameworks (see Table 12.2, pgs. 8 ff. in Chorus 2012). The parameters analyzed in the different monitoring procedures are mostly cyanobacterial cell density, biovolume, or foam formation but are combined with other measures such as dissolved microcystin-LR (equivalent) concentration in a range from 10 $\mu g\ l^{-1}$ to >25 $\mu g\ l^{-1}$ (Uruguay, Canada, partly Denmark, Portugal, Spain, France, Germany, Hungary, Italy, New Zealand, and Turkey) or chlorophyll-a concentration (the Czech Republic, Denmark, Germany, Hungary, the Netherlands, Turkey, and Singapore). Similarly, in the different US states, there are separate regulations for recreational waters that establish levels by cell number but also delimit the concentrations of various cyanotoxins: microcystins (California, Florida, Indiana, Iowa, Kansas, Maryland, Massachusetts, Nebraska, Ohio, Oklahoma, Oregon, Rhode Island, Vermont, and Washington), cylindrospermopsin (California and Oregon), anatoxin-a (California, Oregon, and Washington), and saxitoxins (Oregon) (Chorus 2012).

12.4 Discussion

Most national regulations are aimed to ensure the safety of drinking water supplies, and, as far as cyanotoxins are concerned, they usually include the minimum concentrations for which health hazard is assumed, but in a very limited spectrum of toxins potentially present in blooms. Thus, it can be stated that, despite the worldwide increasing occurrence of cyanotoxins, they are still underestimated in regulations, with microcystin-LR being the only one considered in most of the cases. There is also a general lack of knowledge on the affected ecosystems and the local scenario regarding the distribution of cyanobacteria and their toxins, which limits the establishment of appropriate indicator variables and action levels. Therefore, it is unlikely that all toxins produced by cyanobacteria have been identified, as toxicity cases of unknown etiology are still occurring, and new analogues continue to be identified through modifications of existing techniques and the emergence of new ones.

Besides, it has often been argued that it is difficult to establish universal risk doses for human health if individual susceptibility is considered (Chorus and Bartram 1999). Moreover, the prerequisite for any risk assessment of chemicals on health (IPCS 1995) is not only the knowledge of the chemical/biochemical characteristics, kinetics, and toxicity/carcinogenicity dynamics of the compound (e.g., the fact that the action of some toxins may intensify the toxicity of others) but also the understanding of individual sensitivity and time of exposure (e.g., absorption through the skin may cause dermatitis, the seriousness of which depends on individual sensitivity and time of exposure) and the main routes of administration or contact (e.g., oral ingestion produces at least one degree less of toxicity than the intraperitoneal injection, but intranasal application was demonstrated to be as toxic as injection). In this regard, a very important aspect to consider in the development of legal provisions is that in natural ecosystems, cyanotoxins are not found individually but as a mixture of toxins, so it is necessary to understand the interactions between them and to develop new protocols and techniques that contemplate these cases of toxic mixtures. Additionally, the true incidence of cyanotoxins in terms of mortality is restricted to the lack of recognition of their symptomatology, reporting, and analytical epidemiology (Codd et al. 2005).

Another general drawback lies in the fact that regulations do not usually recommend any analytical method of measurement, which makes their application and intercalibration difficult. Finally, an issue of great concern is that, in some regions, the necessary monitoring plans or the established control values and adequate levels cannot always be implemented with the available financial, technical, and institutional resources.

12.5 Conclusions

The global picture shows a clear regional inequality in the definition of levels of a wide range of cyanotoxins, the representativeness of indicator variables, levels of action, and existing local knowledge of affected ecosystems, all of which limit the development of legal regulations on their control. There is a need to stimulate research to improve our understanding of toxin interactions and to develop new protocols and techniques, facilitate risk management capacity, and broaden the awareness and dissemination of relevant expertise to harmonize cyanotoxin control methods and practices internationally.

References

Baker TC, Tymm FJ, Murch SJ (2018) Assessing environmental exposure to β-N-methylamino-l-alanine (BMAA) in complex sample matrices: a comparison of the three most popular LC-MS/MS methods. Neurotox Res 33(1):43–54

Bláha L, Babica P, Maršálek B (2009) Toxins produced in cyanobacterial water blooms – toxicity and risks interdisciplinary. Toxicology 2(2):36–41

Burch MD (2008) Effective doses, guidelines & regulations. Adv Exp Med Biol 619:831–853

Chorus I (2005) Current approaches to cyanotoxin risk assessment, risk management and regulations in different countries. 117 pp, Federal Environment Agency (Unweltbundesamt), Berlin

Chorus I (2012) Introduction. In: Chorus I (ed) Current approaches to cyanotoxin risk assessment, risk management and regulations in different countries. 147 pp. Section II drinking water and swimming pool water hygiene. Federal Environment Agency (Unweltbundesamt), Berlin

Chorus I, Bartram J (1999) Toxic cyanobacteria in water. A guide to their public health consequences, monitoring and management. World Health Organization, London, p 416

Chorus I, Welker M (2021) Toxic cyanobacteria in water: a guide to their public health consequences, monitoring and management. Taylor & Francis, 858 pp

Chorus I, Welker M (2021) A guide to their public health consequences, monitoring and management. In: Chorus I, Welker M (eds) CRC Press, London, 8587 pp
Chorus I, Falconer IR, Salas HJ, Bartram J (2000) Health risks caused by freshwater cyanobacteria in recreational waters. J Toxicol Environ Health, Part B 3:323–347
Cobo F (2015) Métodos de control de las floraciones de cianobacterias en aguas continentales. Limnetica 34(1):247–268
Codd GA (2000) Cyanobacterial toxins, the perception of water quality, and the prioritisation of eutrophication control. Ecol Eng 16:51–60
Codd GA, Bell SG, Brooks WP (1989) Cyanobacterial toxins in water. Water Sci Technol 21(3): 1–13
Codd GA, Lindsay J, Young FM, Morrison LF, Metcalf JS (2005) Harmful cyanobacteria: risk management for health protection. Toxicol Appl Pharmacol 203:264–272
Cottingham KL, Weathers KC, Ewing HA, Greer ML, Carey CC (2021) Predicting the effects of climate change on freshwater cyanobacterial blooms requires consideration of the complete cyanobacterial life cycle. J Plankton Res 43(1):10–19
Cox PA, Banack SA, Murch SJ, Rasmussen U, Tien G, Bidigare RR, Metcalf JS, Morrison LF, Codd GA, Bergman B (2005) Diverse taxa cyanobacteria produce β-N-methylamino-L-alanine, a neurotoxic amino acid. Proc Natl Acad Sci USA 102(1):5074–5078
Dawson RM (1998) Review article: the toxicology of microcystins. Toxicon 36(7):953–962
Dietrich D, Hoeger S (2005) Guidance values for microcystins in water and cyanobacterial supplement products (blue-green algal supplements): a reasonable or misguided approach? Toxicol Appl Pharmacol 203:273–289
Douma M, Ouahid Y, del Campo FF, Loudiki M, Mouhri KH, Oudra B (2010) Identification and quantification of cyanobacterial toxins (microcystins) in two Moroccan drinking-water reservoirs (Mansour Eddahbi Almassira). Environ Monit Assess 160:439–450
Dunn J (1996) Algae kills dialysis patients in Brazil. Br Med J 312:1183–1184
Falconer IR (1996) Potential impact on human health of toxic cyanobacteria. Phycologia 35(Suppl 6):6–11
Falconer I, Bartram J, Chorus I, Kuiper-Goodman T, Utkilen H, Burch M, Codd GA (1999) Safe levels and safe practices. In: Chorus I, Bartram J (eds) Toxic cyanobacteria in water: a guide to public health significance, monitoring and management. E & FN Spon, London
Farrer D, Counter M, Hillwig R, Cude C (2015) Health-based cyanotoxin guideline values allow for cyanotoxin-based monitoring and efficient public health response to cyanobacterial blooms. Toxins 7(2):457–477
Fitzgerald D, Cunliffe D, Burch M (1999) Development of health alerts for cyanobacteria and related toxins in drinking water in South Australia. Environ Toxicol 14(1):203–207
Francis G (1878) Poisonous Australian lake. Nature 18:11–12
Giddings M, Aranda-Rodriguez R, Yasvinski G, Watson SB, Zurawell R (2012) Canada: cyanobacterial toxins: drinking and recreational water quality guidelines. In: Chorus I (ed) Current approaches to cyanotoxin risk assessment, risk management and regulations in different countries. 147 pp. Section II drinking water and swimming pool water hygiene. Federal Environment Agency (Unweltbundesamt), Berlin
Gkelis S, Papadimitriou T, Zaoutsos N, Leonardos I (2014) Anthropogenic and climate-induced change favors toxic cyanobacteria blooms: evidence from monitoring a highly eutrophic, urban Mediterranean lake. Harmful Algae 39:322–333
Graham JL (2021) Cyanotoxin occurrence in the United States: a 20 year retrospective. Lakeline 41: 8–11
Grupta S (1998) Cyanobacterial toxins: microcystin-LR. Health criteria and other supporting information. In: Organización Mundial de la Salud (OMS) (ed) Guidelines for drinking-water quality. Génova, pp 95–110
Hotto AM, Satchwell MF, Boyer GL (2007) Molecular characterization of potential microcystin-producing cyanobacteria in Lake Ontario embayments and nearshore waters. Appl Environ Microbiol 73(14):4570–4578

Hudnell HK (2010) The state of US freshwater harmful algal blooms assessments, policy and legislation. Toxicon 55(5):1024–1034
Hudnell HK, Backer LC, Andersen J, Dionysiou DD (2012) United States of America: historical review and current policy addressing cyanobacteria. In: Chorus I (ed) Current approaches to cyanotoxin risk assessment, risk management and regulations in different countries. 147 pp. section II drinking water and swimming Pool water hygiene, Federal Environment Agency (Unweltbundesamt), Berlin
Huisman J, Codd GA, Paerl HW, Ibelings BW, Verspagen JM, Visser PM (2018) Cyanobacterial blooms. Nat Rev Microbiol 16(8):471–483
Humpage A (2008) Toxin types, toxicokinetics and toxicodynamics. In: Hudnell HK (eds) Cyanobacterial harmful algal blooms: state of the science and research needs. Advances in experimental medicine and biology, vol 619. Springer, New York, NY. https://doi.org/10.1007/978-0-387-75865-7_16
Humpage AR, Falconer IR (2003) Oral toxicity of the cyanobacterial toxin cylindrospermopsin in male swiss albino mice: determination of no observed adverse effect level for deriving a drinking water guideline value. Environ Toxicol 18:94–103
Ibelings BW, Backer LC, Kardinaal WEA, Chorus I (2014) Current approaches to cyanotoxin risk assessment and risk management around the globe. Harmful Algae 40:63–74
IPCS (1995) Inorganic lead. Geneva, World Health Organization, International Programme on Chemical Safety (Environmental Health Criteria) 165
Jeppesen E, Kronvang B, Meerhoff M, Søndergaard M, Hansen KM, Andersen HE, Olesen JE et al (2009) Climate change effects on runoff, catchment phosphorus loading and lake ecological state, and potential adaptations. J Environ Qual 38(5):1930–1941
Kapsalis VC, Kalavrouziotis IK (2021) Eutrophication—a worldwide water quality issue. In: Zamparas MG, Kirapoulus GL (eds) Chemical Lake restoration. Technologies, innovation and economic perspective. Springer, Cham, pp 1–21
Llewellyn LE (2009) Sodium channel inhibiting marine toxins. In: Fusetani N, Kem W (eds) Marine toxins as research tools, vol 46. Springer, Berlin, pp 67–97
Mankiewicz-Boczek J, Jurczak T, Izydorczyk K, Zalewski M (2005) Poland: regulation on cyanotoxins in legislation. Current approaches to cyanotoxin risk assessment, risk management and regulations in different countries. WaBoLu-Hefte 99–101
Mazur-Marzec H (2006) Characterization of phycotoxins produced by cyanobacteria. Oceanol Hydrobiol Stud 35(1):85–109
Mcbarron EJ, May V (1966) Poisoning of sheep in new south wales by the blue-green alga anacystis cyanea (kuetz.) Dr. and Dail. Aust Vet J 42:449–453
Merel S, Walker D, Chicana R, Snyder S, Baurès E, Thomas O (2013) State of knowledge and concerns on cyanobacterial blooms and cyanotoxins. Environ Int 59:303–327
Metcalf JS, Codd GA (2009) Chapter 8. The status and potential of cyanobacteria and their toxins as agents of bioterrorism. In: Gault PM, Marler HJ (eds) Handbook on cyanobacteria: biochemistry, biotechnology and applications. Nova Publishers, New York, pp 259–281
Metcalf JS, Codd GA (2012) Chapter 24. Cyanotoxins. In: Whitton BA (ed) Ecology of cyanobacteria II: their diversity in space and time. Springer Science + Busisness Media B.V, pp 651–675
Moe SJ, Haande S, Couture RM (2016) Climate change, cyanobacteria blooms and ecological status of lakes: a Bayesian network approach. Ecol Model 337:330–347
Molica R, Azevedo S (2009) Ecofisiologia de cianobacterias productoras de cianotoxinas. Oecologia brasiliensis 13(2):229–246
Niamien-Ebrottie JE, Bhattacharyya S, Deep PR, Nayak B (2015) Cyanobacteria and cyanotoxins in the world: review. Int J Appl Res 1(8):563–569
WHO (World Health Organization) (2003) Guidelines for safe recreational water environments, vol 1: Coastal and fresh waters, Geneva
Orr PT, Jones GJ (1998) Relationship between microcystin production and cell division rates in nitrogen-limited *Microcystis aeruginosa* cultures. Limnol Oceanogr 43(7):1604–1614

Paerl HW, Huisman J (2009) Climate change: a catalyst for global expansion of harmful cyanobacterial blooms. Environ Microbiologu Rep 1(1):27–37

Paerl HW, Otten TG (2013) Harmful cyanobacterial blooms: causes, consequences, and controls. Microb Ecol 65(4):995–1010

Pérez DS, Soraci AL, Tapia MO (2008) Cianobacterias y cianotoxinas: rol de las microcistinas en la salud humana y animal y su detección en las muestras de agua. Analecta Veterinaria 28(1): 48–56

Plaas HE, Paerl HW (2020) Toxic cyanobacteria: a growing threat to water and air quality. Environ Sci Technol 55(1):44–64

Quesada A, Moreno E, Carrasco D, Paniagua T, Wörmer L, de Hoyos C, Sukenik A (2006) Toxicity of *Aphanizomenon ovalisporum* (Cyanobacteria) in a Spanish water reservoir. Eur J Phycol 41(1):39–45

Rohrlack T, Hyenstrand P (2007) Fate of intracellular microcystins in the cyanobacterium *Microcystis aeruginosa* (Chroococcales, Cyanophyceae). Phycologia 46:277–283

Roset J, Aguayo S, Muñoz MJ (2001) Detección de cianobacterias y sus toxinas. Una revisión Toxicología 18:65–71

Rutkowska M, Płotka-Wasylka J, Majchrzak T, Wojnowski W, Mazur-Marzec H, Namieśnik J (2019) Recent trends in determination of neurotoxins in aquatic environmental samples. TrAC Trends Anal Chem 112:112–122

Sellner KG, Doucette GJ, Kirkpatrick GJ (2003) Harmful algal blooms: causes, impacts and detection. J Ind Microbiol Biotechnol 30:383–406

Sha J, Xiong H, Chengjun LI, Zhiying LU, Zhang J, Zhong H, Zhang W, Yan B (2021) Harmful algal blooms and their eco-environmental indication. Chemosphere 129912

Shaw G, Seawright A, Shahin M, Senogles P, Mueller J, Moore M (2000) The cyanobacterial toxin, cylindrospermopsin: human health risk assessment. In: Harmful algal blooms. 9th international conference on harmful algal blooms. Hobart. 56. National Research. Center for Environmental Toxicology, Australia

Spencer PS, Palmer VS, Kisby GE (2016) Seeking environmental causes of neurodegenerative disease and envisioning primary prevention. Neurotoxicology 56:269–283

Svirčev Z, Lalić D, Savić GB, Tokodi N, Backović DD, Chen L, Meriluoto J, Codd GA (2019) Global geographical and historical overview of cyanotoxin distribution and cyanobacterial poisonings. Arch Toxicol 93(9):2429–2481

van Appeldoorn ME, van Egmond HP, Speijers GJ, Bakker GJ (2007) Review. Toxins of cyanobacteria. Mol Nutr Food Res 51:7–60

Vergara Larrrayad Y, Moya Jaraba A, Peleato Sánchez ML, Sevilla Miguel E, S. López gomollón. (2005) Nuevos riesgos para el agua potable: microcistina. Mejora de la calidad del agua de consumo por eliminación de toxinas. Consejo económico y social de Aragón, Spain 16

Wang Z, Akbar S, Sun Y, Gu L, Zhang L, Lyu K, Huang Y, Yang Z (2021) Cyanobacterial dominance and succession: factors, mechanisms, predictions, and managements. J Environ Manag 297:113281

Wood S, Williamson W, Kouzminov A (2012) New Zealand: regulation and management of cyanobacteria. In: Chorus I (ed) Current approaches to cyanotoxin risk assessment, risk management and regulations in different countries. 147 pp. Section II drinking water and swimming pool water hygiene. Federal Environment Agency (Unweltbundesamt), Berlin

Xia R, Zhang Y, Critto A, Wu J, Fan J, Zheng Z, Zang Y (2016) The potential impacts of climate change factors on freshwater eutrophication: implications for research and countermeasures of water management in China. Sustainability 8:229

Yadav S, Sinha RP, Tyagi MB, Kumar A (2011) Cyanobacterial secondary metabolites. Int J Farma Bio Sci 2(2):B144–B167

Yu SZ (1995) Primary prevention of hepatocellular carcinoma. J Gastroenterol Hepatol 10:674–682

Therapeutic Potential of Cyanobacteria as a Producer of Novel Bioactive Compounds

13

Priya Yadav, Rahul Prasad Singh, Ajay Kumar, Prashant Kumar Singh, and Rajan Kumar Gupta

Abstract

Cyanobacteria are photosynthetic prokaryotes that have a variety of biological activities and can be used as dietary supplements. Because of its high nutritional, protein, and digestibility content, it is utilized as a meal supplement. *Spirulina* sp., *Anabaena* sp., *Microcystis* sp., *Nostoc* sp., *Lyngbya* sp., *Oscillatoria* sp., and other cyanobacteria create a wide range of secondary bioactive metabolites with diverse biological functions. Cyanobacteria produce various physiologically active and chemically diverse compounds such as saccharides, fatty acid amides, lipopeptides, cyclic and acyclic peptides, and alkaloids. More than half of marine cyanobacteria could be used to extract bioactive compounds that can destroy cancer cells by causing apoptosis. Their effectiveness as an antiviral, antitumor, antibacterial, anti-HIV, and dietary additive is extensively documented. However, such drugs are in various phases of clinical studies, and only a few have made it to the market. This chapter examines the bioactive chemicals in cyanobacteria with antiprotozoal, antibacterial, anti-inflammatory, anticancer, antioxidant, and antiviral properties.

P. Yadav · R. P. Singh · R. K. Gupta (✉)
Laboratory of Algal Research, Centre of Advanced Study in Botany, Institute of Science, Banaras Hindu University, Varanasi, India
e-mail: rajang.bot@bhu.ac.in

A. Kumar
Department of Postharvest Science, Agriculture Research Organization, Volcani Center, Rishon LeZion, Israel

P. K. Singh
Department of Biotechnology, Mizoram University (A Central University), Pachhunga University College Campus, Aizawl, India

B. Neilan et al. (eds.), *Cyanobacterial Biotechnology in the 21st Century*,
https://doi.org/10.1007/978-981-99-0181-4_13

Keywords

Bioactive compounds · Cyanobacteria · Human health · Metabolites · Therapeutic application

13.1 Introduction

The emergence and development of antibiotic resistance in the pathogenic organism is a severe worldwide problem developing at an alarming rate (Bin et al. 2017). Depending on the drug resistance criteria, bacteria have been categorized into three categories: critical, high, and medium. The World Health Organization has expressed concern about the need to conduct an urgent discovery for new antibiotics that target prioritized bacteria, such as critical (*Pseudomonas aeruginosa*, *Acinetobacter baumannii*, etc.), high (*Staphylococcus aureus* methicillin-resistant strain (MRS), *Enterococcus faecium*, etc.), and medium (*Streptococcus pneumoniae*) (World Health Organization report 2020). The antimicrobial competence of natural compounds from various sources has acquired prominence as an alternative to antibiotics due to antibiotic resistance. Antibiotics were once thought to be magic bullets specifically targeting disease-causing germs (Bin et al. 2017). However, microorganisms from marine habitats, particularly cyanobacteria, are diverse sources of putative bioactive chemicals that have received little attention. Cyanobacteria are a primordial, varied, and widespread genus of oxygen-evolving prokaryotes that have similarities with green plants and gram-negative bacteria in oxygenic photosynthesis and cellular organization (Stanier Cohen-Bazire 1988).

On the other hand, cyanobacteria colonize a wide range of environments, including saline water, freshwater, hot springs, agriculture fields, and the bark of trees. Secondary metabolites are metabolites that are created to withstand extreme environmental conditions. As a result, these metabolites are crucial in adaptation strategies. Cyanobacteria are natural sources of secondary bioactive compounds, and they've been used in a variety of pharmaceutical industries due to their abundance of primary and secondary metabolites (Volk 2008). Bioactive substances are physiologically active molecules that positively or negatively affect a living creature, tissue, or cell at low concentrations (Estela Silva-Stenico et al. 2013). They produce a diverse range of bioactive chemicals because they have ribosomal and non-ribosomal machinery (Cane and Walsh 1999). Antibacterial, antialgal, antifungal, antioxidative, and anticancer activities of algal extracts comprising bioactive components such as lipids, polysaccharides, proteins, terpenes, and polyphenols, esters, vitamins, and carotenoids are critical in the creation of novel medications (Michalak and Chojnacka 2015). Immune modulators, anticancer drugs, and antioxidants have been successfully employed with cyanobacterial glycoproteins, polysaccharides, and compounds like carotenoid, phycocyanin, butylated hydroxytoluene, and vitamin C (Kini et al. 2020). Essential fatty acids from cyanobacteria, including linoleic acid and omega-3 fatty acid, have been shown to help with chronic obstructive pulmonary disease, hypertension, diabetes type II, renal disease, and coronary heart disease (Simopoulos 1997). Clinical studies of isolated bioactive chemicals are being conducted to determine their pharmacokinetics, effectiveness, bioavailability, and safety to produce a novel medicine in various forms (Sasidharan et al. 2011). Cyanobacterial phytochemical investigations and systemic screening identified a diversity of biologically active compounds that can be used to treat a variety of human illnesses (Nowruzi et al. 2018). This chapter discusses cyanobacterial bioactive chemicals with antibacterial, anti-inflammatory, antioxidant, anticancer, antiprotozoal, and antiviral actions and other biological properties.

13.2 Nutritional Supplements

Cyanobacteria are an excellent source of food and nutritional supplements, and they are rich in proteins, glucose, vital fatty acids, active enzymes, beta-carotene, minerals, and vitamins. For ages, humans have consumed cyanobacteria such as *Spirulina maxima*, *Aphanizomenon flos-aquae*, *Spirulina platensis*, *S. fusiformis*, *Nostoc commune*, and *Nostoc flagelliforme*. *Spirulina* sp. is high in protein, vital fatty acids, minerals, vitamins, and other nutrients and contains around 70% of its dry weight in protein (Vonshak 1997). Dried *spirulina* has a carbohydrate content of 24%, a protein content of 51–71%, a fat content of 8%, and a water content of 5%. It is an affluent source of cyanocobalamin (vitamin B12), iron, and other critical nutrients and dietary minerals (Zahra et al. 2020). It is utilized in fodder, food, and medicinal agents. *Spirulina* protein has less cysteine, lysine, and methionine than meat or milk but is better than other plant- and animal-based proteins (Ciferri 1983). *Spirulina* has been shown to help with various disorders, including hyperglycemia and hypercholesterolemia, and with cardiovascular, inflammatory, viral, and cancer diseases. *Spirulina* sp. reduced triacylglycerols and total plasma cholesterol at 4.5 g daily for 6 weeks by 28% and 10%, respectively (Torres-Duran et al. 2007). Several studies of *Spirulina* extracts have exhibited anti-inflammatory, cytotoxic, and antioxidant properties. In an α-synuclein model of Parkinson's disease, a *Spirulina*-enriched diet reduces microglial activation by enhancing the expression of the chemokine receptor CX3CR1 (Pabon et al. 2012). It has been seen that diet supplementation with *Spirulina fusiformis* induces reduction in leukoplakia and suppresses tumours and buccal pouch cancer. From ancient times, humans consume *Nostoc commune* for medical purposes such as digestion, inflammation, chronic weariness, and night blindness (Qiu et al. 2002). *Nostoc* has atheroprotective properties and could be used as an alternative to statin therapy for hypercholesterolemia. Lipids extracted from *Nostoc* inhibit cholesterol production and lipogenesis, lowering the total cholesterol and triglyceride levels of plasma considerably (Rasmussen et al. 2008, 2009).

The oral dose of *Aphanizomenon flos-aquae* in healthy men resulted in a modest decrease in polymorph nucleated cell phagocytic activity (Jensen et al. 2000). Furthermore, when administered with or before a possible antigen danger, cyanobacteria increased IgA levels to prevent food allergies (Hayashi et al. 1998).

This characteristic could effectively cure disorders like chronic inflammatory, allergies, and autoimmune diseases that include a suppressed immune system and reactions. In addition, a diet supplemented with cyanobacteria was proven to alleviate adopohepatosis induced by a high-cholesterol diet. This was attributable to the regulation of lipoprotein lipase, a triglyceride-rich lipoprotein processing enzyme.

13.3 Vitamins

Cyanobacteria are an excellent source of vitamins. *Spirulina* sp. is an affluent source of tocopherol (vit. E) and cyanocobalamin (vit. B12) among different food sources (Watanabe et al. 2002). The results of an accretion study on humans showed that the effect of a symptom of retinol (vit. A) paucity, Bitøt's spot, declined from 10% to 80% after 1 g of *Spirulina* per day was provided for 5 months to 5000 children below the age of 4 years in Madras, India (Seshadri 1993). *Spirulina* is rich in tocopherol, with the reported content (190 mg kg^{-1}) being several times more than wheat grain (Challem et al. 1981). B-complex vitamins like pantothenic acid and nicotinic acid are also found in *Anabaena hassali*, *Aphanizomenon flos-aquae*, *Nostoc punctiforme*, *Phormidium bijugatum*, and *Microcystis pulverana* (Sharma et al. 2011).

13.4 Gamma-Linolenic Acid (GLA)

GLA is an uncommon and distinctive polyunsaturated fatty acid (PUFA) with a wide range of health benefits. The very important GLA is found in cyanobacteria, specifically in *Spirulina platensis*, which is easily metabolized into arachidonic acid and then prostaglandin E2 in the human body (Euler and Eliassen 1967). GLA is applicable for the cure of premenstrual syndrome, atopic eczema, Parkinson's disease, multiple sclerosis, and heart diseases (Horrobin 2000). GLA decreases low-density lipoprotein in hypercholesterolemic patients and is a precursor for prostaglandins, which are master hormones that regulate various bodily activities. For example, prostaglandin E1 regulates blood pressure, lipid production, inflammation, and cell proliferation, and prostaglandin E2 lowers blood pressure and regulates smooth muscle contractions; hence, it plays a vital role in the metabolism of the body (Kernoff et al. 1977). *Aphanizomenon flos-aquae* is high in 1–3-a-linoleic acid, which prevents the synthesis of arachidonic acid and inflammatory prostaglandins (Kushak et al. 1999).

Spirulina platensis is a possible alternative for GLA production (Qiang et al. 1997). Nitrogen status, growth phase, light intensity, and temperature of the medium influence GLA production in cyanobacteria (Cohen et al. 1993). The dry cell mass of the blue-green alga can accumulate 1% GLA. For GLA production in crops, the D6-desaturase gene from cyanobacteria (*Synechocystis* sp.) was cloned in tobacco. Transgenic tobacco plants expressing the cyanobacterial D6-desaturase also produced octadecatetraenoic acid, a highly unsaturated fatty acid, and this was the first time that transgenic plants were used to enhance the production of PUFA (Reddy and Thomas 1996).

13.5 Pigments

13.5.1 Cyanobacterial Phycocyanin (CPC)

Cyanobacterial pigments like phycobiliproteins, chlorophyll-a, and carotenoids are applicable in the bioindustry. CPC can compose up to 15% of the total dry weight of cyanobacteria and contribute to the neuroprotective, antioxidant, hepatoprotective, and anti-inflammatory effects (Eriksen 2008). Extracted and purified CPC has nutraceutical and pharmaceutical potentials (Belay 2002), and it has set a milestone in developing fluorescent probes for several immunodiagnostics (Kronick and Grossman 1983). Phycobiliproteins such as phycocyanin, phycoerythrin, and allophycocyanin have potent anti-inflammatory, cytotoxic, and antioxidant properties (Bhaskar et al. 2005). CPC was estimated as an antioxidant in vitro as it could scavenge peroxy, alkoxyl, and hydroxyl radicals and inhibit lipid peroxidation (Romay et al. 2003). CPC can restrain cell proliferation, invoke apoptosis in cancerogenic cell lines, and affect gene regulation in mammalian cell lines (Liu et al. 2000; Cherng et al. 2007). CPC inhibits cyclooxygenase (COX), particularly COX-2, which is over-expressed in breast cancer cells but has no or minimal effect on COX-1 (Rimbau et al. 2001); it also works as an inhibitor of cytochrome oxidase II (Reddy et al. 2000). It is applicable in curing diseases like Parkinson's and Alzheimer's and averts cataracts, constipation, degenerative diseases, pancreatitis, and experimental oral and skin cancers (Rimbau et al. 2001). CPC can be potentially applicable for treating lipopolysaccharide (LPS)-related acute lung infection by inhibiting responses and increasing apoptosis in infected lung tissues (Leung et al. 2013). These developments have sparked an interest in CPC as a pharmaceutical or nutraceutical with anticarcinogenic, antitumor, and other possible health effects.

13.5.2 Scytonemin

As an essential photoprotective mechanism, cyanobacteria generate "sunscreen" substances such as scytonemin and mycosporine to shield themselves from UV radiation. It is a yellow-brown symmetrical indole-alkaloid with 544 Da hydrophobic pigments deposited in the cyanobacteria's exopolysaccharide sheaths, where it absorbs approximately 85–90% of incident UV radiation. In addition, it is a pharmacological compound with antiproliferative and anti-inflammatory properties (Stevenson et al. 2002a, b). More than 300 cyanobacterial species have been identified, including *Nostoc punctiforme*, *Scytonema* sp., *Rivularia* sp., and *Leptolyngbya* sp. (Rastogi et al. 2013).

13.5.3 Mycosporine-like Amino Acids (MAAs)

Cyanobacteria synthesize UV-absorbing water-soluble pigments called mycosporine-like amino acids (MAAs), including porphyra-334 and shinorine (Rastogi and Incharoensakdi 2014a). In cyanobacteria like *Gloeocapsa* sp., *Nostoc commune*, *Lyngbya* sp., *Scytonema* sp., etc., the production of these natural sunscreen chemicals has been studied (Rastogi and Incharoensakdi 2014b). Recently, a rare novel MAA was isolated from the unicellular cyanobacterium *Euhalothece* sp. containing the amino acid alanine [2-(e)-2,3-dihydroxipro-1-enylimino-mycosporine-alanine] (Volkmann et al. 2006). *Synechocystis* sp. produced a different novel MAA that was tentatively identified as dehydroxyl-usujirene (Zhang et al. 2007). MAAs have been shown to protect human fibroblast cells (Oyamada et al. 2008) and mice against UV-induced damage in vivo (De la Coba et al. 2009). MAAs from macroalgae, such as shinorine and porphyra-334, are already employed to defend against UV-A radiation in commercial sunscreens like Helioguard 365 and Helionori (Singh et al. 2021).

13.6 Cyanobacterial Peptides

Peptides are protein fragments that have beneficial effects on human health (Kitts and Weiler 2003). Peptides are usually inactive in their natural form, but they can perform various physiological functions (Sarmadi et al. 2011; Ghanbari 2019). Peptides or peptidic substructures make up a large portion of cyanobacterial secondary metabolites. Approximately 600 cyanobacterial peptides have been identified to date. Many of these oligopeptides are thought to be produced by polyketide synthase (PKS) or non-ribosomal peptide synthetase (NRPS), or NRPS/PKS hybrid pathways, which result in a variety of structural classes of metabolites, including acyclic peptides, cyclic and linear peptides, linear alkynoic lipopeptides, and decadepsipeptides, Microorganisms manufacture macrolides using modular polyketide synthases (PKS) similar to NRPS in terms of modularity. Several carboxylic acids are combined, activated, and optionally changed, unlike peptide-synthesizing enzymes. The acyltransferase, ketoreductase, ketosynthase, dehydratase, enoyl reductase, and acyl carrier protein domains of an individual PKS module are identical to those of animal fatty acid synthase (Jenke-Kodama et al. 2005). The hybrid nature of cyanobacterial pathways, i.e., the frequent combination of PKS and NRPS modules, is a distinguishing feature (Staunton and Weissman 2001). This class of peptides has a lot of diversity and many biosynthetic and bioactive potentials.

Cyanobacteria have produced linear peptides like aeruginosins (Aer). *Planktothrix* sp., *Microcystis* sp., *Oscillatoria* sp., and other bacteria generate aeruginosins and serine protease inhibitors. Lukesova 30/93 is the first aeruginosin-type peptide discovered in the genus *Nostoc*, containing both a carbohydrate and a fatty acid moiety (Murakami et al. 1994). Aeruginosins have been demonstrated to inhibit plasmin and thrombin in most cases. Aeruginosin 298A and

Aer-865 were isolated from *Microcystis aeruginosa* and oscillarin from *Oscillatoria agardhii*; they have an anti-inflammatory impact linked to the suppression of NF-kB translocation to the nucleus, although they had no cytotoxic effect (Kapuścik et al. 2013). Linear peptides extracted from *Nostoc* sp., *Planktothrix*, *Microcystis aeruginosa*, and *Oscillatoria* are called microginins. Exopeptidases have been demonstrated to be inhibited by microginins. Four microginins have been shown to inhibit angiotensin-converting enzymes (ACE), a target for congestive heart failure treatment and hypertension (Okino et al. 1993). Analogs of dragonamide were extracted from *Lyngbya polychrox* and *Lyngbya majuscule*. Dragonamide A, carmabin A, and dragomabin show antimalarial activity. Dragonamide A is a potent antileishmanial (McPhail et al. 2007). Almiramides B and C, two further antileishmanial lipopeptides derived from *Lyngbya majuscula*, had higher potencies, with IC50 of 2.4 and 1.9 mM, respectively (Sanchez et al. 2010). Grassystatin A–C possessing γ-amino β-hydroxybutyric acid was extracted from *Lyngbya confervoides*, and it is a very effective inhibitor of metalloprotease tumor necrosis factor-a-converting enzyme (TACE) (Kwan et al. 2009).

13.7 Alkaloids

The anatoxins, saxitoxins, L-alanine, cyanobacterial indole alkaloids, and cylindrospermopsins with cytotoxic, antiviral, and antifungal activities are the major groups of cyanobacterial alkaloids (Vasas et al. 2010). Cyanobacteria like *Fischerella* sp., *Hapalosiphon* sp., *Westiella* sp., *Westiellopsis* sp., etc., produced isonitrile-containing indole alkaloids such as fischerindoles, ambiguine isonitriles, hapalindoles, and wetwitindolinones. They all have polycyclic carbon skeletons produced from geraniol pyrophosphate and L-tryptophan, and they have cytotoxic, antibacterial, and antifungal properties (Baran and Richter 2005). Calothrixins are phenanthridine-containing alkaloids isolated from *Calothrix* sp. Human cervical cancer cells (HeLa) showed considerable cytotoxicity to both alkaloids (Abd El-Hack et al. 2019). Demay et al. evaluated a wide range of bioactive cyanobacterial metabolites, including *Westiella intricate*, *Fischerella musicola*, *Fischerella ambigua*, *Hapalosiphon fontinalis*, and *Hapalosiphon welwitschii*, which contain cytotoxic compounds such as hapalindole-like alkaloids and ambigols (Demay et al. 2019).

13.8 Polyketides

Borophycin, a polyketide extract from *Nostoc linckia* and *Nostoc spongiaeforme*, had a remarkable anticancer impact on human colon carcinoma cell lines (Davidson 1995).

13.9 Application of Cyanobacterial Metabolites

13.9.1 Antitumoral and Anticancerous Activity

Tumor cells are acquiring resistance to currently available medications, and there is an urgent need for new anticancer drugs due to failure in the chemotherapeutic treatment of malignancies. By interfering with the function of lysophosphatidic acid, *Spirulina* intake as a protein source may have slowed the growth and spread of ovarian cancer (McCarty 2015). In addition, bilirubin-like tetrapyrrolic compounds in a cyanobacterium showed a substantial antiproliferative effect on human pancreatic cancer (Koníckova et al. 2014).

Tolyporphin—It is isolated from *Tolypothrix nodosa* and has photosensitizing activity against tumor cells, and it is a thousand times more efficient than photodynamic therapy (Morliere et al. 1998).

Borophycin—It is a boron-containing metabolite extracted from *Nostoc spongiaeforme* and *Nostoc linckia* that exhibited potent cytotoxic activity against human colorectal adenocarcinoma and epidermoid carcinoma cell lines (Burja et al. 2001).

Curacin A—It is a lipid component isolated from *Lyngbya majuscule*, which inhibits tubulin polymerization, has potent cytotoxicity towards leukemic cell lines, and inhibits the growth of breast-, colon-, and renal-derived cell lines (Jones et al. 2009). Furthermore, in silico molecular docking between the cyanobacterial bioactive chemicals and lung cancer-causing receptor, *Lyngbya majuscula*, had the greatest docking score against the human lung cancer cell line A549 (Muniaraj et al. 2018).

Dolastatin 10—It is a pentapeptide with four amino acids—dolaproline, dolaisoleucine, dolaphenine, and dolavaline—isolated from *Symploca* sp. Dolastatin is a potent antiproliferative drug that binds with tubulin at the rhizoxin-binding site and inhibits microtubule assembly, causing the cell to enter the G2/M phase (Luesch et al. 2002).

Symplostatin 1—Extracted from *Symploca hydnoides*, it is an analog of dolastatin 10 that is efficacious against solid tumors (Harrigan et al. 1998).

Aeruginosin 103-A—Extracted from *Microcystis viridis* (NIES-103), it is a linear peptide that inhibits thrombin (Kodani et al. 1998).

Somocystinamide A—Extracted from cyanobacterium *Lyngbya majuscule*, it is a pluripotent inhibitor of tumor cell proliferation and angiogenesis (Stevenson et al. 2002a). In addition, extracts of *Oscillatoria terebriformis* have been shown to cause apoptosis in human lung cancer A549 cells, indicating that they may have anticancer properties (Mukund and Sivasubramanian 2014).

Nostocarboline—It is a potent cholinesterase inhibitor obtained from *Nostoc* 78-12A, which may be helpful in Alzheimer's treatment (Becher et al. 2005).

13.9.2 Antiviral Activity

Since ancient times, viruses have been responsible for many severe and debilitating illnesses globally. They are the world's largest cause of human sickness and death. Cyanobacterial polysaccharides are nontoxic, inexpensive, biodegradable, and biocompatible natural polymers. Sulfated polysaccharides are a promising viral treatment approach. They've been studied for antiviral activity against various influenza, HIV, and dengue viruses. As a result, they've gained traction in the pharmaceutical and biomedical industries, and they can be used to generate therapeutic compounds that target SARS-CoV-2 (Prabhakar et al. 2013).

Nostoflan—It is an acidic polysaccharide extracted from *Nostoc flagelliforme* (Whitton and Potts 2007) that has antiviral properties against viruses that employ carbohydrates as receptors. It has antiviral properties against influenza A, human cytomegalovirus, Herpes simplex virus (HSV) 1 and 2. In addition, it prevents virus infection from starting, including the binding and internalization steps (Kanekiyo et al. 2007).

Calcium spirulan (Ca-SP)—It is a sulfated polysaccharide obtained from *Spirulina platensis* hot water extract that has antiviral efficacy against HSV, measles, mumps, human immunodeficiency viruses (HIV), influenza A, and Henoch-Schonlein Purpura (HSP) (Hernández-Corona et al. 2002). It inhibits the virus from entering the host cell and lowers melanoma lung metastasis by preventing tumor cells from invading the basal membrane (Khandia et al. 2019).

Cyanovirin-N—It is a protein obtained from *Nostoc ellipsosporum*; it effectively binds with envelope glycoprotein and inhibits a variety of viruses such as HIV-1 and HIV-2, SIV (simian immunodeficiency virus), and FIV (feline immunodeficiency virus) (Mori and Boyd 2001). *Scytonema* sp. produced cyanovirin-N, scytovirin, and sulfoglycolipids with a significant antiviral activity against HIV, which will aid in the treatment of AIDS (Bokesch et al. 2003).

Ichthyopeptins—The cyclic depsipeptides extracted from *Microcystis ichthyoblabe* have antiviral actions against influenza virus (Zainuddin et al. 2007).

13.9.3 Antibacterial Activity

Multidrug-resistant bacteria like *Enterococci*, *Enterobacteria*, and *Staphylococcus aureus* become resistant to antibiotics like Vancomycin, AmpC b-lactamase, and Methicillin; thus, there is an urgent need to discover new antibiotics (Reinert et al. 2007). Noscomin, isolated from *Nostoc commune*, has a broad antibacterial activity for several bacteria like *Bacillus cereus*, *Staphylococcus epidermidis*, and pathogenic *Escherichia coli* comparable to several standard drugs (Mundt et al. 2003). *Westiellopsis prolifica* ARM 365, *Fischerella* sp. extract ARM 354, *Nostoc muscorum* extract ARM 221, and *Hapalosiphon hibernicus* extract ARM 178 showed potent antibacterial activity against several nonpathogenic bacteria such as *Bacillus subtilis*, *Escherichia coli*, *Bradyrhizobium* sp., and *Pseudomonas striata* (Tyagi et al. 2014), whereas *Anabaena* sp. extract showed antibacterial activity against the pathogenic strains of *E. coli*, *Klebsiella pneumonia*, *Pseudomonas aeruginosa*, *Salmonella typhi*, *and Staphylococcus aureus* (Chauhan et al. 2010). Methanolic extract of *Fischerella* sp. was active against *E. coli*, *Enterobacter aerogenes*, *P. aeruginosa*, *Mycobacterium tuberculosis*, *S. aureus*, and *Salmonella typhi* as well as multidrug-resistant (MDR) pathogenic species of *E. coli* strains in in vitro assays (Asthana et al. 2006).

13.9.4 Antiprotozoal and Insecticidal Activity

According to the World Health Organization (WHO), *Plasmodium*, *Trypanosoma*, *Leishmania*, *Schistosoma*, and other protozoal illnesses affect more than 100 million people worldwide. Malaria and leishmaniasis treatments have failed because these parasites have developed resistance to chemicals and pharmaceuticals (Fernández-Prada et al. 2019).

Nostocarboline—It is a protease and alkaloid inhibitor extracted from *Nostoc* sp. which is effective against *Trypanosoma brucei*, *Leishmania donovani*, *Plasmodium falciparum*, and *Trypanosoma cruzi* (Harrigan et al. 1998).

Aerucyclamide C68 and aerucyclamide B—It is extracted from *Microcystis aeruginosa* PCC 7806, they were effective against *Trypanosoma brucei* and *Plasmodium falciparum*. The acyl proline derivatives tumonoic acids D–I were isolated from *Blennothrix cantharidosmum*, showing antimalarial activity against *Trypanosoma* (Clark et al. 2008).

Calothrixins A and B—It is an indole phenanthridine extract from *Calothrix* sp. that inhibits *Plasmodium falciparum*'s proliferation (Rickards et al. 1999). *Synechococcus* PCC 6301 was transformed with mosquitocidal toxin genes from *Bacillus sphaericus* and *Bacillus thuringiensis*; transformed cells showed antiplasmodial and anti-larval activity (Sangthongpitag et al. 1996).

13.10 Future Prospects

Cyanobacteria are a special type of photosynthetic, oxygen-evolving gram-negative bacteria found in all possible environments around the world. Several bioactive chemicals produced from cyanobacteria display a broad spectrum of actions, including antibacterial, anticancer, antiviral, antifungal, and antiplasmodial properties, demonstrating their potential as an excellent source of novel therapeutic lead compounds over the last few decades. Another advantage of cyanobacteria as a microbiological source for drug development is the lower cost of cultivation than other organisms. The former requires simple inorganic nutrients and light for growth. As a result, it appears that cyanobacteria have the potential to be used more extensively in drug discovery.

Furthermore, because of their enormous microbial variety, cyanobacterial secondary metabolites could be a rich source of novel entities, perhaps leading to the development of new medications. However, cyanophycean species exploitation has been limited by various challenges connected to their care. With the majority of these issues now addressed, cyanobacteria have the potential to broaden the range of natural products derived from microbes. Furthermore, cyanobacteria have become intriguing sources of innovative pharmaceuticals for application in various therapeutic areas due to their relative disregard in the past compared to other microbial sources of natural products and the tremendous chemical diversity and biological activity of their products. As a result, the medicinal potential of cyanobacteria merits increased scientific attention and interdisciplinary research, with cyanobacterial strains from yet-to-be-discovered and extreme habitats serving as ideal possibilities.

Acknowledgments The authors are thankful to the Head of Centre of Advanced Study in Botany, Institute of Science, Banaras Hindu University, Varanasi, and DST-FIST for providing the necessary research facilities. The University Grants Commission, Government of India, New Delhi, is also acknowledged for providing Senior Research Fellowship (SRF) for financial assistance and ISLS, ISC, and BHU for instrumental facilities.

References

Abd El-Hack ME, Abdelnour S, Alagawany M, Abdo M, Sakr MA, Khafaga AF, Gebriel MG (2019) Microalgae in modern cancer therapy: current knowledge. Biomed Pharmacother 111: 42–50

Asthana RK, Srivastava A, Kayastha AM, Nath G, Singh SP (2006) Antibacterial potential of γ-linolenic acid from Fischerella sp. colonizing neem tree bark. World J Microbiol Biotechnol 22:443–448

Baran PS, Richter JM (2005) Enantioselective total syntheses of welwitindolinone A and fischerindoles I and G. J Am Chem Soc 127:15394–15396

Becher PG, Beuchat J, Gademann K, Jüttner F (2005) Nostocarboline: isolation and synthesis of a new cholinesterase inhibitor from *Nostoc* 78-12A. J Nat Prod 68:1793–1795

Belay A (2002) The potential application of *Spirulina* (*Arthrospira*) as a nutritional and therapeutic supplement in health management. J Am Nutraceutical Assoc 5:27–48

Bhaskar SU, Gopalaswamy G, Raghu R (2005) A simple method for efficient extraction and purification of C-phycocyanin from *Spirulina platensis* Geitler

Bin ZS, Hussain MA, Nye R, Mehta V, Mamun KT, Hossain N (2017) A review on antibiotic resistance: alarm bells are ringing. Cureus. Available from: https://doi.org/10.7759/cureus.1403

Bokesch HR, O'Keefe BR, McKee TC, Pannell LK, Patterson GM, Gardella RS, Boyd MR (2003) A potent novel anti-HIV protein from the cultured cyanobacterium *Scytonema varium*. Biochemistry 42:2578–2584

Burja AM, Banaigs B, Abou-Mansour E, Burgess JG, Wright PC (2001) Marine cyanobacteria—a prolific source of natural products. Tetrahedron 57:9347–9377

Cane DE, Walsh CT (1999) The parallel and convergent universes of polyketide synthases and nonribosomal peptide synthetases. Chem Biol 6:R319–R325

Challem JJ, Passwater RA, Mindell EM (1981) Spirulina. Keats Publishing, New Canaan, CT

Chauhan A, Chauhan G, Gupta PC, Goyal P, Kaushik P (2010) In vitro antibacterial evaluation of Anabaena sp. against several clinically significant microflora and HPTL Canalysis of its active crude extracts. Indian J Pharmacol 42:105–107

Cherng SC, Cheng SN, Tarn A, Chou TC (2007) Anti-inflammatory activity of c-phycocyanin in lipopolysaccharide-stimulated RAW 264.7 macrophages. Life Sci 81:1431–1435

Ciferri O (1983) *Spirulina*, the edible microorganism. Microbiol Rev 47:551–578

Clark BR, Engene N, Teasdale ME, Rowley DC, Matainaho T, Valeriote FA, Gerwick WH (2008) Natural products chemistry and taxonomy of the marine cyanobacterium *Blennothrix cantharidosmum*. J Nat Prod 71:1530–1537

Cohen Z, Reungjitchachawali M, Siangdung W, Tanticharoen M (1993) Production and partial purification of γ-linolenic acid and some pigments from *Spirulina platensis*. J Appl Phycol 5: 109–115

Davidson BS (1995) New dimensions in natural products research: cultured marine microorganisms. Curr Opin Biotechnol 6:284–291

De la Coba F, Aguilera J, De Galvez MV, Alvarez M, Gallego E, Figueroa FL, Herrera E (2009) Prevention of the ultraviolet effects on clinical and histopathological changes, as well as the heat shock protein-70 expression in mouse skin by topical application of algal UV-absorbing compounds. J Dermatol Sci 55:161–169

Demay J, Bernard C, Reinhardt A, Marie B (2019) Natural products from cyanobacteria: focus on beneficial activities. Mar Drugs 17:320

Eriksen NT (2008) Production of phycocyanin-a pigment with applications in biology, biotechnology, foods and medicine. Appl Microbiol Biotechnol 80:1–14

Estela Silva-Stenico M, Kaneno R, Zambuzi F, Vaz M, Alvarenga D, Fiore M (2013) Natural products from cyanobacteria with antimicrobial and antitumor activity. Curr Pharm Biotechnol 14:820–828

Euler US, Eliassen R (1967) Prostaglandins. Academic Press, New York

Fernández-Prada C, Douanne N, Minguez-Menendez A, Pena J, Tunes LG, Pires DE, Monte-Neto RL (2019) Repurposed molecules: a new hope in tackling neglected infectious diseases. In: Silico drug design, pp 119–160

Ghanbari R (2019) Review on the bioactive peptides from marine sources: indication for health effects. Int J Pept Res Ther 25:1187–1199

Harrigan GG, Yoshida WY, Moore RE, Nagle DG, Park PU, Biggs J, Valeriote FA (1998) Isolation, structure determination, and biological activity of dolastatin 12 lyngbyastatin 1 from *Lyngbya majuscula/Schizothrix calcicola* cyanobacterial assemblages. J Nat Prod 61: 1221–1225

Hayashi O, Hirahashi T, Katoh T, Miyajima H, Hirano T, Okuwaki Y (1998) Class-specific influence of dietary *Spirulina platensis* on antibody production in mice. J Nutr Sci Vitaminol 44:841–851

Hernández-Corona A, Nieves I, Meckes M, Chamorro G, Barron BL (2002) Antiviral activity of *Spirulina maxima* against herpes simplex virus type 2. Antivir Res 56:279–285

Horrobin DF (2000) Essential fatty acid metabolism and modification in atopic eczema. Am J Clin Nutr 71:367–372

Jenke-Kodama H, Sandmann A, Müller R, Dittmann E (2005) Evolutionary implications of bacterial polyketide synthases. Mol Biol Evol 22:2027–2039

Jensen GS, Ginsberg DI, Huerta P, Citton M, Drapeau C (2000) Consumption of *Aphanizomenon flos-aquae* has rapid effects on the circulation and function of immune cells in humans. JANA 2: 50–58

Jones AC, Gu L, Sorrels CM, Sherman DH, Gerwick WH (2009) New tricks from ancient algae: natural products biosynthesis in marine cyanobacteria. Curr Opin Chem Biol 13:216–223

Kanekiyo K, Hayashi K, Takenaka H, Lee JB, Hayashi T (2007) Anti-herpes simplex virus target of an acidic polysaccharide, nostoflan, from the edible blue-green alga *Nostoc flagelliforme*. Biol Pharm Bull 30:1573–1575

Kapuścik A, Hrouzek P, Kuzma M, Bártová S, Novák P, Jokela J, Kopecký J (2013) Novel Aeruginosin-865 from *Nostoc* sp. as a potent anti-inflammatory agent. Chem Bio Chem 14: 2329–2337

Kernoff PB, Willis A, Stone KJ, Davies JA, McNicol GP (1977) Antithrombotic potential of dihomogamma-linolenic acid in man. Br Med J 2:1441–1444
Khandia R, Dadar M, Munjal A, Dhama K, Karthik K, Tiwari R, Chaicumpa W (2019) A comprehensive review of autophagy and its various roles in infectious, non-infectious, and lifestyle diseases: current knowledge and prospects for disease prevention, novel drug design, and therapy. Cell 8:674
Kini S, Divyashree M, Mani MK, Mamatha BS (2020) Algae and cyanobacteria as a source of novel bioactive compounds for biomedical applications. In: Advances in cyanobacterial biology, pp 173–194
Kitts DD, Weiler K (2003) Bioactive proteins and peptides from food sources. Applications of bioprocesses used in isolation and recovery. Curr Pharm 9:1309–1323
Kodani S, Ishida K, Murakami M (1998) Aeruginosin 103-A, a thrombin inhibitor from the cyanobacterium *Microcystis viridis*. J Nat Prod 61:1046–1048
Koníčková R, Vanková K, Vaníková J, Vánová K, Muchová L, Subhanová I, Vítek L (2014) Anti-cancer effects of blue-green alga *Spirulina platensis*, a natural source of bilirubin-like tetrapyrrolic compounds. Ann Hepatol 13:273–283
Kronick MN, Grossman PD (1983) Immunoassay techniques with fluorescent phycobiliprotein conjugates. Clin Chem 29:1582–1586
Kushak R, VanCott E, Drapeau C, Winter H (1999) Effect of algae *Aphanizomenon flos aquae* on digestive enzyme activity and polyunsaturated fatty acids level in blood plasma. Gastroenterology 4:A559–A559
Kwan JC, Eksioglu EA, Liu C, Paul VJ, Luesch H (2009) Grassystatins A-C from marine cyanobacteria, potent cathepsin E inhibitors that reduce antigen presentation. J Med Chem 52: 5732–5747
Leung PO, Lee HH, Kung YC, Tsai MF, Chou TC (2013) Therapeutic effect of C-phycocyanin extracted from blue-green algae in a rat model of acute lung injury induced by lipopolysaccharide. eCAM
Liu Y, Xu L, Cheng N, Lin L, Zhang C (2000) Inhibitory effect of phycocyanin from *Spirulina platensis* on the growth of human leukaemia K562 cells. J Appl Phycol 12:125–130
Luesch H, Yoshida WY, Moore RE, Paul VJ, Mooberry SL, Corbett TH (2002) Symplostatin 3, a new dolastatin 10 analogue from the marine cyanobacterium *Symploca* sp. VP452. J Nat Prod 65:16–20
McCarty MF (2015) Hypothesis: *Spirulina* may slow the growth and spread of ovarian cancer by interfering with the growth factor activity of lysophosphatidic acid. J Mol Genet Med 9:184
McPhail KL, Correa J, Linington RG, González J, Ortega-Barría E, Capson TL, Gerwick WH (2007) Antimalarial linear lipopeptides from a Panamanian strain of the marine cyanobacterium *Lyngbya majuscula*. J Nat Prod 70:984–988
Michalak I, Chojnacka K (2015) Algae as production systems of bioactive compounds. Eng Life Sci 15:160–176
Mori T, Boyd MR (2001) Cyanovirin-N, a potent human immunodeficiency virus-inactivating protein, blocks both CD4-dependent and CD4-independent binding of soluble gp120 (sgp120) to target cells, inhibits sCD4-induced binding of sgp120 to cell-associated CXCR4 and dissociates bound sgp120 from target cells. Antimicrob Agents Chemother 45:664–672
Morliere P, Maziere JC, Santus R (1998) Tolyporphin: a natural product from cyanobacteria with potent photosensitizing activity against tumour cells in vitro and in vivo. Cancer Res 58:3571–3578
Mukund S, Sivasubramanian V (2014) Anticancer activity of *Oscillatoria terebriformis* cyanobacteria in human lung cancer cell line a:549. Int J Appl Biol Pharm 5:2
Mundt S, Kreitlow S, Jansen R (2003) Fatty acids with antibacterial activity from the cyanobacterium Oscillatoria redekei HUB 051. J Appl Phycol 15:263–267
Muniaraj S, Subramanian V, Srinivasan P, Palani M (2018) In silico and in vitro studies on *Lyngbya majuscula* using against lung cancer cell line (A549). Pharm J 10:421

Murakami M, Okita Y, Matsuda H, Okino T, Yamaguchi K (1994) Aeruginosin 298-A, a thrombin and trypsin inhibitor from the blue-green alga *Microcystis aeruginosa* (NIES-298). Tetrahedron 35:3129–3132

Nowruzi B, Haghighat S, Fahimi H, Mohammadi E (2018) *Nostoc* cyanobacteria species: a new and rich source of novel bioactive compounds with pharmaceutical potential. J Pharm Health Serv Res 9:5–12

Okino T, Matsuda H, Murakami M, Yamaguchi K (1993) Microginin, an angiotensin-converting enzyme inhibitor from the blue-green alga *Microcystis aeruginosa*. Tetrahedron 34:501–504

Oyamada C, Kaneniwa M, Ebitani K, Murata M, Ishihara K (2008) Mycosporine-like amino acids extracted from scallop (*Patinopecten yessoensis*) ovaries: UV protection and growth stimulation activities on human cells. Mar Biotechnol 10:141–150

Pabon MM, Jernberg JN, Morganti J, Contreras J, Hudson CE, Klein RL, Bickford PC (2012) A spirulina-enhanced diet provides neuroprotection in an α-synuclein model of Parkinson's disease

Prabhakar U, Maeda H, Jain RK, Sevick-Muraca EM, Zamboni W, Farokhzad OC (2013) Challenges and key considerations of the enhanced permeability and retention effect for nanomedicine drug delivery in oncology. Cancer Res 73:2412–2417

Qiang HU, Zheungu H, Cohen Z, Richond A (1997) Enhancement of eicosapentaenoic acid (EPA) and γ-linolenic acid (GLA) production by manipulating the algal density of outdoor cultures of *Monodus subterraneus* (Eustigmatophyta) and *Spirulina platensis* (Cyanobacteria). Eur J Phycol 32:81–86

Qiu B, Liu J, Liu Z, Liu S (2002) Distribution and ecology of the edible cyanobacterium Ge-Xian-Mi (*Nostoc*) in rice fields of Hefeng County in China. J Appl Phycol 14:423–429

Rasmussen HE, Blobaum KR, Park YK, Ehlers SJ, Lu F, Lee JY (2008) Lipid extract of *Nostoc commune var. sphaeroides* Kutzing, a blue-green alga, inhibits the activation of sterol regulatory element-binding proteins in HepG2 cells. Nutr J 138:476–481

Rasmussen HE, Blobaum KR, Jesch ED, Ku CS, Park YK, Lu F, Lee JY (2009) Hypocholesterolemic effect of *Nostoc commune var. sphaeroides* Kützing, an edible blue-green alga. Eur J Nutr 48:387–394

Rastogi RP, Incharoensakdi A (2014a) UV radiation-induced biosynthesis, stability and antioxidant activity of mycosporine-like amino acids (MAAs) in a unicellular cyanobacterium *Gloeocapsa* sp. CU2556. J Photochem Photobio B Bio 130:287–292

Rastogi RP, Incharoensakdi A (2014b) Characterization of UV-screening compounds, mycosporine-like amino acids, and scytonemin in the cyanobacterium *Lyngbya* sp. CU2555. FEMS Microbiol Ecol 87:244–256

Rastogi RP, Sinha RP, Incharoensakdi A (2013) Partial characterization, UV-induction and photoprotective function of sunscreen pigment, scytonemin from *Rivularia* sp. HKAR-4. Chemosphere 93:1874–1878

Reddy AS, Thomas TL (1996) Expression of a cyanobacterial Δ6-desaturase gene results in γ-linolenic acid production in transgenic plants. Nat Biotechnol 14:639–642

Reddy CM, Bhat VB, Kiranmai G, Reddy MN, Reddanna P, Madyastha KM (2000) Selective inhibition of cyclooxygenase-2 by C-phycocyanin, a biliprotein from *Spirulina platensis*. Biochem Biophys Res Commun 277:599–603

Reinert RR, Low DE, Rossi F, Zhang X, Wattal C, Dowzicky MJ (2007) Antimicrobial susceptibility among organisms from the Asia/Pacific rim, Europe and Latin and North America collected as part of TEST and the in vitro activity of tigecycline. J Antimicrob Chemother 60: 1018–1029

Rickards RW, Rothschild JM, Willis AC, de Chazal NM, Kirk J, Kirk K, Smith GD (1999) Calothrixins A and B, novel pentacyclic metabolites from *Calothrix* cyanobacteria with potent activity against malaria parasites and human cancer cells. Tetrahedron 55:13513–13520

Rimbau V, Camins A, Pubill D, Sureda FX, Romay C, González R, Pallàs M (2001) C-phycocyanin protects cerebellargranule cells from low potassium/serum deprivation-induced apoptosis. Naunyn Schmiedeberg's Arch Pharmacol 364:96–104

Romay CH, Gonzalez R, Ledon N, Remirez D, Rimbau V (2003) C-phycocyanin: a biliprotein with antioxidant, anti-inflammatory and neuroprotective effects. Curr Protein Pept Sci 4:207–216

Sanchez LM, Lopez D, Vesely BA, Della Togna G, Gerwick WH, Kyle DE, Linington RG (2010) Almiramides A-C: Discovery and development of a new class of leishmaniasis lead compounds. J Med Chem 53:4187–4197

Sangthongpitag K, Delaney SF, Rogers PL (1996) Evaluation of four fresh-water unicellular cyanobacteria as potential hosts for mosquitocidal toxins. Biotechnol Lett 18:175–180

Sarmadi B, Ismail A, Hamid M (2011) Antioxidant and angiotensin-converting enzyme (ACE) inhibitory activities of cocoa (*Theobroma cacao* L.) autolysates. Food Res Int 44:290–296

Sasidharan S, Chen Y, Saravanan D, Sundram KM, Latha LY (2011) Extraction, isolation and characterization of bioactive compounds from plants' extracts. Afr J Tradit Complement Altern Med 8

Seshadri CV (1993) All India coordinated project on *Spirulina*. Shri Amm Murugappa Chettiar Research Center (MCRC), Chennai

Sharma NK, Tiwari SP, Tripathi K, Rai AK (2011) Sustainability and cyanobacteria (blue-green algae): facts and challenges. J Appl Phycol 23:1059–1081

Simopoulos AP (1997) Nutrition tid-bites: essential fatty acids in health and chronic disease. Food Rev Int 13:623–631

Singh A, Čížková M, Bišová K, Vítová M (2021) Exploring mycosporine-like amino acids (MAAs) as safe and natural protective agents against UV-induced skin damage. Antioxid 10:683

Stanier G (1988) Fine structure of cyanobacteria. In Meth Enzymol 167:157–172

Staunton J, Weissman KJ (2001) Polyketide biosynthesis: a millennium review. Nat Prod Rep 18: 380–416

Stevenson CS, Capper EA, Roshak AK, Marquez B, Eichman C, Jackson JR, Marshall LA (2002a) The identification and characterization of the marine natural product scytonemin as a novel antiproliferative pharmacophore. J Pharmacol Exp Th 303:858–866

Stevenson CS, Capper EA, Roshak AK (2002b) Scytonemin-a marine natural product inhibitor of kinases key in hyperproliferative inflammatory diseases. Inflamm Res 51:112–114

Torres-Duran PV, Ferreira-Hermosillo A, Juarez-Oropeza MA (2007) Antihyperlipemic and anti-hypertensive effects of *Spirulina maxima* in an open sample of Mexican population: a preliminary report. Lipids Health Dis 6:1–8

Tyagi R, Kaushik BD, Kumar J (2014) Antimicrobial activity of some cyanobacteria. In: Microbial diversity and biotechnology in food security. Springer, New Delhi, pp 463–470

Vasas G, Borbely G, Nanasi P, Nanasi PP (2010) Alkaloids from cyanobacteria with diverse, potent bioactivities. Mini Rev Med Chem 10:946–955

Volk RB (2008) A newly developed assay for the quantitative determination of antimicrobial (anticyanobacterial) activity of both hydrophilic and lipophilic test compounds without any restriction. Microbiol Res 163:161–167

Volkmann M, Gorbushina AA, Kedar L, Oren A (2006) Structure of euhalothece-362, a novel red-shifted mycosporine-like amino acid, from a halophilic cyanobacterium (*Euhalothece* sp.). FEMS Microbiol Lett 258:50–54

Vonshak A (1997) Spirulina platensis arthrospira: physiology, cell-biology and biotechnology. CRC

Watanabe F, Takenaka S, Kittaka-Katsura H, Ebara S, Miyamoto E (2002) Characterization and bioavailability of vitamin B12-compounds from edible algae. J Nutr Sci Vitaminol 48:325–331

Whitton BA, Potts M (2007) The ecology of cyanobacteria: their diversity in time and space. Springer Science & Business Media

World Health Organization (2020) WHO report on cancer: setting priorities, investing wisely and providing care for all. World Health Organization, Geneva

Zahra Z, Choo DH, Lee H, Parveen A (2020) Cyanobacteria: review of current potentials and applications. Environments 7:13

Zainuddin EN, Mentel R, Wray V, Jansen R, Nimtz M, Lalk M, Mundt S (2007) Cyclic depsipeptides, ichthyopeptins A and B, from *Microcystis ichthyoblabe*. J Nat Prod 70:1084–1088

Zhang L, Li L, Wu Q (2007) Protective effects of mycosporine-like amino acids of *Synechocystis* sp. PCC 6803 and their partial characterization. J Photochem Photobiol 86:240–245

Bioactivity Potential of Cyanobacterial Species Inhabitant of Southwestern India

14

Kodandoor Sharathchandra and Kandikere Ramaiah Sridhar

Abstract

Cyanobacteria possess potent bioactive compounds of nutritional and medicinal significance. This study evaluates the antioxidant compounds, antioxidant enzymes, and antioxidant potentials of 12 cyanobacterial species isolated from different unpolluted and polluted ecosystems (lentic water, reservoir, thermal spring, dairy effluents, domestic sewage effluents, pharmaceutical effluents, and iron ore mine effluents). The top producers of antioxidant compounds (total phenolics, flavonoids, tannins, and reduced glutathione), vitamins (C and E), and enzymes (catalase, glutathione peroxidase, glutathione reductase, glutathione S-reductase, and superoxide dismutase) showed a high antioxidant potential (total antioxidant activity, DPPH radical-scavenging activity, nitric oxide-scavenging activity, and ferric-reducing power). Methanol extracts possess consistently higher quantities of total phenolics, flavonoids, and tannins and antioxidant activities than ethanol extracts. Among the 12 species, *Leptolyngbya fragilis*, *Phormidium corium*, and *Planktolyngbya limnetica* were common among the top 5–6 species. The quantities of reduced glutathione, catalase, and superoxide dismutase of these species showed a strong positive correlation with all the antioxidant activities assessed, justifying their effective protective roles by their antioxidant potential. This study recommends three cyanobacteria (*L. fragilis*, *Ph. corium*, and *Pl. limnetica*) based on their antioxidant compounds, antioxidant enzymes, and antioxidant activities for future progress in cyanobacterial biotechnology.

K. Sharathchandra · K. R. Sridhar (✉)
Department of Biosciences, Mangalore University, Mangalore, Karnataka, India

B. Neilan et al. (eds.), *Cyanobacterial Biotechnology in the 21st Century*,
https://doi.org/10.1007/978-981-99-0181-4_14

Keywords

Antioxidants · Antioxidant activity · Antioxidant enzymes · *Cyanophyceae* · Phenolics · Reduced glutathione · Vitamins

14.1 Introduction

The production of free radicals leads to about 200 oxidative stress-related diseases in humans (e.g., atherosclerosis, cancer, cardiovascular diseases, diabetes, hyperoxaluria, hypertension, and neurological ailments) (Rop et al. 2012). To combat such diseases, a diet consisting of natural antioxidants is highly preferred against synthetic antioxidants owing to their carcinogenic effects (Vaz et al. 2010; Agyei et al. 2015). Natural antioxidants in the diet (e.g., fruits, vegetables, and seaweeds) include phenolic compounds (tannins, phenolic acids, and flavonoids), vitamins (carotenoids, vitamin C, and vitamin E), and enzymes (ascorbate peroxidase, catalase, and superoxide dismutase) (Regelsberger et al. 2002; Vellard 2003; Li et al. 2007; Zeeshan and Prasad 2009; Freile-Pelegrín and Robledo 2013; El-Aty et al. 2014; Kunur and Singh 2020).

In addition to several value-added plant products, cyanobacteria are also known for their richness in a wide range of secondary metabolites and antioxidants of substantial nutritional and medicinal significance (Nunnery et al. 2010; Goiris et al. 2012; Guedes et al. 2013; Hossain et al. 2016). Cyanobacteria are highly diverse, representing up to 8000 species (150 genera) worldwide with widespread occurrence in different habitats (terrestrial, freshwater, and marine), providing an ample opportunity for exploration (Dvořák et al. 2015, 2017). Recent studies examined the bioactive compounds of freshwater cyanobacteria possessing antioxidant, anti-inflammatory, anticancer, antimicrobial, and pharmacological potential (Kaushik et al. 2009; Guedes et al. 2013). Among the cyanobacteria, *Arthrospira* is well known for its commercial applications in the food and feed industry industries owing to its wide disease-preventive ability and rich antioxidants, pigments, polysaccharides, and fatty acids (Kepekçi and Saygideger 2012; Belay 2013; Shalaby and Shanab 2013; Singh et al. 2017). Besides *Arthrospira*, other cyanobacteria such as *Anabaena* and *Nostoc* are known for considerable health-promoting attributes, especially in Asian countries (Li et al. 2007; Yu et al. 2009; Hajimahmoodi et al. 2010; Ma et al. 2015). In addition, some cyanobacterial strains are applied in the cosmetic and cosmeceutical industries (Morone et al. 2019).

Another novelty of cyanobacteria is their potential to produce antioxidant enzymes (e.g., catalase and superoxide dismutase) (Zeeshan and Prasad 2009; Kunur and Singh 2020). These enzymes produced by cyanobacteria are useful in antioxidant therapies, food industries (processing and storage), and medicinal requirements (Tiwari and Pandey 2012; Gacheva et al. 2013). The biological systems' antioxidant defense mechanisms are mainly nonenzymatic and enzymatic (Latifi et al. 2008; Zeeshan and Prasad 2009). The nonenzymatic antioxidants include phenolic compounds and vitamins, while enzymatic antioxidants include catalase, peroxidase, reductase, and superoxide dismutase (Luximon-Ramma et al. 2002; Hong et al. 2008). Cyanobacteria are well equipped with nonenzymatic and enzymatic antioxidants to stimulate antioxidant functions. Although cyanobacteria have been studied from the pristine and contaminated aquatic habitats in southern India, their antioxidant compounds and antioxidant potential have not been well focused (Albertano et al. 1999; Raghavendra and Hosmani 2002; Vijayakumar et al. 2007; Rajeshwari and Rajashekhar 2012). Hence, the current study aims to evaluate 12 different species of cyanobacteria isolated from varied freshwaters and stressed habitats (tank, reservoir, thermal springs, and polluted habitats) to evaluate the antioxidant functions of antioxidant compounds and enzymes.

14.2 Cyanobacterial Strains and Their Processing

Twelve cyanobacterial species (non-heterocystous forms, ten spp.; heterocystous forms, two spp.) isolated from the natural and polluted habitats in southern India are used in this study (Table 14.1). Such habitats include lentic water (*Oxynema acuminatum*), reservoir (*Phormidium lucidum*), thermal spring (*Calothrix fusca*, *Leptolyngbya fragilis*, *Planktolyngbya limnetica*, and *Scytonema bohnerii*), dairy effluents (*Geitlerinema calcuttense*), domestic sewage effluents (*Jaaginema pseudogeminatum* and *Phormidium chlorinum*), pharmaceutical effluents (*Leptolyngbya tenuis* and *Phormidium chalybeum*), and iron ore mine effluents (*Phormidium corium*).

Processing of cyanobacterial cultures and assessment have been schematically represented in Fig. 14.1. First, the cyanobacterial masses obtained from different habitats were rinsed repeatedly in distilled water to remove the impurities. Next, the samples were diluted and inoculated into the BG11 broth amended with neomycin (Stainer et al. 1976; Rippka et al. 1979). After 2 weeks, the cultures were microscopically examined for the growth of cyanobacteria. Micropipettes picked up the individual colonies and plated them onto the antibiotic-amended BG11 agar medium (Dor 1987). To avoid contamination, the isolated cyanobacteria were subcultured in antibiotic-free BG11 broth up to five times. Cultures in antibiotic-free BG11 broth were grown in the Erlenmeyer flasks incubated under fluorescent light (2000 lux) at $28 \pm 2°$ with 14:10 h light and dark regimes. The antibiotic-amended BG11 broth and BG11 agar medium without nitrogen source served as isolation of heterocystous cyanobacteria (*Calothrix fusca* and *Scytonema bohnerii*).

As detailed above, the axenic culture of each species was grown in a 250 ml Erlenmeyer flask with 150 ml BG11 broth without antibiotics. At the mid-stationary growth phase (25 days), the cultures were harvested, and the biomass was separated by filtration (Millipore; porosity, 0.45 μm). The cultures were rinsed in distilled water repeatedly to remove the traces of medium and inorganic components. The biomass was transferred to an aluminum foil followed by lyophilization, transferred to airtight vials, and stored in a deep freezer. The lyophilized biomass of the cyanobacterial samples was used to determine various bioactive constituents. The cyanobacterial extracts (ethanol, methanol, and aqueous) were prepared by

Table 14.1 Cyanobacterial strains used to evaluate bioactive potential ([a], heterocystous form)

	Species code in the figures	Name used in the text	Geographic origin	Geographic coordinates
[a]*Calothrix fusca* (Kutz.) B. and F.	Cafu	*C. fusca*	Thermal spring, Panekal	12°54′N, 75°17.5′E
Geitlerinema calcuttense (Biswas) Anagnostidis	Geca	*G. calcuttense*	Dairy effluent, Mangalore	12°53′N, 74°53′E
Jaaginema pseudogeminatum Anagnostidis and Komarek	Japs	*J. pseudogeminatum*	Domestic sewage, Mangalore	12°52′N, 74°51′E
Leptolyngbya fragilis (Gomont) Anagnostidis and Komarek	Lefr	*L. fragilis*	Thermal spring, Panekal	12°54′ N, 75°17.5′E
Leptolyngbya tenuis (Gomont) Anagnostidis and Komarek	Lete	*L. tenuis*	Pharmaceutical effluent, Mangalore	12°51′N, 74°51′E
Oxynema acuminatum (Gomont) Chatchawan, Komarek, Strunecky, Smarda and Peerapornpisal	Oxac	*O. acuminatum*	Tank, Malavalli	12° 22′N, 77°4′E
Phormidium chalybeum (Martens ex Gomont) Anagnostidia and Komarek	Phcha	*Ph. chalybeum*	Pharmaceutical effluent, Mangalore	12° 51′N, 74°51′E
Phormidium chlorinum (Kutzing ex Gomont) Umezaki and Watanabe	Phchl	*Ph. chlorinum*	Domestic sewage, Mangalore	12°52′N, 74°51′E
Phormidium corium (C. Agardh) Kutzing ex Gomont	Phco	*Ph. corium*	Iron ore mine effluent, Kudremukh	13°13′N, 75°13′E
Phormidium lucidum Kutzing ex Gomont	Phlu	***Ph. lucidum***	Bhadra reservoir, Shimoga	13°42′N, 75°38′E
Planktolyngbya limnetica (Lemmermann) J. Komarkova-Legnerova and G. Cronberg	Plli	*Pl. limnetica*	Thermal spring, Panekal	12°54′N, 75°17.5′E
[a]*Scytonema bohnerii* Schmidle	Scbo	*S. bohnerii*	Thermal spring, Panekal	12°54′N, 75°17.5′E

following the methods by Kaushik et al. (2009). The lyophilized biomass was subjected to extraction by mixing with solvents. The mixture was kept overnight (4 °C), brought to laboratory temperature (2–3 h), and followed by sonication (10 min) (Branson ultrasonic corporation, 1510, USA). The extraction mixture was centrifuged (4000 rpm, 10 min), and the supernatant was transferred into

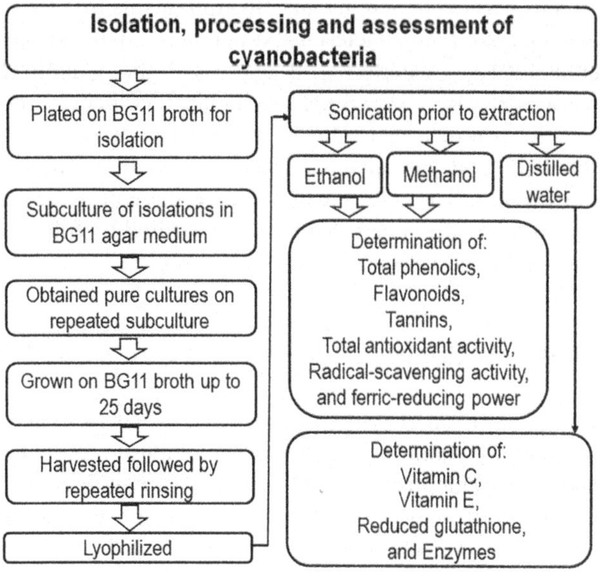

Fig. 14.1 Schematic presentation of isolation, processing, and assessment of cyanobacteria

pre-weighed test tubes and concentrated by nitrogen gas until dried. Test tubes were reweighed, and the dried sample was suspended in ethanol or methanol to prepare defined concentrations for different assays. The aqueous extracts were prepared by suspending lyophilized cultures in deionized water. On mixing, the homogenized mixture was kept at the laboratory (28 ± 2 °C, 30 min) followed by sonication. The suspension was centrifuged (2500 rpm, 6 min), and the supernatant was transferred in pre-weighed test tubes and concentrated with nitrogen until complete drying. The test tubes were reweighed, and the dried sample was suspended in deionized water to prepare solutions possessing known concentrations.

14.3 Antioxidant Compound Extracts from Cyanobacterial Strains

Six antioxidant compounds, i.e., total phenolics, flavonoids, tannins, vitamin C, vitamin E, and reduced glutathione, of processed cultures of 12 cyanobacteria were estimated following the standard protocols. The total phenolics of ethanol and methanol extracts were estimated using spectrophotometer by Ciocalteu's method with gallic acid as the standard. The quantity was expressed in gallic acid equivalents (mg GAE/g) (Mc Donald et al. 2001). The aluminum chloride spectrophotometric method was employed for flavonoid determination in ethanol and methanol extracts (Chang et al. 2002). Quercetin served as the standard, and the flavonoid content was expressed in microgram equivalents to quercetin (μg QE/g). The tannin content of ethanol and methanol extracts was determined by the spectrometric method by Okwu (2005). Catechin served as the standard, and the quantity of tannins was expressed in catechin equivalents (mg CEs/g). The vitamin C content in the aqueous cyanobacterial extract was analyzed spectrophotometrically (Kapur et al. 2012). Ascorbic acid expressed the vitamin C content in ascorbic acid equivalents (mg AAEs/g). The vitamin E content in aqueous extracts was estimated spectrophotometrically (Jayashree and Kamat 1985). The α-tocopherol served as the standard, and the quantity was expressed in milligram tocopherol equivalents (mg TEs/g). The reduced glutathione content in aqueous extracts of cultures was measured by the spectrophotometric method by Silber et al. (1992), and its quantity was expressed in nanomoles (nM/g protein).

14.4 Antioxidant Enzymes Assay from Cyanobacterial Strains

Five antioxidant enzymes in aqueous extracts were assayed using standard methods (catalase, glutathione peroxidase, glutathione reductase, glutathione S-reductase, and superoxide dismutase). Catalase was measured according to Rao et al. (1996), and its content was expressed in enzyme units (units/mg protein) (1 unit is the quantity of enzyme required to oxidize 1 μM of hydrogen peroxide per min). The glutathione peroxidase activity was assayed based on the procedure by Bernroitner et al. (2009). The specific activity was evaluated by the extinction coefficient (6.22/mM/cm), and it was expressed in units/mg protein. The glutathione reductase activity was evaluated by spectrophotometric method (Carlberg and Mannervik 1985). The enzyme activity was measured by extinction coefficient (6.224/mM/cm) and expressed in μM oxidized NADPH/min/mg protein (1 unit is the quantity of enzyme required to oxidize 1 μM of NADPH/min). The glutathione S-transferase was measured based on the method by Campbell et al. (2008). Its activity is expressed in μM conjugated 1-chloro-2, 4-dinitrobenzene/min/mg protein using its extinction coefficient ($\varepsilon = 9.6$/mM/cm). The superoxide dismutase assay was based on the spectrophotometric method (Beauchamp and Fridovich 1971; Paya et al. 1992). The activity was expressed in units/mg (1 unit is the quantity of enzyme required to inhibit 50% reduction of nitro blue tetrazolium chloride).

Four antioxidant assays (total antioxidant, DPPH radical-scavenging, nitric oxide-scavenging, and ferric-reducing activities) of ethanol and methanol extracts were assessed by standard protocols and the total antioxidant assay by reduction of Mo (VI) to Mo (V) by the sample and the formation of green phosphate/Mo (V) complex at the acidic pH using a spectrophotometer (Prieto et al. 1999). The tannic acid served as the standard, and the total antioxidant activity was expressed in tannic acid equivalents (mg TAE/g). The DPPH radical-scavenging activity assessment was by a spectrophotometric method using butylated hydroxyanisole (BHA) as the standard to express the activity in percentage (Barreira et al. 2008). The nitric oxide-scavenging activity assessment was carried out with ethanol as a control to express the activity in percentage (Tenpe et al. 2008). Finally, the ferric-reducing power was assessed using the spectrophotometric method and expressed absorbance at 700 nm using BHA as a standard antioxidant (Benzie and Strain 1996). The data on antioxidant compounds (total phenolics, flavonoids, and tannins) and antioxidant activity between ethanol and methanol extracts were compared by paired t-test (StatSoft Inc. 2008). The Pearson correlation coefficient was adopted to follow the relationship between the antioxidant compounds vs antioxidant activity and antioxidant enzymes vs antioxidant activity of pooled data of three common species (*L. fragilis*, *Ph. corium*, and *Pl. limnetica*) (*p* value, two-tailed; confidence interval, 95%) (SPSS 6.0 Windows Student Version # 3.5).

14.5 Bioactivity Potential of Cyanobacterial Isolates

14.5.1 Antioxidant Compounds

The total phenolic, flavonoid, and tannin contents were higher in methanol extract than in ethanol extract (Fig. 14.2a–c). All these compounds were the highest in *J. pseudogeminatum*, followed by either *Pl. limnetica* or *Ph. corium* or *Ph. chalybeum*. *Ph. lucidum* showed the least activity in the total phenolic and flavonoid contents, while tannins were the least in *O. occuminatus*. The total phenolics of methanol extract of *J. pseudogeminatum*, *Pl. limnetica*, *Ph. corium*, *Ph. chalybeum*, and *Ph. chalybeum* were significantly higher than in ethanol extract ($p < 0.05$). Flavonoids in methanol extract of *J. pseudogeminatum*, *Ph. chalybeum*, *Pl. limnetica*, *Ph. chalybeum*, *L. fragilis*, and *O. acuminatum* were significantly higher than in ethanol extract ($p < 0.05$). Tannins in methanol extract of *Ph. chalybeum*, *L. fragilis*, and *Ph. chalybeum* were markedly higher than in ethanol extract ($p < 0.05$). Our results partially resemble the higher quantities of the total phenolics and flavonoids in methanol extract of cyanobacterial strains studied by Guerreiro et al. (2020).

There is a strong notion that the phenolic compounds in plants and microbes are accountable for their antioxidant potential. However, the phenolic compounds in cyanobacteria have been less studied than in higher plants. Blagojević et al. (2018) studied the total phenolics in 18 strains of cyanobacteria, but its quantity was lower than the 12 species in our study. Miranda et al. (1998) evaluated the antioxidant activity of the total phenolics from *Spirulina maxima* and reported that they were accountable for the antioxidant potential. Asan-Ozusaglam et al. (2013) assessed the total phenolic and flavonoid contents in *Anabaenopsis* sp. and reported the total phenolic contents in different extracts (dichloromethane, water, and methanol), with the highest quantity in aqueous extract.

Similarly, the flavonoid content was the highest in dichloromethane. Pant et al. (2012) studied the methanolic extract of *Anabaena* sp. and found a high quantity of total phenolics. Some of these reports align with our study that the total phenolics, flavonoids, and tannins were higher in methanol extracts than in ethanol extracts. The quantity of total phenolics, flavonoids, and tannins in two species among the top five species (*Pl. limnetica* and *Ph. chalybeum*) coincides with their high antioxidant activities (total antioxidant, DPPH radical-scavenging, nitric oxide-scavenging, and ferric-reducing activities).

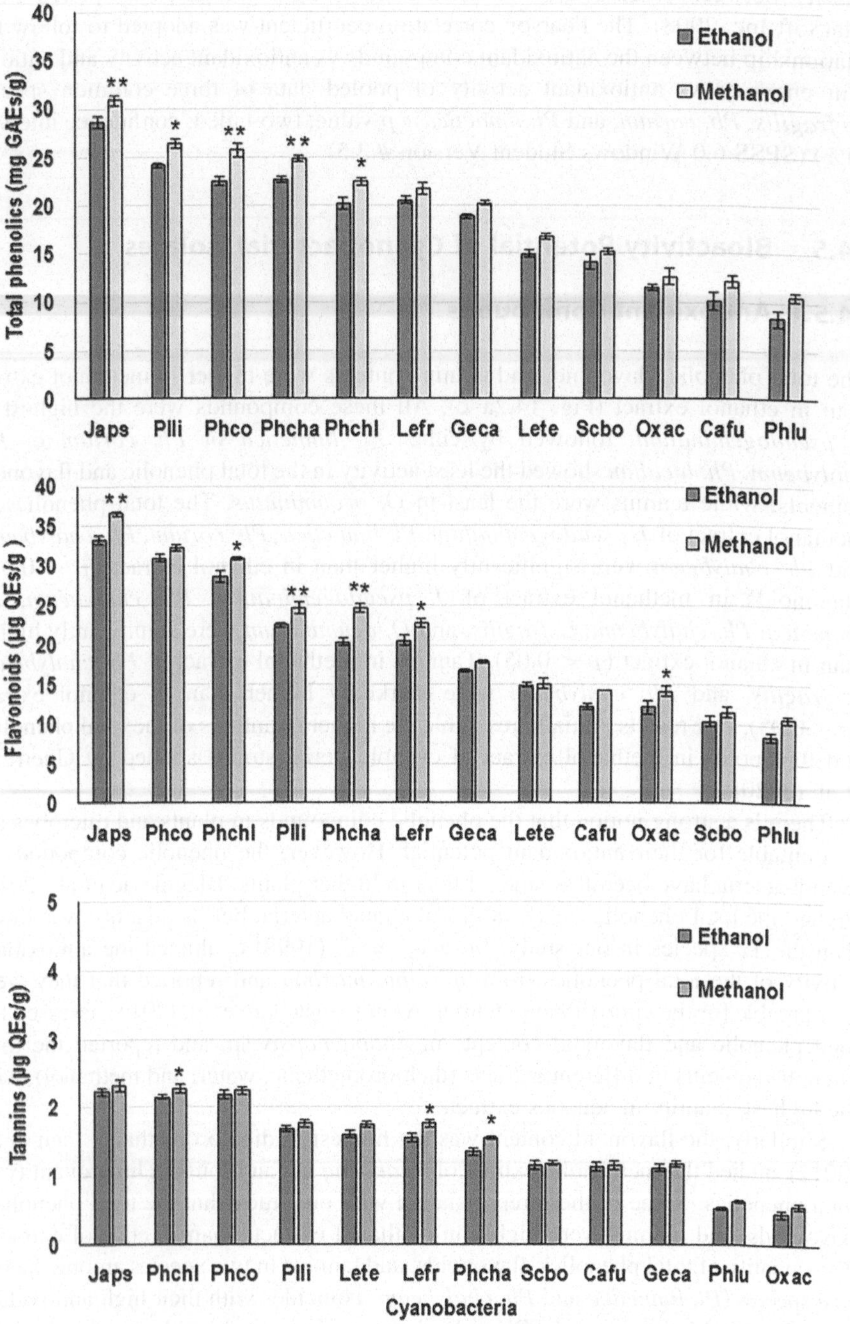

Fig. 14.2 The total phenolics, flavonoids, and tannins of 12 species of cyanobacteria ($n = 3$, mean ± SD; paired t-test: * $p < 0.05$, ** $p < 0.01$; see Table 14.1 for culture codes)

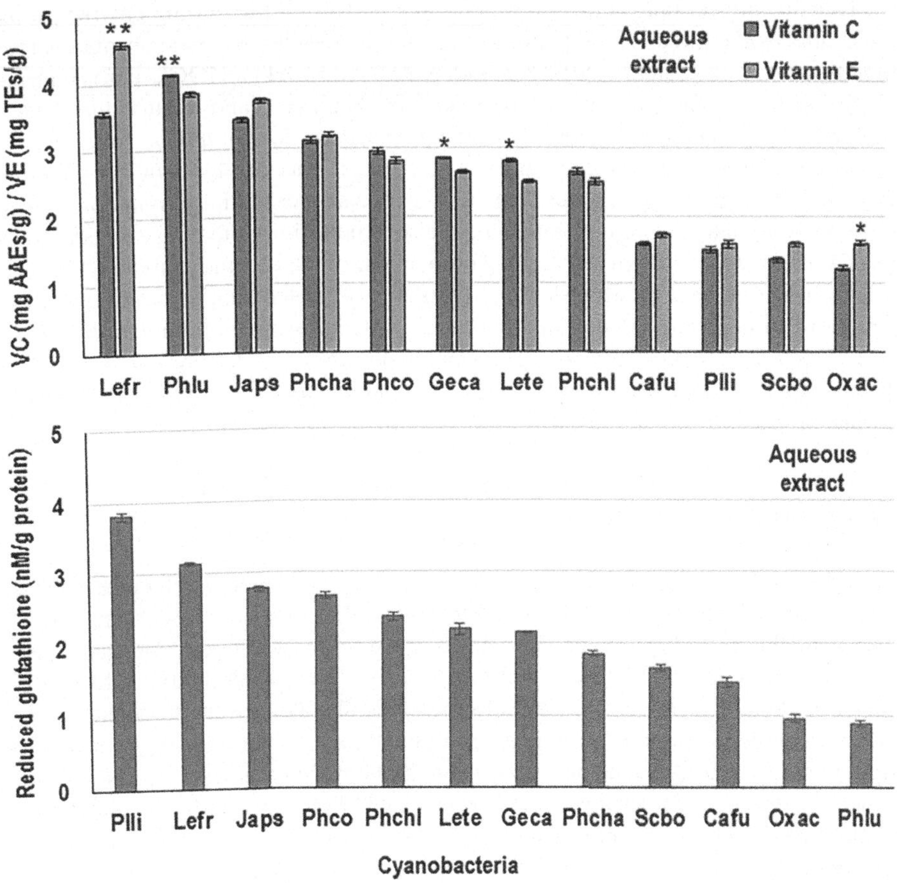

Fig. 14.3 Vitamin C, vitamin E, and reduced glutathione in 12 species of cyanobacteria ($n = 3$, mean ± SD; paired *t*-test: * $p < 0.05$, ** $p < 0.01$; see Table 14.1 for culture codes)

Similar to phenolics, some of the vitamins are responsible for antioxidant potential. The vitamin C content was the highest in *Ph. lucidum* followed by *L. fragilis*, *J. pseudogeminatum*, and *Ph. chalybeum*, while it was the least in *O. acuminatum* (Fig. 14.3a). The vitamin E content was the highest in *L. fragilis*, followed by *Ph. lucidum*, *J. pseudogeminatum*, and *Ph. chalybeum*, while it was the least in *Pl. limnetica* and *S. bohnerii* (Fig. 14.3a). The quantities of vitamin C and E are special for their stress-combating properties. Miranda et al. (1998) assessed the antioxidant activity of vitamin E from *Spirulina maxima*. The vitamin C content was significantly higher than vitamin E in *Ph. lucidum*, *G. calcuttense*, and *L. tenuis*, while vitamin E in *L. fragilis* and *O. acuminatum* was substantially higher than vitamin C ($p < 0.05$).

Besides phenolics and vitamins, reduced glutathione is also responsible for antioxidant activity. It is known as the most abundant thiol present in cyanobacteria involved in intracellular antioxidant activities with multiple biological functions, either direct or indirect participation, to reduce hydrogen peroxide. It also serves directly as a free radical scavenger and reacts with OH^-, HOCl, peroxynitrite, reactive oxygen, and carbon-centered radicals (Halliwell and Gutteridge 1990). The reduced glutathione content was the highest in *Pl. limnetica*, followed by *L. fragilis*, *J. pseudogeminatum*, and *Ph. corium*; it was the least in *Ph. lucidum* (Fig. 14.3b). Furthermore, reduced glutathione plays an essential role in conserving the oxidation state of protein thiols. It is also crucial in protecting the cell from several threats (e.g., low pH, chlorine compounds, oxidative and osmotic stresses). Finally, it serves as a posttranslational regulator of protein under oxidative stress conditions through the direct modification of proteins by glutathionylation (Masip et al. 2006). In our present study, a moderate amount of glutathione was found in species isolated from the thermal spring (*Pl. limnetica* and *L. fragilis*), domestic sewage (*J. pseudogeminatum*), and iron ore effluent (*Ph. corium*).

14.5.2 Antioxidant Enzymes

According to Wiktelius and Stenberg (2007), each cyanobacterial cell consists of an array of antioxidant defense enzymes, which consists of catalase, ascorbate peroxidase, glutathione reductase, glutathione S-transferase, and superoxide dismutase. Among these, 4 enzymes were tested for the 12 isolates in our study (Table 14.2). Catalase, glutathione S-transferase, and superoxide dismutase were the highest in *Pl. limnetica* while glutathione peroxidase and glutathione reductase in *L. fragilis*. The top four catalase, glutathione peroxidase, and glutathione reductase species were the same (*L. fragilis*, *L. tenuis*, *Ph. corium*, and *Pl. limnetica*). In glutathione S-transferase and superoxide dismutase, three species were common among the top four species (*L. fragilis*, *Pl. limnetica*, and *Ph. corium*). *Ph. lucidum* showed consistently low quantities of all enzymes except for glutathione peroxidase, while *O. acuminatum* showed the least glutathione peroxidase.

Catalase was the highest in *Pl. limnetica*, followed by *L. fragilis*, *L. tenuis*, and *Ph. corium*; it was the least in *Ph. lucidum*. The glutathione peroxidase was the highest in *L. fragilis*, followed by *Pl. limnetica*, *L. tenuis*, and *Ph. corium*; it was the least in *O. acuminatum*. The glutathione reductase was the highest in *L. fragilis*, followed by *Pl. limnetica*, *Ph. corium*, and *L. tenuis*; it was below the detectable limit in *C. fusca*, *Ph. lucidum*, and *S. bohnerii*. The glutathione S-transferase was the highest in *Pl. limnetica*, followed by *Ph. corium*, *L. fragilis*, and *L. tenuis*. It was below the detectable limit in *C. fusca*, *O. acuminatum*, and *P. lucidum*. The superoxide dismutase was the highest in *Pl. limnetica*, followed by *L. fragilis*, *J. pseudogeminatum*, and *Ph. corium*; it was the least in *Ph. lucidum*. The activities of catalase, ascorbate peroxidase, and glutathione peroxidase have been reported in many cyanobacteria (Tel-Or et al. 1986; Miyake et al. 1991). Despite the low level of ascorbate, the ascorbate peroxidase-like activities are reported in *Nostoc muscorum* (PCC 7119) as well as in two species of *Synechococcus* spp. (PCC 6311 and PCC 7942), while dehydroascorbate and glutathione reductases were responsible for

Table 14.2 Antioxidant enzyme activity of 12 cyanobacterial strains ($n = 3$, mean ± SD; BDL, below the detectable limit)

	Catalase (U/mg protein)	Glutathione peroxidase (U/mg protein)	Glutathione reductase (nM NADPH oxidized/min/mg protein)	Glutathione S-transferase (nM/mg protein)	Superoxide dismutase (U/mg protein)
Calothrix fusca	6.18 ± 0.45	1.35 ± 0.06	BDL	BDL	1.93 ± 0.20
Geitlerinema calcuttense	15.56 ± 0.06	5.49 ± 0.46	5.36 ± 0.35	4.65 ± 0.14	4.75 ± 0.22
Jaaginema pseudogeminatum	18.65 ± 0.24	5.75 ± 0.06	4.62 ± 0.15	3.19 ± 0.15	7.33 ± 0.18
Leptolyngbya fragilis	26.82 ± 0.40	10.51 ± 0.25	8.17 ± 0.20	5.21 ± 0.15	9.43 ± 0.35
Leptolyngbya tenuis	24.80 ± 0.70	6.64 ± 0.82	6.10 ± 0.35	5.12 ± 0.35	5.30 ± 0.22
Oxynema acuminatum	8.20 ± 0.30	0.52 ± 0.10	0.35 ± 0.02	BDL	1.65 ± 0.02
Phormidium chalybeum	18.35 ± 0.30	5.18 ± 0.22	4.20 ± 0.15	2.70 ± 0.05	4.10 ± 0.10
Phormidium chlorinum	12.80 ± 0.40	3.58 ± 0.06	3.05 ± 0.22	2.80 ± 0.41	2.40 ± 0.10
Phormidium corium	20.52 ± 0.25	6.13 ± 0.40	6.36 ± 0.60	5.45 ± 0.16	6.34 ± 0.40
Phormidium lucidum	5.40 ± 0.06	0.85 ± 0.06	BDL	BDL	0.88 ± 0.10
Planktolyngbya limnetica	32.41 ± 0.75	8.72 ± 0.32	7.80 ± 0.10	6.65 ± 0.15	12.52 ± 0.50
Scytonema bohnerii	8.32 ± 0.50	3.23 ± 0.05	BDL	1.06 ± 0.02	2.72 ± 0.14

producing ascorbate and glutathione in *Synechococcus* spp., respectively (Tel-Or et al. 1986). The production of catalase, glutathione peroxidase, and superoxide dismutase in cyanobacteria was enhanced under oxidative stress conditions (Li et al. 2004). In *Nostoc muscorum* (7119) and *Synechococcus* sp. (6311), enzymes responsible for antioxidant activity as well as cellular antioxidant functions (glutathione and ascorbate) have been evaluated for hydrogen peroxide removal. In *Nostoc muscorum*, elevated activities of catalase and ascorbate peroxidase were found in vegetative and heterocyst cells (Tel-Or et al. 1986). However, except for catalase, the rest of the enzymes were not in appreciable quantities in two heterocystous forms in our study (*C. fusca* and *S. bohnerii*).

Fluctuations in antioxidant enzyme activity towards protection from pollution are essential in an antioxidant system (Zhao et al. 2008). A widespread distribution of cyanobacteria indicates that they are capable of coping with a broad-ranging spectrum of climate change and environmental stresses (e.g., temperature, desiccation, cold, salinity, depletion of nitrogen, anaerobiosis, photo-oxidation, and metal stress) (Fay 1992; Sinha and Hader 1996; Tripathi and Srivastava 2001). Key enzymes that cope with environmental stresses include catalases, peroxidases, and superoxide dismutase (Canini et al. 2001). This outcome agrees with our study on those species isolated from thermal springs (*Pl. limnetica* and *L. fragilis*) as they showed good activity antioxidant enzymes. Likewise, species isolated from the sewage effluents (*J. pseudogeminatum* and *Ph. chlorinum*) possess moderate levels of antioxidant enzymes.

Antioxidant enzymatic responses in *Gloeocapsa* sp. against temperature and irradiance have been studied (Gacheva et al. 2013). In *Synechocystis* sp. (PCC 6803), Rady et al. (1994) showed a heat-induced decline in the activity of catalase, glutathione peroxidase, and superoxide dismutase, resulting in increased lipid peroxidation. In *Anabaena doliolum*, elevated activities of catalase, glutathione reductase, superoxide dismutase, and ascorbate peroxidase were responsible for protection against temperature stress (43 °C) (Mishra et al. 2005). Similarly, the high acclimation ability of *Arthrospira platensis* to low temperature (15 °C) is linked to the increased activity of catalase, glutathione reductase, and superoxide dismutase (Chien and Vonshak 2011). These results are in order with two thermal spring isolations in our study (*L. fragilis* and *Pl. limnetica*). However, the other two thermal spring species did not show such activities (*C. fusca* and *S. bohnerii*).

Kumar et al. (2013) evaluated the impact of chromium on the antioxidant enzymes in *Dolichospermum flos-aquae* (NTMS07) and showed a significant increase in catalase and superoxide dismutase. Similarly, the antioxidant enzyme concentration in *Anabaena variabilis* also increased by treatment with metals (Padmapriya and Anand 2010). The species isolated from the iron ore mine (*Ph. corium*) also showed increased antioxidant enzymes in our study.

14.5.3 Antioxidant Potential

The total antioxidant, DPPH radical-scavenging, nitric oxide-scavenging, and ferric-reducing activities of 12 species were higher in methanol extract than in ethanol extract (Fig. 14.4a–d). These antioxidant activities are the highest in *Pl. limnetica*, followed by *Ph. chlorinum* or *L. fragilis*. It was the least in *Ph. lucidum* in total antioxidant activity, DPPH radical-scavenging activity, and ferric-reducing power, whereas *O. occuminatus* exhibited the least nitric oxide-scavenging activity. The top three species were the same in all antioxidant activities tested (*L. fragilis*, *Pl. limnetica*, and *Ph. chlorinum*).

Methanol extract of *Pl. limnetica*, *Ph. chalybeum*, *Ph. corium*, *S. bohnerii*, and *O. occuminatus* showed a significantly higher antioxidant activity than ethanol extract ($p < 0.05$). Methanol extract of *Pl. limnetica*, *Ph. chalybeum*, and *S. bohnerii* displayed a significantly higher DPPH radical-scavenging activity than ethanol extract ($p < 0.05$). This result corroborates with the DPPH radical-scavenging activity in the eight strains of cyanobacteria studied by Guerreiro et al. (2020). Methanol extract of *L. fragilis* and *C. fusca* showed a significantly higher nitric oxide-scavenging activity than ethanol extract ($p < 0.05$). The antioxidant activity of ethanolic extracts of 18 strains of cyanobacteria (range, 8.4–21 mg AAEs/g) was lower than the 12 species evaluated in our study (10.3–44.4 mg AAEs/g) (Blagojević et al. 2018). Interestingly, the heterocystous forms showed a lower antioxidant activity than the non-heterocystous forms (Blagojević et al. 2018), which corroborates in our study as the antioxidant activity was relatively lower in heterocystous forms (*C. fusca* and *S. bohnerii*), possibly owing to the low production of antioxidant compounds as well as antioxidant enzymes.

Blagojević et al. (2018) showed a correlation of phenolic contents against the DPPH radical-scavenging activity and ferric-reducing power of 18 strains of cyanobacteria. This study corroborates that the top producers of antioxidant compounds and enzymes also showed a high antioxidant potential. Blagojević et al. (2018) indicated that the antioxidant potential of cyanobacteria is not dependent on the cultivation conditions; possibly, the genetic makeup of stress-tolerant cyanobacterial species or strains has been fine-tuned to face the environmental changes and stresses.

Methanolic extracts of seven cyanobacteria showed a lower DPPH radical-scavenging activity compared to the top ten species in our study (0–14.6 vs. 16.4–32.6%) (Jerez-Martel et al. 2017). However, *Nostoc* sp. showed up to 27.9% radical-scavenging activity, lower than *Pl. limnetica* and *L. fragilis* (30.5–32.6%). However, the radical-scavenging and nitric oxide-scavenging activities and ferric-reducing power of the 12 species in our study were not equivalent to the BHA. However, the top two species (*Pl. limnetica* and *L. fragilis*) fulfilled up to 50% DPPH radical-scavenging and nitric oxide-scavenging activities shown by the BHA. Our study revealed that two thermal spring isolates (*L. fragilis* and *Pl. limnetica*) and two domestic sewage isolates (*J. pseudogeminatum* and *Ph. chlorinum*) possess good antioxidant activity in a methanolic extract with notable quantities of antioxidant compounds as well as enzymes.

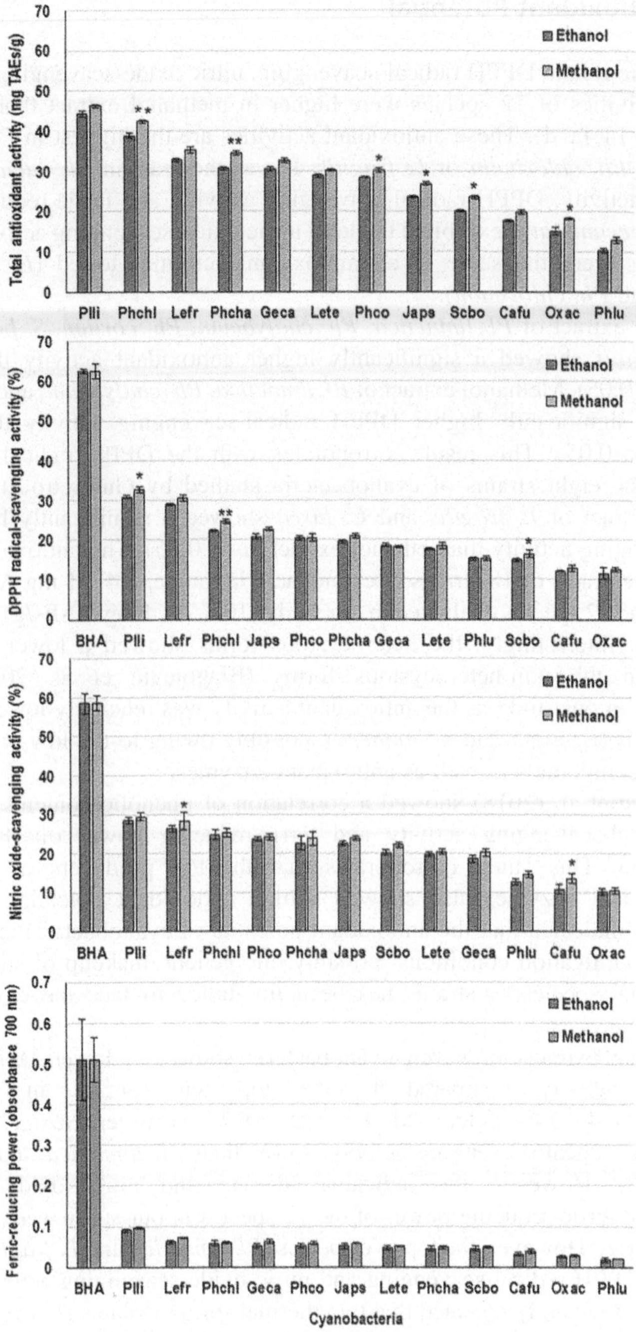

Fig. 14.4 The total antioxidant activity, DPPH radical-scavenging activity, nitric oxide-scavenging activity, and ferric-reducing power of 12 species of cyanobacteria ($n = 3$, mean ± SD; paired t-test: * $p < 0.05$, ** $p < 0.01$; BHA, butylated hydroxyanisole; see Table 14.1 for culture codes)

Currently employed synthetic antioxidants butylated hydroxyanisole (BHA) and butylated hydroxytoluene (BHT) are toxic as well as pose potential health risks (Jayalakshmi and Sharma 1986; Felter et al. 2021). Therefore, the development of suitable natural antioxidant molecules has recently gained much significance. It shows a vital role in preventing or suspending or delaying pathological conditions (e.g., cardiovascular diseases, hepatotoxicity, and cancer) (Gryglewski et al. 1987). Shazia et al. (2011) studied the antioxidant capacity of the methanol extracts of various cyanobacteria. Among them, extracts obtained from *Plectonema boryanum* and *Scytonema* sp. showed antioxidant activity (30%) and DPPH radical-scavenging activity (27%) than the ascorbic acid control (25%).

The nitric oxide-scavenging potential of cyanobacteria has potential health significance owing to its vital role in pathological conditions (e.g., septic shock, atherosclerosis, ischemia-reperfusion, neurodegenerative disorders, cancer, and diabetes) (Wojcik et al. 2010). In our study, two species showed suitable nitric oxide-scavenging activities (*L. fragilis* and *Pl. limnetica*). Furthermore, the highest antioxidant activity was exhibited by methane extract compared to ethanol extract (Hong et al. 2009; Asan-Ozusaglam et al. 2013). These studies corroborate our study as almost all species' antioxidant compounds and antioxidant activities were higher in methanol extract than in ethanol extract.

14.5.4 Antioxidant Compounds and Enzymes Versus Antioxidant Activities

Pooled data of three common species (*L. fragilis, Ph. corium*, and *Pl. limnetica*) among the top 5–6 were evaluated by Pearson correlation to follow the relationship among antioxidant compounds and enzymes against antioxidant activities. The total phenolics and vitamin E did not significantly correlate with any of the antioxidant activities assessed ($p > 0.05$). At the same time, the flavonoids and tannins were positively correlated with DPPH radical-scavenging and nitric oxide-scavenging activities ($p < 0.01$) (Table 14.3). Reduced glutathione was positively correlated with four antioxidant activities assessed ($p < 0.05$). Among the antioxidant enzymes, catalase and superoxide dismutase were positively correlated with four antioxidant activities assessed ($p < 0.01$) (Table 14.4). Glutathione peroxidase and glutathione reductase were positively correlated with DPPH radical-scavenging as well as nitric oxide-scavenging activities ($p < 0.05$). The glutathione S-reductase was positively correlated with the total antioxidant activity as well as ferric-reducing power ($p < 0.05$).

The total phenolics were not the major contributor to the DPPH radical-scavenging potential in the ten cyanobacterial strains tested by Blagojević et al. (2018). Our study corroborates that the total phenolics in three major species have no significant correlation with any of the four antioxidant activities assessed. However, individual phenolic compounds will positively correlate with antioxidant activity (Blagojević et al. (2018)). Cepoi et al. (2009) predicted that the high antioxidant activity in *Nostoc linckia* related to the quantities of carotenoids and tocopherols.

Table 14.3 Correlation matrix of antioxidant compounds vs antioxidant activities in cyanobacteria (*Leptolyngbya fragilis*, *Phormidium corium*, and *Planktolyngbya limnetica*)

	TP	F	T	VC	VE	RG	TAA	FSA	NOS	FRP
TP	1									
F	0.5491	1								
T	0.3829	0.9823**	1							
VC	−0.831	0.0068	0.1940	1						
VE	−0.957*	−0.284	−0.099	0.956**	1					
RG	0.2295	−0.687	−0.811*	−0.731	−0.501	1				
TAA	0.3327	0.605	0.743	**0.801***	0.591	**0.994****	1			
FSA	−0.230	**0.939****	**0.987****	0.347	0.060	**0.894***	0.841*	1		
NOS	−0.169	**0.916****	**0.975****	0.402	0.123	**0.921****	0.873*	0.998**	1	
FRP	0.2668	0.658	0.788	0.756	0.534	**0.999****	0.997**	0.876*	0.905*	1

TP, Total phenolics; F, flavonoids; T, tannins; VC, vitamin C; VE, vitamin E; RG, reduced glutathione; TAA, total antioxidant activity; FSA, DPPH free radical-scavenging activity; NOS, nitric oxide-scavenging activity; FRP, ferric-reducing power; * $p < 0.05$; ** $p < 0.01$

Table 14.4 Correlation matrix of antioxidant enzymes vs antioxidant activities in cyanobacteria (*Leptolyngbya fragilis*, *Phormidium corium*, and *Planktolyngbya limnetica*)

	CAT	GLP	GLR	GST	SOD	TAA	FSA	NOS	FRP
CAT	1								
GLP	0.6155	1							
GLR	0.7751	0.97501	1						
GST	0.7555	−0.051	0.1718	1					
SOD	0.999**	0.5880	0.7529	0.7777	1				
TAA	**0.972****	0.4078	0.6003	**0.891***	**0.9783****	1			
FSA	**0.946****	**0.837***	**0.938****	0.5033	**0.9346****	0.841*	1		
NOS	**0.965****	**0.801***	**0.914***	0.557	**0.955****	0.873*	0.998**	1	
FRP	**0.985****	0.4698	0.654	**0.858***	**0.992****	0.997**	0.876*	0.904**	1

CAT, catalase; GLP, glutathione peroxidase; GLR, glutathione reductase; GST, glutathione S-transferase; SOD, superoxide dismutase; TAA, total antioxidant activity; FSA, DPPH free radical-scavenging activity; NOS, nitric oxide-scavenging activity; FRP, ferric-reducing power; * $p < 0.05$; ** $p < 0.01$

Similarly, in our study, only vitamin C (but not vitamin E) was positively correlated with the total antioxidant potential ($p < 0.05$). Among the antioxidant compounds, reduced glutathione and catalase and superoxide dismutase among the antioxidant enzymes showed a strong positive correlation with all the antioxidant activities assessed, justifying their strong protective role. Further assessment of *L. fragilis*, *Ph. corium*, and *Pl. limnetica* will pave the way to adapt them in future biotechnological innovations.

14.6 Conclusions

Interest in natural antioxidants than in synthetic antioxidants is gaining significance in the health and pharmaceutical industries. Owing to adaptation to several extreme habitats, cyanobacteria serve as suitable representatives to exploit their antioxidant potential. Several cyanobacteria showed their high quantities of antioxidant compounds, antioxidant enzymes, and antioxidant activities. Considering those properties, *Leptolyngbya fragilis*, *Phormidium corium*, and *Planktolyngbya limnetica* could be recommended as potential candidates for future progress in therapeutic research (nutraceuticals, cosmeceuticals, and pharmaceuticals). Besides their significance in industry applications, they serve as good models to evaluate the consequences of climate change-related stresses as they have been adapted to various extreme habitats.

Acknowledgments The authors are grateful to Mangalore University and the Department of Biosciences for the necessary facilities. One of us (KS) acknowledges the University Grants Commission, Government of India, New Delhi, for financial assistance. Finally, the authors are thankful to Dr. S. Mahadevakumar, Department of Microbiology, Karnataka State Open University, Mysore, for the constructive suggestions.

References

Agyei D, Danquah MK, Sarethy IP, Pan S (2015) Antioxidative peptides derived from food proteins. In: Rani V, Yadav U (eds) Free radicals in human health and disease. Springer, New Delhi, pp 417–430

Albertano PR, Congestri, Shubert LE (1999) Cyanobacterial biofilms in sewage treatment plants along the Thyrranian coast (Mediterranean Sea), Italy. Arch Hydrobiol 94:13–24

Asan-Ozusaglam M, Cakmak YS, Kaya M (2013) Bioactivity and antioxidant capacity of *Anabaenopsis* sp. (cyanobacteria) extracts. J Algal Biomass Utln 4:50–58

Barreira JCM, Ferreira ICFR, Oliveira MBPP, Pereira JA (2008) Antioxidant activity and bioactive compounds of ten Portuguese regional and commercial almond cultivars. Food Chem Toxicol 46:2230–2235

Beauchamp CO, Fridovich I (1971) Superoxide dismutase: improved assays and an assay applicable to acrylamide gels. Anal Biochem 44:276–287

Belay A (2013) Biology and industrial production of *Arthrospira* (*Spirulina*). In: Richmond A, Hu Q (eds) Handbook of microalgal culture: applied phycology and biotechnology. Blackwell, Oxford, pp 339–358

Benzie IFF, Strain JJ (1996) The ferric reducing ability of plasma (FRAP) as a measure of antioxidant power: the FRAP assay. Anal Biochem 239:70–76

Bernroitner M, Zamocky M, Furtmuller PG, Peschek GA, Obinger C (2009) Occurrence, phylogeny, structure and function of catalases and peroxidases in cyanobacteria. J Exp Bot 60:423–440

Blagojević DB, Babić O, Rašeta M, Šibul F, Janjušević L, Simeunović J (2018) Antioxidant activity and phenolic profile in filamentous cyanobacteria: the impact of nitrogen. J Appl Phycol 30: 2337–2346

Campbell A, Holstege D, Swezey R, Medina-Cleghorn D (2008) Detoxification of molinate sulfoxide: comparison of spontaneous and enzymatic glutathione conjugation using human and rat liver cytosol. J Toxicol Environ Health A 71:1338–1347

Canini A, Leonardi D, Caiola MG (2001) Superoxide dismutase activity in the cyanobacterium *Microcystis aeruginosa* after surface bloom formation. New Phytol 152:107–116

Carlberg I, Mannervik B (1985) Glutathione reductase. Methods Enzymol 113:484–490

Cepoi L, Rudi L, Miscu V, Cojocari A, Chiriac T, Sadovnic D (2009) Antioxidative activity of ethanol extracts from Spirulina platensis and *Nostoc linckia* measured by various methods. Analele Universității din Oradea, Fascicula Biologie 16:43–48

Chang C, Yang M, Wen H, Chern J (2002) Estimation of total flavonoid content in propolis by two complementary colorimetric methods. J Food Drug Anal 10:178–182

Chien LF, Vonshak A (2011) Enzymatic antioxidant response to low-temperature acclimation in the cyanobacterium *Arthrospira platensis*. J Appl Phycol 23:887–894

Dor I (1987) Preservation and microscopy of blue-green algae (cyanobacteria) on dry agar. Bot Mar 30:507–510

Dvořák P, Poulı́ˇcková A, Hašler P, Belli M, Casamatta DA, Papini A (2015) Species concepts and speciation factors in cyanobacteria, with connection to the problems of diversity and classification. Biodivers Conserv 24:739–757

Dvořák P, Casamatta DA, Hašler P, Jahodářová E, Norwich AR, Poulíčková A (2017) Diversity of the cyanobacteria. In: Hallenbeck P (ed) Modern topics in the phototrophic prokaryotes. Springer International, Cham, pp 3–46

El-Aty AMA, Mohamed AA, Samhan FA (2014) In vitro antioxidant and antibacterial activities of two fresh water cyanobacterial species, *Oscillatoria agardhii* and *Anabaena sphaerica*. J Appl Pharm Sci 4:69–75

Fay P (1992) Oxygen relationships of nitrogen fixation in cyanobacteria. Microbiol Rev 56:340–373

Felter SP, Zhang X, Thompson C (2021) Butylated hydroxyanisole: carcinogenic food additive to be avoided or harmless antioxidant important to protect food supply? Regul Toxicol Pharmocol 121:104887. https://doi.org/10.1016/j.yrtph.2021.104887

Freile-Pelegrín Y, Robledo D (2013) Bioactive phenolic compounds from algae. In: Hernández-Ledesma B, Herrero M (eds) Bioactive compounds from marine foods: plant and animal sources. Wiley, Chichester, pp 113–129

Gacheva GV, Gigova LG, Ivanova NY, Pilarski PS, Lukavsky J (2013) Growth, biochemical and enzymatic responses of thermal cyanobacterium *Gloeocapsa* sp. (Cyanophyceae) to temperature and irradiance. Phycol Res 1–11

Goiris K, Muylaert K, Fraeye I, Foubert I, De Brabanter J, De Cooman L (2012) Antioxidant potential of microalgae in relation to their phenolic and carotenoid content. J Appl Phycol 24: 1477–1486

Gryglewski RJ, Korbut R, Robak J (1987) On the mechanism of antithrombotic action of flavonoids. Biochem Pharmacol 36:317–321

Guedes AC, Gião MS, Seabra R, Ferreira AC, Tamagnini P et al (2013) Evaluation of the antioxidant activity of cell extracts from microalgae. Mar Drugs 11:1256–1270

Guerreiro A, Andrade MA, Menezer C, Vilarinho F, Dias E (2020) Antioxidant and cytoprotective properties of cyanobacteria: potential for biotechnological applications. Toxins 12:548. https://doi.org/10.3390/toxins12090548

Hajimahmoodi M, Faramarzi MA, Mohammadi N, Soltani N, Oveisi MR, Nafissi-Varcheh N (2010) Evaluation of antioxidant properties and total phenolic contents of some strains of microalgae. J Appl Phycol 22:43–50

Halliwell B, Gutteridge JMC (1990) Role of free radicals and catalytic metal ions in human disease: an overview. Met Enzymol 186:1–85

Hong H, Takahashi K, Ichisaka T, Aoi T, Kanagawa O et al (2008) Suppression of induced pluripotent stem cell generation by the p53-p21 pathway. Nature 460:1132–1135

Hong Y, Hu H-Y, Xie X, Li F-M (2009) Responses of enzymatic antioxidants and non-enzymatic antioxidants in the cyanobacterium *Microcystis aeruginosa* to the allelochemical ethyl 2-methyl acetoacetate (EMA) isolated from reed (*Phragmites communis*). J Pl Physiol 165:1264–1273

Hossain MF, Ratnayake RR, Meerajini K, Kumara KLW (2016) Antioxidant properties in some selected cyanobacteria isolated from freshwater bodies of Sri Lanka. Food Sci Nutr 4:753–758

Jayalakshmi CP, Sharma JD (1986) Effect of butylated hydroxyanisole (BHA) and butylated hydroxytoluene (BHT) on rat erythrocytes. Environ Res 41:235–238

Jayashree VS, Kamat SY (1985) Distribution of tocopherol (vitamin E) in marine algae from Goa, west coast of India. Ind J Mar Sci 14:228–229

Jerez-Martel I, Garcia-Poza S, Rodriguez-Martel G, Rico M, Afonso-Olivares C, Gómez-Inchetti JL (2017) Phenolic profile and antioxidant activity of crude extracts from microalgae and cyanobacteria strains. J Food Qual 2924508:1. https://doi.org/10.1155/2017/2924508

Kapur A, Haskovic A, Copra-Janicijevic A, Klepo L, Topcagic A et al (2012) Spectrophotometric analysis of total ascorbic acid content in various fruits and vegetables. Bull Chem Technol Bosnia Herzegovina 38:39–42

Kaushik P, Chauhan A, Chauhan G, Goyal P (2009) Antibacterial potential and V-HPLC analysis of laboratory-grown culture of *Anabaena variabilis*. Int J Food Safety 11:11–18

Kepekçi RA, Saygideger SD (2012) Enhancement of phenolic compound production in *Spirulina platensis* by two-step batch mode cultivation. J Appl Phycol 24:897–905

Kumar JIN, Bora A, Kumar RN, Amb MK, Khan S (2013) Toxicity analysis of pesticides on cyanobacterial species by 16S rDNA molecular characterization. Proc Int Acad Ecol Environ Sci 3:101–132

Kunur K, Singh SS (2020) Protective role of antioxidative enzymes and antioxidants against iron-induced oxidative stress in the cyanobacterium *Anabaena sphaerica* isolated from iron rich paddy field of Chattisgarh, India. Ind J Exp Biol 58:617–630

Latifi A, Ruiz M, Zhang C-C (2008) Oxidative stress in cyanobacteria. FEMS Microbiol Rev 33(2009):258–278

Li H, Singh AK, McIntyre LM, Sherman LA (2004) Differential gene expression in response to hydrogen peroxide and the putative PerR regulon of *Synechocystis* sp. strain PCC 6803. J Bacteriol 186:3331–3345

Li H-B, Cheng K-W, Wong C-C, Fan K-W, Chen F, Jiang Y (2007) Evaluation of antioxidant capacity and total phenolic content of different fractions of selected microalgae. Food Chem 102:771–776

Luximon-Ramma A, Bahorun T, Soobrattee M, Aruoma OI (2002) Antioxidant activities of phenolic, proanthocyanidin and flavonoid components in extracts of *Cassia fistula*. J Agric Food Chem 50:5042–5047

Ma R, Lu F, Bi Y, Hu Z (2015) Effects of light intensity and quality on phycobiliprotein accumulation in the cyanobacterium *Nostoc sphaeroides* Kützing. Biotechnol Lett 37:1663–1669

Masip L, Veeravalli K, Georgiou G (2006) The many faces of glutathione in bacteria. Antiox Redox Signal 8:753–762

Mc Donald S, Prenzler PD, Autolovich M, Robards K (2001) Phenolic content and antioxidant activity of olive extracts. Food Chem 73:73–84

Miranda MS, Cintra RG, Barros SBM, Filho JM (1998) Antioxidant activity of the microalga *Spirulina maxima*. Braz J Med Biol Res 31:1075–1079

Mishra Y, Bhargava P, Rai LC (2005) Differential induction of enzymes and antioxidants of the antioxidative defense system in *Anabaena doliolum* exposed to heat stress. J Therm Biol 30: 524–531

Miyake C, Michihata F, Asada K (1991) Scavenging of hydrogen peroxide in prokaryotic and eukaryotic algae: acquisition of ascorbate peroxidase during the evolution of cyanobacteria. Pl Cell Physiol 32:33–43

Morone J, Alfeus A, Vasconcelos V, Martins R (2019) Revealing the potential of cyanobacteria in cosmetics and cosmeceuticals - a new bioactive approach. Algal Res 41:101541

Nunnery JK, Mevers E, Gerwick WH (2010) Biologically active secondary metabolites from marine cyanobacteria. Curr Opin Biotechnol 21:787–793

Okwu DE (2005) Phytochemicals, vitamins and mineral contents of two Nigerian medicinal plants. Int J Mol Med Adv Sci 1:375–381

Padmapriya V, Anand N (2010) The influences of metals on the antioxidant enzyme, superoxide dismutase, present in the cyanobacterium, *Anabaena variabilis* KÜTZ. ARPN J Agric Biol Sci 5:4–9

Pant G, Kumar G, Karthik L, Prasuna RG, Vankata K, Rao B (2012) Effect of electric treatment on total phenolic content and antioxidant activity of *Anabaena variabilis*. Int J Pharm Pharmaceut Sci 4:128–131

Paya M, Halliwell B, Hoult JR (1992) Interactions of a series of coumarins with reactive oxygen species. Biochem Pharmacol 44:205–214

Prieto P, Pineda M, Aguilar M (1999) Spectrophotometric quantitation of antioxidant capacity through the formation of a phosphomolybdenum complex: specific application to the determination of vitamin E. Anal Biochem 269:337–341

Rady AA, El-Scheekh MM, Matkovics B (1994) Temperature shift induced changes in the antioxidant enzyme system of the cyanobacterium, *Synechocystis* PCC 6803. Int J Biochem 26:433–435

Raghavendra, Hosmani SP (2002) Hydrobiological study of Mandakally Lake, a polluted water body at Mysore. Nat Environ Poll Technol 1:291–293

Rajeshwari KR, Rajashekhar M (2012) Diversity of cyanobacteria in some polluted aquatic habitats and a sulfur spring in the Western Ghats of India. Algol Stud 138:37–56

Rao MV, Paliyath G, Ormrod DP (1996) Ultraviolet-B and ozone-induced biochemical changes in antioxidant enzymes of *Arabidopsis thaliana*. Plant Physiol 110:125–136

Regelsberger G, Jakopitsch C, Plasser L, Schwaiger H, Furtmüller PG et al (2002) Occurrence and biochemistry of hydroperoxidases in oxygenic phototrophic prokaryotes (cyanobacteria). Pl Physiol Biochem 40:479–490

Rippka R, Dereulles J, Waterbury JB, Herdman M, Stanier RY (1979) Generic assignments, strain histories and properties of pure cultures of cyanobacteria. J Gen Microbiol 111:1–61

Rop O, Mlcek J, Jurikova T, Neugebauerova J, Vabkova J (2012) Edible flowers—a new promising source of mineral elements in human nutrition. Molecules 17:6672–6683

Shalaby EA, Shanab SMM (2013) Comparison of DPPH and ABTS assays for determining antioxidant potential of water and methanol extracts of *Spirulina platensis*. Ind J Geomar Sci 42:556–564

Shazia S, Deboshree B, Alvina F, Arif JM, Zeeshan M (2011) Antibacterial and free radical scavenging potential of some cyanobacterial strains and their growth characteristics. J Chem Pharm Res 3:472–478

Silber R, Farber M, Papopoulos E, Nervla D, Liebes L et al (1992) Glutathione depletion in chronic lymphocytic leukemia B-lymphocytes. Blood 80:2038–2040

Singh R, Parihar P, Singh M, Bajguz A, Kumar J et al (2017) Uncovering potential applications of cyanobacteria and algal metabolites in biology, agriculture and medicine: current status and future prospects. Front Microbiol 8:515. https://doi.org/10.3389/fmicb.2017.00515

Sinha RP, Hader DP (1996) UV-induced DNA damage and repair: a review. Photochem Photobiol Sci 1:225–236

Stainer RY, Kunisawa R, Mandel M, Cohen-Bazire G (1976) Purification and properties of unicellular blue-green algae (order Chroococcales). Bacteriol Rev 32:171–205
StatSoft Inc (2008) Statistica, Version # 8. StatSoft, Tulsa
Tel-Or E, Huflejt ME, Packer L (1986) Hydrogen peroxide metabolism in cyanobacteria. Arch Biochem Biophys 246:396–402
Tenpe CR, Upaganlawar AB, Thakre A, Yeole PG (2008) In vitro antioxidant properties of leaves of *Terminalia catappa* Linn Ex. Roxb Indian Drugs 45:132–137
Tiwari A, Pandey A (2012) Role of antioxidative enzymes in toxic bloom forming cyanobacteria. J Appl Nat Sci 4:228–233
Tripathi SN, Srivastava P (2001) Presence of stable oxygen scavenging enzymes superoxide dismutase, ascorbate peroxidase and catalase in a desiccation tolerant cyanobacterium *Lyngbya arbicola* under dry state. Curr Sci 81:197–200
Vaz JA, Heleno SA, Martins A, Almeida GM, Vasconcelos MH, Ferreira IC (2010) Wild mushrooms *Clitocybe alexandri* and *Lepista inversa*: in vitro antioxidant activity and growth inhibition of human tumour cell lines. Food Chem Toxicol 48:2881–2884
Vellard M (2003) The enzyme as drug: application of enzymes as pharmaceuticals. Curr Op Biotech 14:1–7
Vijayakumar S, Thajuddin N, Manoharan C (2007) Biodiversity of cyanobacteria in industrial effluents. Acta Bot Malacitana 32:27–34
Wiktelius E, Stenberg G (2007) Novel class of glutathione transferases from cyanobacteria exhibit high catalytic activities towards naturally occurring isothiocyanates. Biochem J 406:115–123
Wojcik M, Burzynska-Pedziwiatr I, Wozniak LA (2010) A review of natural and synthetic antioxidants important for health and longevity. Curr Med Chem 17:3262–3288
Yu H, Jia S, Dai Y (2009) Growth characteristics of the cyanobacterium *Nostoc flagelliforme* in photoautotrophic, mixotrophic and heterotrophic cultivation. J Appl Phycol 21:127–133
Zeeshan M, Prasad SM (2009) Differential response of growth, photosynthesis, antioxidant enzymes and lipid peroxidation to UV-B radiation in three cyanobacteria. South Afr J Bot 75: 466–474
Zhao L, He J, Wang X, Zhang L (2008) Nitric oxide protects against polyethyleneglycol-induced oxidative damage in two ecotypes of reed suspension cultures. J Pl Physiol 165:182–191

Cyanobacteria-Mediated Heavy Metal and Xenobiotic Bioremediation

15

Lalrokimi, Gajanan Mehetre, Zothanpuia, Bhim Pratap Singh, Mukesh Kumar Yadav, and Esther Lalnunmawii

Abstract

Overpopulation and industrialization are the leading cause of environmental pollution. However, environmental pollutants such as xenobiotics and heavy metals are of more concern because of their toxicity; in addition, the degradation of both compounds is impossible and persists in the environment for a longer time. Photoautotrophic microorganism such as cyanobacteria is considered to be the potential organism in the process of bioremediation because of its ability to photosynthetically break down toxic heavy metals and xenobiotic compounds into less toxic forms. It is also known to be conventional, eco-friendly, and cost-effective. This chapter highlighted the capability of different genera of cyanobacteria in the biodegradation of heavy metals and xenobiotics.

Keywords

Bioremediation · Blue-green algae · Heavy metal · Xenobiotics

Lalrokimi · G. Mehetre · E. Lalnunmawii
Department of Biotechnology, Mizoram University, Tanhril, Mizoram, India

Zothanpuia (✉)
Department of Biotechnology/Life Sciences, Pachhunga University College, Mizoram University, Aizawl, Mizoram, India

B. P. Singh
Department of Agriculture and Environmental Sciences (AES), National Institute of Food Technology Entrepreneurship and Management (NIFTEM), Sonepat, Haryana, India

M. K. Yadav
Department of Biotechnology/Life Sciences, Pachhunga University College, Mizoram University, Aizawl, Mizoram, India

Department of Microbiology, Central University of Punjab, Bhatinda, Punjab, India

B. Neilan et al. (eds.), *Cyanobacterial Biotechnology in the 21st Century*,
https://doi.org/10.1007/978-981-99-0181-4_15

15.1 Introduction

Advances in anthropogenic activities such as industrial, medicinal, and synthetic materials lead to the exploitation of nature immensely. Typically, these include heavy metal pollutants such as arsenic, cadmium, lead, mercury, nickel, silver, zinc, etc. and xenobiotic compounds such as polyaromatic hydrocarbons (PAHs), pesticides, etc. Xenobiotic compounds and heavy metal pollutants are considered hazardous since these are not easily degraded and present in the environment for a longer time (Ughy et al. 2015). Therefore, the decontamination or the degradation of these pollutants from the terrestrial and aquatic environments has been a great challenge for the past few decades and is getting massive attention in research. The bioremediation technique is one of the safest and most economical approaches for degrading toxic heavy metals and xenobiotics to less toxic forms (Ojuederie and Babalola 2017). Cyanobacteria, also known as blue-green algae, have attracted attention due to their availability; ability to obtain energy from photosynthesis, carbon and nitrogen fixation, and oxygen production; and ability to transport electrons (Wilson et al. 2001; Thajamanbi et al. 2016). Different organisms like bacteria, plants, and fungi are commonly used for the bioremediation of toxic pollutants such as heavy metal and xenobiotic compounds. Still, the unique properties of cyanobacteria have attracted the attention of many researchers. It is a promising microorganism for the bioremediation of heavy metals and xenobiotics due to its capability for bioaccumulation, biotransformation, and degradation (Chaillan et al. 2004; El-Sheekh and Hamouda 2014). In addition to this, the viability and the metabolic activities are not affected by the breaking down of biodegradable pollutants (Megharaj et al. 1987). Despite the vast availability of higher plants, fungi, or bacteria, etc., cyanobacteria is considered as an alternative for heavy metal remediation due to its high metal sorption ability and its unique structural properties (Singh 2020). Cyanobacteria are prokaryotes under the phylum of bacteria; they receive energy by photosynthesis and are considered to be one of the oldest organisms present on Earth (Mazard et al. 2016). Cyanobacteria exist as free-living, colonial, or forming a symbiotic relationship with other living forms such as plants and fungi (Al-Thani 2015). They are found in diverse habitats that include aquatic (freshwater, marine), terrestrial (soil), and extreme environments such as volcanic vents, geothermal hot springs, frozen ponds, etc. (Panosyan 2015). Most cyanobacteria produce phycocyanin that is responsible for the cells to give a bluish color; hence, it explains the name blue-green algae. Sometimes a red pigment known as phycoerythrin is produced. Cyanobacteria have a thick gelatinous cell wall, but flagella are absent, and unlike green plants, they lack chloroplast but can perform photosynthesis in distinctive folds. Cyanobacteria have a unique cell wall which comprises a greater mucilage volume with a high binding affinity; the cell has simple nutrient requirements with a large surface area (Micheletti et al. 2008a, b). The availability of light controls the metabolism of cyanobacteria (Stal 1995). Bioremediation is the process of mineralization and detoxification of natural or manufactured contaminants present in the environment into less toxic forms with the help of living things such as microorganisms and plants or indirectly by the enzymes produced by plants or microbes (Rani et al. 2007). Bioremediation utilizes the reactions in the metabolic process of microorganisms to transform contaminant compounds into nontoxic compounds. The microorganisms could be indigenous or isolated from an additional source brought to the contaminated areas.

15.2 Xenobiotics and Heavy Metal Pollutants

Heavy metals are metallic elements with a comparatively high density and are poisonous or toxic even at low concentrations (Lenntech 2004). The term is also used as a collective term for a collection of metals and metalloids with an atomic density greater than water or which is greater than 4 g/cm^3 (Hutton and Symon 1986; Nriagu and Pacyna 1988; Hawkes 1997). Xenobiotics are unnatural and foreign chemical compounds found in environments where complete biodegradation or biotransformation is impossible. Due to these, xenobiotics can persist in the background for a longer duration of time. Xenobiotics have a recalcitrant nature which is due to the nonphysiological chemical bonds or substituents in their structure; due to this it blocks the attack of catabolic microbial enzymes (Fetzner 1999). Xenobiotics include synthetic azo dyes, fuels, pesticides, solvents, alkanes, antibiotics, pollutants, polycyclic hydrocarbons (PAHs), and polyaromatic, chlorinated, and nitroaromatic compounds (Sinha et al. 2009).

15.3 Heavy Metal- and Xenobiotic-Mediated Toxicity

Heavy metals are generally known to be hazardous to the environment, known mainly as environmental pollutants due to their toxicity and prevalence; the adverse contribution to public health needs to be taken care of (Jaishankar et al. 2014). Heavy metal pollutants are different from organic pollutants on the persistence level (Orta-García et al. 2016). Organic pollutants are easily degradable, whereas heavy metal pollutants cannot be degraded. Heavy metals and their compounds are separated into "black list" and "gray list" according to their toxicities by the European Economic Community (EEC) (McEldowney et al. 1993). Among the heavy metals, Cd, Hg, and their compounds are the most dangerous toxic compounds which are categorized as black list, and Cu, Cr, Co, Pb, Zn, etc. are placed under the gray list, which shows the less hazardous pollutants (Morin-Crini et al. 2013). Different pollutants, including heavy metals, have been included in the EPA (Environmental Protection Agency) list of "priority pollutants" (McEldowney et al. 1993). Heavy metal exists as hydrated ionic species in the aquatic and terrestrial environment. Sometimes heavy metals can form a complex with inorganic and organic ligands or associated with colloids and suspended particulate matter. The toxicity of heavy metal against living organism depends upon its concentration, chemical form, bioavailability, chemical form, and duration of exposure.

On the other hand, the pH of water or soil is proven to control the mobility, bioavailability, and toxicity of heavy metals. For example, acidification of aquatic bodies by acid rain worsens the problem of heavy metal toxicity (McEldowney et al. 1993). There is a possibility that heavy metals present in the environment of both aquatic and terrestrial may enter into the food chain and bioaccumulate in living organisms (Grimanis et al. 1978; Rayms-Keller et al. 1998). Exposure to heavy metals, especially for humans, can occur through food, air, or water contamination. Occupational exposure, such as industry, etc., is another everyday exposure for humans (Pauls et al. 2003). Heavy metal pollutants are systematic toxicants and therefore can cause a variety of pathophysiological conditions and may induce organ damage, such as liver failure, renal dysfunction, bronchitis, arthritis, osteoporosis, pulmonary edema, and neurological damage, and may even cause cancer (Hayes 1997; Costa and Klein 2006). In addition, the impact of xenobiotics is dangerous to nature as well as human beings. Due to its recalcitrant property, it is not degraded in the environment. Therefore, it remains in the wilderness for a more extended period which eventually results in bioaccumulation and biomagnification. In this way, the concentration of xenobiotics will increase as they move up to the trophic level of the ecosystem. The constant exposure of xenobiotics to humans can result in many complications such as skin diseases and cancers (Sinha et al. 2009).

Heavy metal is a natural constituent of the Earth's crust, and it is also present as an environmental contaminant since they cannot be degraded or destroyed. Sources of heavy metal pollution could be both natural and anthropogenic. Naturally occurring heavy metals are present in insoluble forms such as mineral structures or in precipitated or complex forms that are not readily available for uptake by plants (Fulekar et al. 2009). Comparatively, the bonding energy between naturally occurring heavy metals and soil is much higher than with anthropogenic sources. Naturally occurring heavy metals in the environment are due to comets, weathering of minerals, erosion, and volcanic eruptions, whereas the anthropogenic sources are due to alloy production, battery production, atmospheric deposition, coating, biosolids, explosive manufacturing, pesticides, leather tanning, mining, phosphate fertilizer, dyes, sewage irrigation, textiles and wood preservation, and printing pigments (Dixit et al. 2015). The accumulation of heavy metal in the food web depends on its sources, concentrations, degree of uptake by plants, soil properties, and the degree of absorption by animals (Bolan et al. 2014). The source of xenobiotic pollution is usually anthropogenic rather than natural and occurs due to the production of pesticides, fuels, solvents, antibiotics, polyaromatics, alkanes, and chlorinated and nitroaromatic compounds (Sinha et al. 2009).

15.4 Cyanobacteria for the Bioremediation of Heavy Metals and Xenobiotics

Cyanobacteria, which are aquatic and photosynthetic prokaryotes, have been present on the Earth's surface for the past 3 billion years, and they are known to have a promising activity in bioremediation processes (Noel and Ranjan 2014). They are suitable organisms in the process of bioremediation since they are found at the bottom of the food chain and therefore consume pollutants which in turn protect the organisms that are at the top of the food chain from harmful pollutants. They can be used in many forms such as wild-type, mutant, or genetically engineered forms (Ananya and Ahmad 2014). Therefore, cyanobacteria have gained attention extensively as a potential bioremediation agent for controlling various pollutants, including heavy metals (Lem and Glick 1985; Radwan and Al-Hasan 2000; Gupta et al. 2013). Cyanobacteria degrade the pollutants by binding with high efficiency to the contaminants with the help of different types of proteins and polysaccharide receptors on their surface (Priyadarshani et al. 2011). The pollutants are trapped by the receptors present on the surface and bind to the cellular structure by the process known as biosorption (Malik 2004).

The advantage of using cyanobacteria to remove heavy metals and xenobiotics is their ability to sequester metals and their abundance in natural environments, especially in soil and water. There are two known metal resistance in cyanobacteria, namely: exogenous metal chelation and endogenous metal chelation. Exogenous metal chelation is done by EPS (exopolysaccharides) present in mucilaginous sheath or capsule (Ozturk and Aslim 2008) and by energy-dependent metal efflux (Verma and Singh 1995), and endogenous metal chelation is done by metallothioneins (MTs) (Turner and Robinson 1995). The separation of metal from contaminated water is a difficult task. Metals such as arsenic, cadmium, chromium, copper, cobalt, lead, lanthanum, nickel, mercury, uranium, and zinc could enter into water directly or indirectly, which leads to water pollution. These metal pollutants present in different water bodies are harmful to live organisms. The separation of such metal pollutants from water is a challenging work (Archarya and Apte 2013). Therefore, conventional techniques such as coprecipitation, cloud-point extraction (CPT), ion exchange, solvent extraction, and electrodeposition could be applied to separate metal pollutants from water bodies.

1. *Coprecipitation*
   It is a process where the metals are treated with a collector or a carrier. The metal and the carrier have the same crystalline structure. Three mechanisms take place in the process of coprecipitation, namely: inclusion, occlusion, and adsorption. When the charge and the ionic radius of the metal and the carrier are similar, it is known as inclusion. When the metal gets substantially trapped inside the crystals, it is known as occlusion, and in adsorption, the adsorbate is weakly bonded to the surface of the precipitate (Elsevier Science Ltd. 2002).
2. *Cloud-point extraction (CPT)*
   CPT extraction is when the nonionic surfactant is transferred from one liquid phase to another by the process of heating. When the temperature increases above the CPT, it forms micelles. It gathers and becomes dehydrated. Therefore, the original surfactant solution gets separated into a surfactant phase preconcentrate. The surfactant phase contains the separated metal, and the preconcentrate is obtained from the diluted aqueous phase (Bezerra et al. 2005).
3. *Ion exchange*
   In ion-exchange chromatography, the metal ions are separated using cation and anion exchange resin. In this separation technique, the functional groups, i.e.,

Table 15.1 Degradation of xenobiotics and heavy metals by cyanobacteria

S. No	Cyanobacteria	Heavy metals	References
1.	*Aphanothece halophytica*	Hg, As, Cd	Laloknam et al. (2009)
2.	*Microcystis aeruginosa*	Cu	Tien et al. (2005)
3.	*Spirulina platensis*	Hg, Cu, Cd	Cain et al. (2008), Fang et al. (2011)
4.	*Anabaena cylindrical*	Cu, Ni	Corder and Reeves (1994), Tien et al. (2005)
5.	*Nostoc muscorum*	Cu, Zn, Pb, Cd	Hazarika et al. (2015), Goswami et al. (2015)
6.	*Gloeomargarita lithophora, Cyanothece* sp.	Sr, Ba	Cam et al. (2016)
7.	*Anabaena azollae*	Lignin	Schoeny et al. (1988)
8.	*Chlorella sorokiniana*	Phenanthrene	Kuritz (1999)
9.	*Chlorella* sp.	Black oil	Perales-Vela et al. (2006)
10.	*Spirulina platensis*	Rhenium	Zinicovscaia et al. (2018a, b)

positive charge or negative charge, are used for identifying cation exchange resin or anion exchange resin (Srogi 2008). Therefore, the metals used ion exchange to bind to cyanobacteria. Metals such as Ba, Al, Fe, Cr, Sr, and Zn are separated using spirulina biomass in the order of Ba > Fe > Sr > Al > Zn > Cr, but the metal-removal efficiency varies from 37% to 100% for chromium and barium, respectively (Zinicovscaia et al. 2018a, b).

There have been several reports on the potential of cyanobacteria in the biodegradation of xenobiotics and heavy metals (Table 15.1). For example, Laloknam et al. (2009) reported the possibility of *Aphanothece halophytica* in degrading Hg, As, and Cd. Similarly, Cu was shown to be degraded by *Microcystis aeruginosa* (Tien et al. 2005); Hg, Cu, and Cd by *Spirulina platensis* (Cain et al. 2008; Fang et al. 2011); Cu and Ni by *Anabaena cylindrica* (Corder and Reeves 1994; Tien et al. 2005); Cu, Zn, Pb, and Cd by *Nostoc muscorum* (Hazarika et al. 2015); Sr and Ba by *Gloeomargarita lithophora* and *Cyanothece* sp. (Cam et al. 2016); lignin by *Anabaena azollae* (Schoeny et al. 1988); phenanthrene by *Chlorella sorokiniana* (Kuritz 1999); black oil by *Chlorella* sp. (Perales-Vela et al. 2006); and rhenium by *Spirulina platensis* (Zinicovscaia et al. 2018a, b).

15.5 Conclusion

Heavy metals and xenobiotic compounds pose a severe threat to the environment because of their persistency and nondegradable nature. Bioremediation using cyanobacteria has shown to be effective in eliminating heavy metals and xenobiotic compounds. Even though several studies have been performed in the field of heavy metal biosorption and microbial biosorbents by different microbes, cyanobacteria could be the most promising area for the degradation of heavy metals and xenobiotics due to their availability and ability to perform photosynthesis.

References

Al-Thani RF (2015) Cyanomatrix and cyanofilm. J Res Dev 3:123

Ananya KA, Ahmad IZ (2014) Cyanobacteria the blue-green algae and its novel applications. Int J Innov Appl Stud 7:251–261

Archarya C, Apte SK (2013) Insights into the interactions of cyanobacteria with uranium. Photosynth Res 118(1–2):83–94

Bezerra MA, Arruda MAZ, Ferreira SLC (2005) Cloud-point extraction as a procedure of separation and pre-concentration for metal determination using spectroanalytical techniques: a review. Appl Spectrosc Rev 40:269–299

Bolan N, Kunhikrishnan A, Thangarajan R, Kumpiene J, Park J, Makino T, Kirkham MB, Scheckel K (2014) Remediation of heavy metal(loid)s contaminated soils—to mobilize or to immobilize? J Hazard Mater 266:141–166

Cain A, Vannela R, Woo LK (2008) Cyanobacteria as a biosorbents for mercuric ion. Bioresour Technol 99:6578–6586

Cam N, Benzerara K, Georgelin T, Jaber M, Lambert J-F, Poinsot M, Panet FS, Cordier L (2016) Selective uptake of alkaline earth metals by cyanaobacteria forming intracellular carbonates. Environ Sci Technol 50(21):11654–11662

Chaillan F, Le Flèche A, Bury E, Phantavong YH, Grimont P, Saliot A, Oudot J (2004) Identification and biodegradation potential of tropical aerobic hydrocarbondegrading microorganisms. Res Microbiol 155(7):587–595

Corder SL, Reeves M (1994) Biosorption of nickel in complex aqueous waste streams by cyanobacteria. Appl Biochem Biotechnol 45(46):847–859

Costa M, Klein C (2006) Toxicity and carcinogenicity of chromium compounds in humans. Crit Rev Toxicol 36:155–163

Dixit R, Malaviya D, Pandiyan K, Singh UB, Sahu A, Shukla R, Singh BP, Rai JP, Sharma PK, Lade H (2015) Bioremediation of heavy metals from soil and aquatic environment: an overview of principles and criteria of fundamental processes. Sustainability 7:2189–2212

Elsevier Science Ltd (2002) Encyclopedia of separation science. Elsevier Science, Amsterdam

El-Sheekh MM, Hamouda RA (2014) Biodegradation of crude oil by some cyanobacteria under heterotrophic conditions. Desalin Water Treat 52(7–9):1448–1454

Fang L, Zhou C, Cai P, Chen W, Rong X, Dai K, Liang W, Gu JD, Huang Q (2011) Binding characteristics of copper and cadmium by cyanobacterium Spirulina platensis. J Hazard Mater 190:810–815

Fetzner S (1999) Bacterial dehalogenation. Appl Microbiol Biotechnol 50:633–657

Fulekar M, Singh A, Bhaduri AM (2009) Genetic engineering strategies for enhancing phytoremediation of heavy metals. Afr J Biotechnol 8:529–535

Goswami S, Diengdoh OL, Syiem MB, Pakshirajan K, Kiran MG (2015) Zn(II) and Cu(II) removal by *Nostoc muscorum*: a cyanobacterium isolated from a coal mining pit in Chiehruphi, Megahalaya, India. Can J Microbiol 61(3):209–215

Grimanis AP, Zafiropoulos D, Vassilaki R, Grimanis M (1978) Trace elements in the flesh and liver of two fish species from polluted and unpolluted areas in the Aegean Sea. Environ Sci Technol 12:723–726

Gupta V, Ratha SK, Sood A, Chaudhary V, Prasanna R (2013) New insights in to the biodiversity and applications of cyanobacteria (blue-green algae)- prospects and challenges. Algal Res 2(2): 79–97

Hawkes JS (1997) Heavy metals. J Chem Educ 74(11):1374

Hayes RB (1997) The carcinogenicity of metals in humans. Cancer Causes Control 8:371–385

Hazarika J, Pakshirajan K, Sinharoy A, Syiem MB (2015) Bioremoval of Cu(II), Zn(II), Pb(II) and Cd(II) by *Nostoc muscorum* isolated from coal mining site. J Appl Phycol 27:1525–1534

Hutton M, Symon C (1986) The quantities of cadmium, Lead, mercury and arsenic entering the U.K. environment from human activities. Sci Total Environ 57:129–150

Jaishankar M, Tseten T, Anbalagan N, Mathew BB, Beeregowda KN (2014) Toxicity, mechanism and health effects of some heavy metals. Interdiscip Toxicol 7(2):60–72. https://doi.org/10.2478/intox-2014-0009

Kuritz T (1999) Cyanobacteria as agents for the control of pollution by pesticides and chlorinated organic compounds. J Appl Microbiol Symp Suppl 85:186–195

Laloknam S, Sirisopana S, Phornphisutthimas S, Takabe T, Incharoensakdi A (2009) Removal of mercury, arsenic and cadmium in synthetic wastewater by cyanobacterium Aphanothece halophytica. NU Sci J 6:96–104

Lem NW, Glick BR (1985) Biotechnological uses of cyanobacteria. Biotechnol Adv 3:195–208

Lenntech Water Treatment and Air Purification (2004) Water treatment, Published by Lenntech, Rotterdamseweg

Malik A (2004) Metal bioremediation through growing cells. Environ Int 30:261–278

Mazard S, Penesyan A, Ostrowski M, Paulsen IT, Egan S (2016) Tiny microbes with a big impact: the role of cyanobacteria and their metabolites in shaping our future. Mar Drugs 14(5):97. https://doi.org/10.3390/md14050097

McEldowney S, Hardman DJ, Waite S (1993) Pollution: ecology and biotreatment. Longman Scientific & Technical, Harlow

Megharaj M, Venkateswarlu K, Rao AS (1987) Metabolism of monocrotophos and quinalphos by algae isolated from soil. Bull Environ Contamin Toxicol 39:251–256

Micheletti E, Colica G, Viti C, Tamagnini P, De Philippis R (2008a) Selectivity in the heavy metal removal by exopolysaccharide-producing cyanobacteria. J Appl Microbiol 105:88–94

Micheletti E, Pereira S, Mannelli F, Moradas-Ferreira P, Tamagnini P, De Philippis R (2008b) Sheathless mutant of cyanobacterium Gloeothece sp. strain PCC 6909 with increased capacity to remove copper ions from aqueous solutions. Appl Environ Microbiol 74:2797–2804

Morin-Crini N, Druart C, Gavoille S, Lagarrigue C, Crini G (2013) Analytical monitoring of the chemicals present in the discharge water generated by the surface treatment industry. J Environ Prot 4(7A):53–60. https://doi.org/10.4236/jep.2013.47A007

Noel SD, Ranjan MR (2014) Cynobacteria as a potential source of phytoremediation from textile industry effluent. J Bioremed Biodegr 5:1–4

Nriagu JO, Pacyna J (1988) Quantitative assessment of worldwide contamination of air, water and soil by trace metals. Nature 333:134–139

Ojuederie OB, Babalola OO (2017) Microbial and plant-assisted bioremediation of heavy metal polluted environments: a review. Int J Environ Res Public Health 14(12):1504. https://doi.org/10.3390/ijerph14121504

Orta-García ST, Ochoa-Martinez AC, Carrizalez-Yáñez L, Varela-Silva JA, Pérez-Vázquez FJ, Pruneda-Álvarez LG, Torres-Dosal A, Guzmán-Mar JL, Pérez-Maldonado IN (2016) Persistent organic pollutants and heavy metal concentrations in soil from the metropolitan area of Monterrey, Nuevo Leon. Mexico Arch Environ Contam Toxicol 70(3):452–463. https://doi.org/10.1007/s00244-015-0239-3

Ozturk S, Aslim B (2008) Relationship between chromium (VI) resistance and extracellular polymeric substances (EPS) concentration by some cyanobacterial isolates. Environ Sci Pollut Res Int 6:478–480

Panosyan H (2015) Thermophiles harbored in Armenian geothermal springs and their potential in biotechnology. Industrial Biotechnology Congress August 10–11, Birmingham

Pauls J, Rydzynski K, Dziubaltowska E, Wyszynska K, Natarajan AT, Nilsson R (2003) Genotoxic effects of occupational exposure to lead and cadmium. Mutat Res 540(1):19–28

Perales-Vela HV, Pena JM, Canizares RO (2006) Heavy metal detoxification in eukaryotic microalgae. Chemosphere 64:1–10

Priyadarshani I, Sahu D, Rath B (2011) Microalgal bioremediation: current practices and perspectives. J Biochem Technol 3:299–304

Radwan SS, Al-Hasan RH (2000) Oil pollution and cyanobacteria. In: Ba W, Potts M (eds) The ecology of cyanobacteria: their diversity in time and space. Kluwer Academic Publishers, Dordrecht, pp 307–319

Rani B, Choopera SL, Maheshwari R (2007) Cleaning up of pollutants by phytoremediation: a novel approach for sustainable development. J Water Land Use Manage 7(1):71–81

Rayms-Keller A, Olson KE, McGaw M, Oray C, Carlson JO, Beaty BJ (1998) Effect of heavy metals on Aedes aegypti (Diptera:Culicidea) larvae. Ecotoxicol Environ Saf 39:41–47

Schoeny R, Cody T, Warshawsky D, Radike M (1988) Metabolism of mutagenic aromatic hydrocarbons by photosynthetic algal species. Mutant Res 197:289–302

Singh S (2020) Biosorption of heavy metals by cyanobacteria: Potential of live and dead cells in bioremediation. https://doi.org/10.1007/978-981-15-1812-6_15

Sinha S, Chattopadhyay P, Pan I, Chatterjee S, Chanda P, Bandyopadhyay D, Das K, Sen SK (2009) Microbial transformation of xenobiotics for environmental bioremediation. Afr J Biotech. 8(22):6016–6027

Srogi K (2008) Developments in the determination of trace elements by atomic spectroscopic techniques. Anal Lett 41(5):677–724

Stal LJ (1995) Physiological ecology of cyanobacteria in microbial mats and other communities. New Phytol 131:1–32

Thajamanbi M, Rout J, Thajuddin N (2016) Blue green algae from rice fields of Karimganj district, Assam, north East India. Int J life Sci Pharma Res 6(4):L25–L34

Tien C-J, Sigee DC, White KN (2005) Copper adsorption kinetics of cultured algal cells and freshwater phytoplankton with emphasis on cell surface characteristics. J Appl Phycol 17:379–389

Turner JS, Robinson NJ (1995) Cyanobacterial metallothioneins: biochemistry and molecular genetics. J Indus Micro 14:119–125

Ughy B, Nagy CI, Kós PB (2015) Biomedical potential of cyanobacteria and algae. Acta Biologica Szegediensis 59(2):203–224

Verma SK, Singh SP (1995) Multiple metal resistance in the cyanobacterium *Nostoc muscorum*. Bull Environ Contam Toxicol 54:614–619

Wilson PE, Nyborg AC, Watt GD (2001) Duplication and extension of the Thorneley and Lowe kinetic model for Klebsiella pneumoniae nitrogenase catalysis using a mathematica software platform. Biophys Chem 91(3):281–304

Zinicovscaia I, Cepoi L, Povar I, Chiriac T, Rodlovskaya E, Culicov OA (2018a) Metal uptake from complex industrial effluent by cyanobacteria *Arthrospira platensis*. Water Air Soil Pollut 22(7): 1–10

Zinicovscaia I, Safonov A, Troshkina I, Demina L, German K (2018b) Biosorption of re(VII) from batch solutions and industrial effluents by cyanobacteria *Spirulina platensis*. CLEAN: soil air. Water 46(7)

The manufacturer's authorised representative in the EU is Springer Nature Customer Service Centre GmbH, Europaplatz 3, 69115 Heidelberg, Germany. If you have any concerns regarding our products, please contact ProductSafety@springernature.com

Printed and bound by CPI Group (UK) Ltd, Croydon, CR0 4YY
30/09/2026
02225318-0010